ARCHITECTURAL INSITU CONCRETE

David Bennett

RIBA ⌗ **Publishing**

Acknowledgements

Thanks for all your help and guidance.

Individuals

Bill Price, Lafarge Cement; John Taylor, Castle Cement; John Anderson and Derek Ballard, Hanson Premix; Peter Stewart, UPM-Kymmene Wood (WISA); Paul Raybone and Derek Smith, A Plant Acrow; Dennis Higgins, Civil and Marine (GGBS); Lindon Sear, UKQAA (PFA); Edwin Trout, Concrete Information Service; Peter Jackson, PT Group Ltd; Dr Adrian Brough, Leeds University; Clara Willet, English Heritage.

Companies

Lanxess GmbH – Bayferrox Pigments; Kronoply GmbH; PERI Ltd UK; SGB Ltd; Creteco Ltd; Trent Concrete Ltd; Wacker (UK) Ltd; Castle Cement.

© **David Bennett, 2007**

Published by RIBA Publishing, 15 Bonhill Street, London EC2P 2EA

ISBN 978 1 85946 259 1

Stock Code 61854

The right of David Bennett to be identified as the Author of this work has been asserted in accordance with the Copyright, Designs and Patents Act 1988.

All rights reserved. No part of this publication may be reproduced, stored in a retrieval system, or transmitted, in any form or by any means, electronic, mechanical, photocopying, recording or otherwise, without prior permission of the copyright owner.

British Library Cataloguing-in-Publication Data
A catalogue record for this book is available from the British Library.

Publisher: Steven Cross
Project Editor: Alasdair Deas
Editor: Andrea Platts
Designed by Paul Gibbs, Kneath Associates
Printed and bound by Cambridge University Press

While every effort has been made to check the accuracy of the information given in thisbook, readers should always make their own checks. Neither the Author nor the Publisher accepts any responsibility for misstatements made in it or misunderstandings arising from it.

RIBA Publishing is part of RIBA Enterprises Ltd

www.ribaenterprises.com

CONTENTS

Introduction

Part One: Technology
Materials and mixes	002
Formwork and practice	041
Concrete workmanship	060

Part Two: Case Studies in Concrete
Thames Barrier Park *Patel Taylor*	075
Persistence Works *Feilden Clegg Bradley*	084
The Art House *Fraser Brown Mackenna*	097
The Anderson House *Jamie Fobert*	107
Aberdeen Lane *Azman Owens*	117
One Centaur Street *deRMM*	127
85 Southwark Street *Allies and Morrison*	137
The Bannerman Centre *Rivington Street Studio*	147
The Brick House *Caruso St John*	157
The Collection *Panter Hudspith*	167
Playgolf, Northwick Park *Charles Mador*	179
E-Innovation Centre *BDP Manchester*	189
The Jones House *Alan Jones*	199
Spedant Works *Greenway and Lee*	209
Central Venture Park *Eger Architects*	219

Glossary of concrete terminology	228
Further reading	230
Useful contacts	230
Picture credits	230
Index	231

INTRODUCTION

Concrete as an architectural and structural material has gone through many changes and evolutions in its development over the years, but probably none more dramatic than in the past decade. There is renewed interest in concrete's plastic and aesthetic qualities in architecture today, helped and encouraged by the expressive way that British architects have exploited its self-finished quality and form with great success. It clearly demonstrates that standard truck-mixed concrete and the right selection of formwork and placement techniques can produce award-winning architecture at affordable prices. What is also a revelation is that many of the architects whose work is highlighted in the case studies in the second part of this book have never designed exposed concrete on such a scale before. What they discovered is that there is a wealth of knowledge within the concrete industry that can be accessed to give them the confidence and encouragement to realise their ambitions.

The first part of this book provides sufficient in-depth technical guidance and practical information on the mechanics and fundamentals of how to achieve a fine concrete finish that there should be no need to make a frantic search through countless books or websites to find solutions.

ARCHITECTURAL INSITU CONCRETE

The techniques of concrete compaction and placement, the surface treatment of cast concrete to prevent dirt ingress and staining, detailing of panel layouts, concrete workmanship and general tips on good practice are well documented and comprehensively covered. But perhaps the two most important aspects are the control of colour in specifying a truck-mixed concrete and the selection of the form face which imparts the character and definition of the surface appearance. Considerable commentary has been developed on how concrete is produced, what controls its colour and the properties and performance of formwork materials commonly available, backed up with good illustrations to give the designer insight and understanding for their selection and specification.

The case studies which make up the second part of the book manifest the integrity and quality of insitu concrete in its many diverse forms, exploiting the grey coolness of concrete, its planar smoothness and textured tonality. I make no apology for mentioning The Collection (aka the Lincoln and City County Museum) designed by architects Panter Hudspith, which represents a significant step forward in the advancement of self-compacting concrete. The quality of the board-marked surface in the café area is as good as the pictures of the walls and as perfect as porcelain. Here is a poured-in-place material that can produce finishes and formed shapes that were once thought to be feasible only with precast concrete. The added value is that it can be built as a monolithic load-carrying structure with built-in thermal mass. New materials, new forms and new developments, like the use of fabric formwork, ultra high strength concrete, recycled waste and steel fibre reinforced concrete, will continue to keep concrete forever inventive and responsive to change.

Whether you prefer the smooth, unmarked surface of film faced ply, the subtle flecks of untreated birch, the heavy graininess of sawn timber, the marble softness of polythene-lined plywood, there is so much variety, choice and opportunity to experiment that there is something here to suit everyone's taste and predilections. The more you discover, the more you want to innovate. The more deeply ingrained your depth of understanding and mastery of the material, the more you realise how little we really know of liquid rock's potential. And yet, on reflection, judging by the 15 schemes highlighted in the case studies, there is a dynamism that echoes the pioneering spirit of the 1930s Modern Movement and something of the concrete finesse of Swiss architecture in what has been achieved. In these projects we can see an increasing number of architects returning to concrete to presage the dawn of a new era of concrete realism.

Infused with such optimism, there is a driving need to take stock and consider climate change, how we can reduce CO_2 emissions and sustain our natural world through responsible design and considerate architecture. What can concrete do to help? This book is not going to provide those solutions, but the examples will certainly make suggestions and show what has been done, and the reading list in the reference section highlights excellent publications on the subject. We can all make a contribution by reducing site wastage, by recycling building materials – especially formwork, by using public transport instead of driving a car and utilising thermal mass to reduce energy consumption. It is equally important to design concrete buildings with durable fine finishes that will last a lifetime, that will not dilapidate or decay or need rebuilding in the near future – another argument for the sustainability of concrete.

There are many people I wish to thank for helping me put this book together. If your name does not appear the acknowledgments, it's not because you have been ignored, it's because I'm forgetful and have been remiss. Thanks to all those wonderful people – the architects, the clients, the engineers, contractors, the cement manufacturers, the concrete and formwork material suppliers – those champions of concrete

fig 1-2
River City Plaza, Chicago

3

4

— for giving of their time so generously and for allowing the publisher to use the images of the buildings, production plant and materials illustrated in the book. Thanks to Ian Cox and The Concrete Centre for generously supporting this publication, to Steven Cross and the publishing team at RIBA Enterprises for making it happen and to Kneath Associates and the book design team for the fantastic visual layouts.

Just when you think you know your subject, you come across a building that takes your breath away and brings you to your knees in admiration. One such is River City Plaza (1985) in Chicago, completed some 20 years ago and still looking great. It is a building from which some of the notes on architectural concrete compaction were derived thanks to Symonds Corporation, the formwork specialists who helped build it. It ranks among the finest cast-in-place concrete structures I have seen and was the last building that Bertrand Goldberg designed. It has curved board-marked load-bearing walls using formliners glued to curved metal formwork and was developed for a social housing programme.

Two other buildings deserve special mention and were important in my early understanding of architectural insitu concrete and just how creative architects can be when they master their art. When researching material for my first book on architectural concrete in 1998, the images of the Social Science Faculty Centre (1999) in Oxford were taken, at a stage when the internal areas were incomplete and there was the odd scaffold tube and scaffold board in the picture. It is a gem of a building by Foster and Partners and a totally exposed concrete structure with fine detailing and well-crafted concrete workmanship. The internal shots of the finished interior, without scaffold tubes this time, are yet another reminder that a good designer and competent concrete contractor can work wonders with concrete – it's not that difficult. If I had to pick one building that is outstanding for sheer bravado, audacity, architectural verve, construction innovation and engineering simplicity (not easy to do) then it would be the extraordinary red-pigmented, gunited, surface skin and structure of the Minnaert Building in Utrecht University. The architect is Nuetlings Riedijke and the structural engineers ABT in Amsterdam. But that was before this book was written and there are now 15 new projects to consider … what a lovely dilemma! I hope you find the technical information rewarding and the concrete case studies an inspiration for your future ambitions.

David Bennett
July 2007

5

fig 3-4
Social Science Faculty, Oxford

fig 5
Minnaert Building, Utrecht University

PART I
TECHNOLOGY

Materials and mixes
Cement 002
Aggregates 014
Concrete production 022
Concrete colour 034
The right mix 039

Formwork and practice
Introduction 041
Untreated timber 042
Film faced plywood 044
Oriented strand board 046
Chipboard (for single usage) 047
Metal 048
Others 048
Formwork practice 050

Concrete workmanship
Introduction 060
Handling and placing 060
Compaction and consolidation 062
Reinforcement and cover 064
Curing 067
Trial panels 067
Clean, dirt-free surface 067
Stains and blemishes 068
Remedies 070
Repairs 070

MATERIALS AND MIXES
CEMENT

AN INTRODUCTION

*With Bill Price,
Lafarge Cement UK*

Concrete is the most widely used construction material on earth and, at its simplest, cement is the glue, or binder, that holds concrete together. Consequently, cement makes a vitally important contribution to the appearance of the world we live in. But what is 'cement'?

While there are a number of different types of cement, the most common is 'Portland' cement. This was named by its inventor, Joseph Aspdin, after its supposed resemblance to Portland stone when set. This type of cement has been produced for over 150 years and is available worldwide.

In terms of its basic chemistry, Portland cement can be represented in terms of four main oxides, which make up more than 95 per cent of its composition. These are CaO (lime), SiO_2 (silica), Al_2O_3 (alumina) and Fe_2O_3 (iron oxide). The remainder consists of various minor constituents, including magnesium, sulphur and alkalis.

HOW IS IT MADE?

The manufacture of Portland cement is a closely controlled process akin to large-scale chemical engineering. The necessary raw materials are principally a source of calcareous material, typically limestone or chalk, which makes up about 80 per cent of the raw ingredients, a source of silica – either clay or shale – and a pinch of sand and iron oxide to optimise the chemistry.

Limestone and chalk are composed of the fossilised remains of millions of micro-organisms that lived in the sea. It is an interesting quirk of prehistory that the shells of the marine creatures of the carboniferous or cretaceous eras are now being recycled to provide us with concrete.

There are a number of different processes used commercially to produce cement, the most common of which are:

> the dry process
> the semi-dry process
> the semi-wet process and
> the wet process.

The energy consumption and the energy efficiency of cement production vary between the different processes (see Table 1). The dry process, particularly when combined with precalcining, is the most energy-efficient process

Table 1
Relative energy performance of different cement production processes

Process	Relative energy consumption for 1 t of clinker
Dry + precalciner	1.00
Dry + preheater	1.09
Semi-dry	1.14
Semi-wet	1.21
Wet	1.95

Dry process

In the dry process, the raw materials are introduced without additional water. These natural ingredients are finely ground and blended (homogenised) in the optimum proportions. The raw materials can be preheated in a flow of hot kiln exhaust gases prior to entering the cement kiln itself. This preheating dries the materials, and in the similar but more efficient process known as 'precalcining' it also decarbonates the majority of the calcium carbonate in the limestone or chalk. The dry materials are then heated in a rotating kiln to a temperature of around 1,500 °C. At this temperature, the homogenised raw materials are only partially melted, although there is a complete transformation of the mineral assemblage. The partial melting combined with the rotation of the kiln produces a granular material known as cement clinker. The raw materials remain in the kiln for about half an hour in a precalcined, dry process system.

Semi-dry process

In the semi-dry process, the raw materials are initially prepared in the same way as for the dry process, but then pelletised (sometimes termed 'nodulised') into spheres of around 15 mm diameter with about 12 per cent moisture. The pellets are gently preheated on a moving grate preheater before entering the kiln.

Semi-wet process

The semi-wet process involves combining the ground raw materials with around 20 per cent moisture into

fig 1
Padeswood cement works

a filter cake, which can either enter the kiln directly or be preheated in a similar way to the semi-dry process prior to entering the kiln.

Wet process

The wet process utilises raw materials that are mixed with around 40 per cent water to form slurry before blending and clinkering. This is a more energy-intensive method and is not found in the more modern cement plants. The residence time of the raw materials and clinker in the kiln is much longer than in an efficient dry–process system – often up to 2 hours. Very few modern cement plants are based on the wet process.

Clinker

Irrespective of the production process, the clinker exits the kiln at a temperature of around 1,200 °C and is cooled before being stored. The clinker now consists of four main minerals:

> tricalcium silicate or 'alite' (C3S)*
> dicalcium silicate or 'belite' C2S)
> tricalcium aluminate or 'aluminate' (C3A)
> tetracalcium aluminoferrite or 'ferrite' (C4AF).

It is these minerals that react with water to cause the setting and subsequent strength development in hardened concrete.

The cooled clinker is ground in a ball mill together with a small quantity of calcium sulphate (gypsum and/or anhydrite) to produce the final Portland cement. The calcium sulphate is a set regulator that controls the setting time of the cement and prevents premature 'flash' setting.

TYPES OF CEMENT

Although Portland cement is by far the most widely used, the European Standard for common cements, BS EN 197-1, lists a total of 11 different groups of factory-produced cement in common use within Europe. All these cements contain a proportion of Portland cement clinker but not all of them are available in the UK. Cements produced commercially throughout the UK are:

> Portland cement (CEM I), sometimes also known as PC or OPC (this group also includes White Portland cement)
> Portland-flyash cement (CEM II/A-V, CEM II/B-V)
> Portland-limestone cement (CEM II/A-LL, CEM II/B-LL).

The following cement types are also produced in some areas of the UK:

> blastfurnace cement (CEM III/A, CEM III/B, CEM III/C)
> Portland-slag cement (CEM II/A-S, CEM II/B-S).

In addition to these groups there is also sulphate-resisting Portland cement (SRPC), which is still covered by a residual British Standard (BS 4027) and masonry cement, covered by BS EN 413.

fig 2
Typical cement works layout

* In cement chemistry, the oxides of calcium, silicon, aluminium and iron are abbreviated to C, S, A and F respectively. Thus, the various cement minerals can be described in shorthand (e.g. C3S for $3CaO.SiO_2$).

fig 3
Selected Lafarge Portland cements
top: Under indoor lighting
below: Under natural light

CEMENT REPLACEMENTS (ALSO KNOWN AS ADDITIONS)

The above classification refers only to factory-produced cements, distinguished by a 'CEM' designation. However, there are cement replacement materials that can be mixed with cement at the 'ready mixed' plant to produce concrete. Examples of such materials are pulverised fuel ash (PFA or fly ash), ground granulated blastfurnace slag (GGBS), metakaolin and silica fume. These materials are formally known as 'additions', but they are often called 'cement replacements'. When additions are mixed with cement at the ready mixed plant the resulting blends are called 'combinations' to distinguish them from factory-produced cements of similar composition.

PROPERTIES OF CEMENT

A full discussion of the properties of cement and concrete is beyond the scope of this book, but some discussion of the factors influencing durability and colour is appropriate.

Durability

Cement and concrete are inherently durable, but exposure to aggressive environments can, in certain circumstances, lead to premature deterioration. Different types of cement perform better in certain environments and guidance on the most appropriate cement types for given situations can be found in BS 8500 and the complementary British Standard to the European Concrete Standard (EN 206-1). However, certain generalisations can be made. No cements are truly resistant to strong acids. For concrete exposed to these or to other highly aggressive chemicals, the quality of the concrete (low water/cement ratio, properly compacted and cured) is of greater significance to durability than the type of cement.

Cement colour

Given the importance of the aesthetic appeal of concrete, the colour of cement and the factors that influence it are worth understanding. The usual perception of cement is that it is grey. Most Portland cements are grey, but they are not the same shade of grey. Portland cement is manufactured in a wide variety of shades of grey, with each cement works producing a unique shade of grey.

Grey cement

The colour of normal grey cement depends, essentially, on the raw materials and the fuels used in the kiln. The key factor is the proportion of iron oxide (Fe_2O_3) in the Portland cement clinker. This is primarily found in the ferrite phase (C4AF). Increasing the amounts of ferrite in the clinker produces a darker colour. The conditions under which the clinker is cooled also has an important influence on cement colour. Conventional cooling in air (oxidising conditions) produces a ferrite that is nearly black in colour and a darker grey cement, whereas cooling in reducing conditions (such as water quenching) results in a paler brown form of ferrite with less impact on the darkness of the cement. There are also increasing environmental pressures on cement producers to burn alternative fuels in their kilns, examples being used tyres, reclaimed solvents and processed sewage pellets. These new fuels can have an effect on the colour of the final cement. Used tyres in particular will generally produce darker coloured cement. The variation in colour of grey cements produced at different works is illustrated in the series of images on page 3.

Colour control

It is also possible to modify the colour of cement by blending Portland cements from different sources. This process offers the possibility of controlling the colour of the cement and minimising any colour variations. Factory-made blends of white and grey cement, with a closely controlled colour range – based on measuring the colour of the blend using a colour meter and adjusting the proportions of the constituents to maintain a constant colour – are already commercially available from one major manufacturer.

Of the commonly available cement replacements added separately to concrete at the ready mixed plant, silica fume tends to darken the colour of concrete – although a white silica fume is available – as does PFA, but to a lesser extent. GGBS, on the other hand, is quite light in colour and will impart a creamy white colour to concrete, particularly when used in higher proportions relative to Portland cement. Due to the nature of the concrete mixing process, control of colour consistency in the ready mixed plant combinations is less precise than when using factory-blended coloured cement.

White cement

The colour of cement depends to a great extent on the raw materials and the fuels used to heat the kiln. To manufacture white cement, the raw materials are specially selected to contain a very low amount of iron and manganese oxides (usually below 0.3 per cent of the weight of the clinker). High purity limestone and chalk and low iron content clays are preferred as raw materials.

The use of oil or gas as a kiln fuel is preferred over coal as this also minimises the potential for contamination. Kiln temperatures for white cement production may be higher than for normal cement manufacture. The kiln will be operated under slightly reducing conditions and the resulting clinker quenched in water. This helps to keep the colour of the iron oxides as pale as possible and maintain the whiteness of the cement.

The whiteness of the cement is also enhanced by finer grinding. The grinding media in the ball mills are also specially selected to be low in iron in order to prevent contamination of the cement during grinding. Although all 'white cements' are white in colour, there are still variations in the colour of white cements from different sources. As with other cements, it is recommended that only white cement from a single source (works) is used for a given project in order to minimise any variation in cement colour.

On a worldwide scale, comparatively low volumes of white cement are produced. This, together with the need for specially selected raw materials, the restrictions on fuels and the more expensive grinding regime, all contribute to the higher cost of white cement when compared to grey cements. Currently, white Portland cement is not manufactured in the UK and supplies are imported from elsewhere in Europe.

Effect of carbonation on colour

When hardened concrete is exposed to the atmosphere, there is a reaction between the carbon dioxide in the air and the hydrated cement compounds (principally calcium hydroxide produced by the hydration of Portland cement). This precipitates calcium carbonate in the pore structure of the concrete surface. This then has the effect of lightening the colour of the concrete surface over time.

PROFILE: KETTON CEMENT WORKS

With John Taylor, Castle Cement

It is important to understand how cement is manufactured in order to know what causes cement to have a particular grey colour. After all, this is the pigment of a standard ready mixed concrete and we should know something about the raw ingredients used in the making of it. What better introduction to the subject than examining what happens at the cement works at Ketton, whose tall, cylindrical chimneys can be seen from Rutland Water and the beautiful Stamford countryside? Modern Ketton is not a dust-laden cement bowl of past decades. There is still a hardened coating of grey on the plant room roofs to be seen and a giant scar cut into the hillside where limestone has been quarried, but the air is surprisingly clean and the surrounding area lush with greenery and landscaping thanks to the stringent pollution regulations with which modern cement factories must comply. Washing may be hung out to dry without any fear of losing its whiteness!

The raw ingredients at Ketton are a buff-coloured oolitic limestone – a calcareous material that contains calcium carbonate compounds – and clay – an argillaceous material rich in aluminate, silicate and iron compounds. You need calcium carbonate and silicate and a certain amount of aluminates and iron compounds to make cement. Oolitic limestone is a chalk-based sedimentary rock of the Jurassic period, which was formed by chemical deposition in shallow lagoons millions of years ago. Ooliths are rounded grains formed by the deposition of calcium carbonate particles, which attach themselves to grains of sand or pieces of shell found in suspension in the sea. These sand and shell particles are rolled to and fro between tides in limey water near limestone coasts and become coated with calcium carbonate deposits. In the course of time the ooliths become cemented together to form a layer of hard oolite and the beginnings of a rock as the layers build up. The overburden pressure compresses the oolite layers to form limestone. The oolitic limestone at Ketton is of similar quality to Bath stone and Portland stone; it has very few shells and an even grain texture that makes it a first-class building stone.

The oolitic limestone lies in horizontal beds quite near the surface and has been extracted for cement manufacture at Ketton since 1926, but it has been quarried for building stone since Elizabethan times. The quarry is over two miles long, it is 25 m deep and 400 m wide and runs in an east–west direction to follow the plane of the most consistent beds of limestone. In areas where the stone and clay have been exhausted, the overburden that was originally removed is brought back to substantially fill the hole, which is then seeded and grassed over and turned into farmland. The ground level is depressed by 7 m overall but you can hardly discern any change between it and the natural gentle contours and wide open spaces.

The method of extraction is to remove the overburden and topsoil down to a depth of 1 m in order to reveal the clay. The overburden and topsoil are stored for later use. The clay is removed by face shovels to a depth of 3 m to reach the top of the limestone and brought to the receiving hopper. The limestone is drilled and charges placed to split the stone, which falls to the quarry floor. The face shovels pick up the limestone boulders and drop them into dump trucks which take them

4

to the receiving hopper. The 3 and 4 tonne stone pieces are fed by conveyor into the jaws of a rotary crushing machine with percussive steel heads that crush the rock down to 75 mm size lumps. The hard clay lumps are also fed into the rotary crusher to reduce them in size and to dry blend them with the limestone.

About four parts of limestone to one part of clay is needed to make cement. The proportion is crudely gauged when the clay and limestone pieces are dumped into the receiving hoppers before being crushed. To assess the correct proportioning of the blended material, samples are taken every 2 hours from the crushed material as it sits on a conveyor transporting it to the cement works. At the sample station the blended material is ground to a fine powder and sent to the central laboratory where it is analysed. A message is relayed back to quarry control to tell the operators to put more or less clay into the receiving hopper. The 2-hour production volume of crushed limestone and clay is stored in a circular mixing bed at the cement works – this has a 45,000 t capacity. Each 2 hours' worth of material is discharged onto a conical mound of blended material already in the mixing bed. The conveyor deposits the material by a radial arm which spreads it across the sloping mound. The stockpile is made up of layers of blended material that have been checked and adjusted for mix proportioning every 2 hours. Generally, the factory will extract about 4,500 t of this crushed blended material, from the mix bed every 24 hours to feed into the kiln to produce cement clinker.

An extractor rakes through the layers of blended material to extract material, which is channelled into a discharge chute that goes to a central storage silo ready for the final stage of milling before it is put into the kiln. A sample is taken from the central storage bin to fine tune the mixture by adding in an amount of pure crushed limestone that has been brought to an adjacent storage hopper. The analysis will determine how much pure limestone to add to the blended material in the weigh batchers before it is fed into the vertical spindle mill to be ground to a fine powder. This method of blending the raw ingredients together is called the 'dry process' and is the most economical method as it requires less plant and equipment than the wet process and is cheaper to operate.

The vertical spindle mill has a rotating table with two grooves in which sit two pairs of grinding wheels. As the raw material is fed from the top of the spindle mill the ingredients are ground to a very fine powder called 'raw meal'. Heated air is fed into the base of the mill and the hot air is used to transport the finely ground dry material to the electrostatic precipitators, which scrub the dust (raw meal) out of the air and pass it to the raw meal storage silo (capacity 2,200 m^3).

The raw meal storage silo has level sensors which relay information back to the vertical spindle mill to instruct it to grind more material when the silo level gets low. The raw meal in the silo is constantly agitated to keep the material well mixed and to prevent it settling. Next, the raw meal is drawn out of the silo and blown to the top of the preheater, precalcining tower. As it free falls down it meets hot gases rising from the bottom of the tower. By the time it reaches the base of the tower it enters the gently sloping kiln as glowing, red-hot embers at 900 °C. The kiln is on a 5° slope and rotates slowly, moving the red-hot material towards the hottest part of the kiln, where the flame temperature is about 2,000 °C, to raise the temperature of the mixture to 1,450 °C. At that temperature all the material in the limestone and clay goes into a molten state and coalesces to form compounds of calcium silicate and aluminates. By the time the molten compounds have reached the end of the kiln, the rotation has nodulised the raw meal into a clinker.

The hot clinker is an unstable compound and seeks out carbon dioxide and water to return it to a stable state as it cools, which must be prevented. If it is allowed to cool slowly the compounds of calcium silicates and aluminates will crystallise with different sized grain structure. Large crystalline structures in the clinker do not make very reactive cement. It is

fig 4
Two 100 t capacity dump trucks and a loading shovel parked in front of limestone boulders blasted from the Ketton Quarry

ARCHITECTURAL INSITU CONCRETE - TECHNOLOGY

6

7

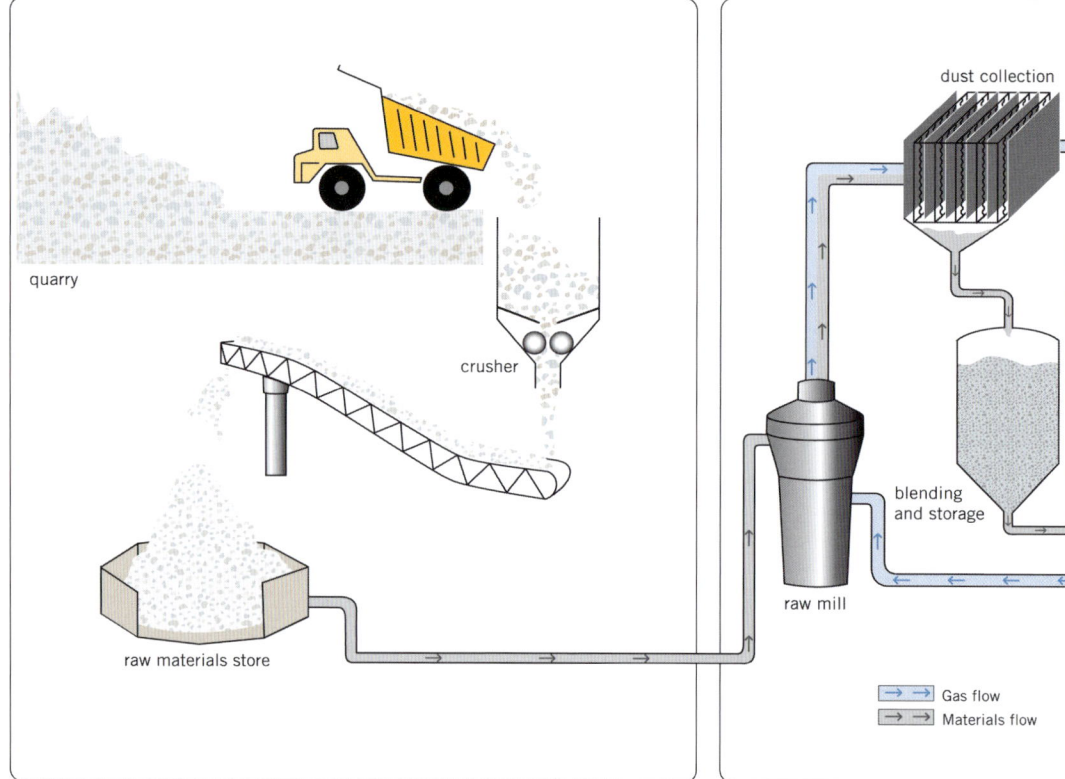
5

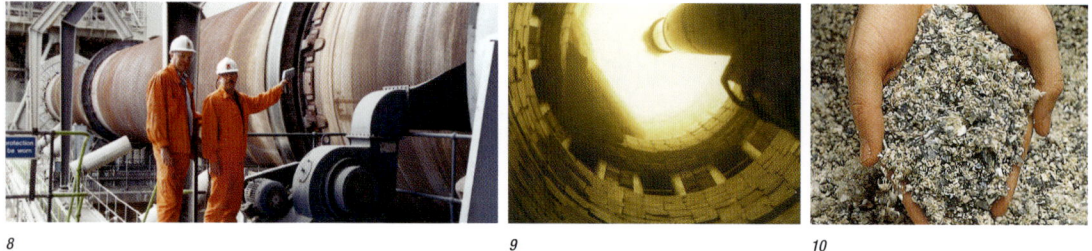
8   9   10

figs 5-14 *The cement production process*
5: Schematic of the process
6: Raw limestone   7: Crushed limestone   8: Kiln   9: Kiln temperature of 1450°C is needed to change the raw material   10: Pro-fuel

MATERIALS AND MIXES

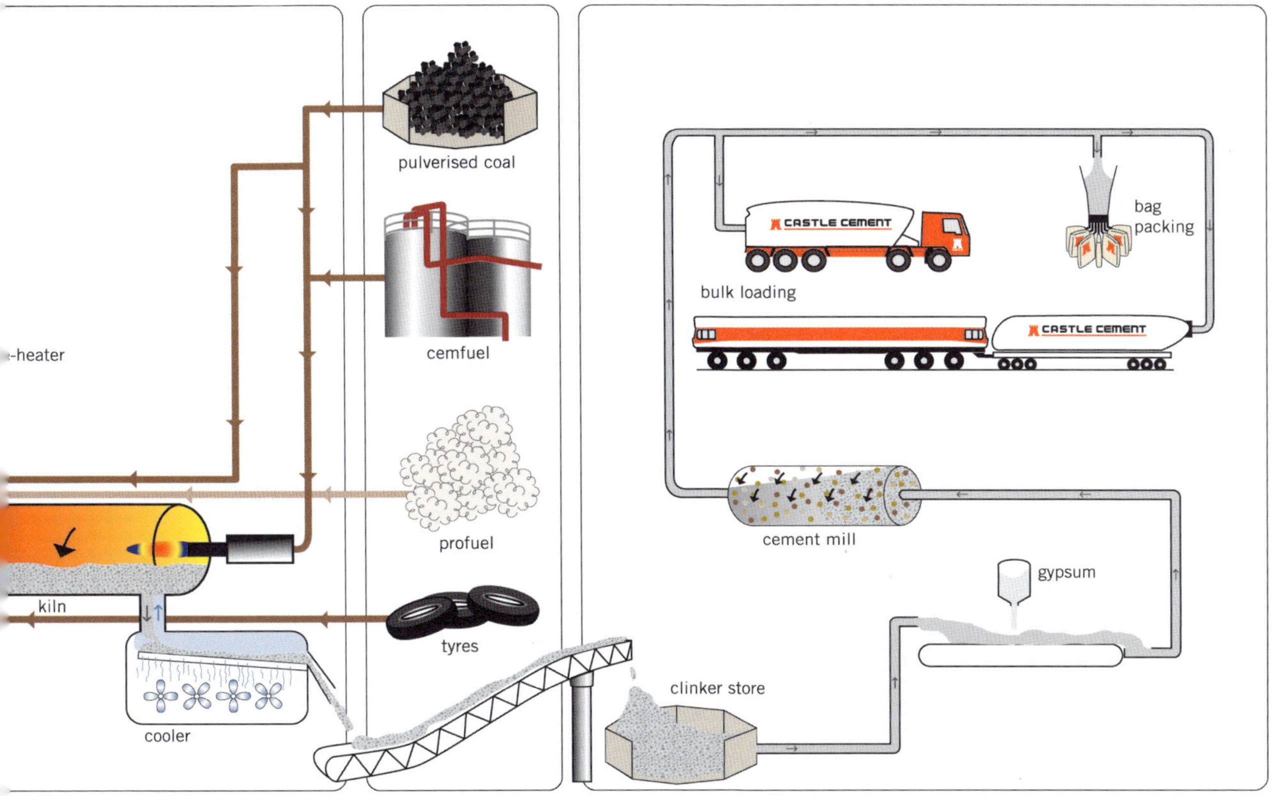

11: Ketton cement works   12: Clinker store   13: Clinker nodules   14: Milled cement

9

critical, therefore, to cool the clinker rapidly to prevent this occurring. The newest kiln at Ketton has a series of large fan coolers that blow ambient air through a grate over which the clinker passes. The cool air blowing through the grate carries away dust particles of clinker, which are collected by electrostatic precipitators, allowing the dust-free, warm air to be used in the kiln to reduce energy consumption before finally being released, when cool and clean, into the atmosphere. The captured dust is returned to the clinker store. The cooled clinker is put into a 100,000 t clinker store.

To manufacture cement, the clinker from the clinker store is transported by conveyor to the cement grinding mill. During the milling process, gypsum is added to the clinker to control the rate of set of the milled cement powder. The cement powder is monitored for gypsum content and fineness; there is not sufficient time to check the chemistry of the cement and no need for it. That control has already been carried out by the earlier sampling of the crushed limestone and clay. The fineness and gypsum content will give the character of the cement.

This is the process for making ordinary Portland cement.

If rapid hardening cement is required the clinker is milled to a much finer grain size. Admixtures and air entraining agent can be added to Portland cement to make air-entrained cement for mortar, concrete and screeds, which will have improved properties both in the plastic state and also in the hardened state. Castle brands this cement 'Multi Cem'– it contains an air-entraining agent for better frost resistance and plasticiser for workability. When Portland limestone cement is manufactured, 10–15 per cent of raw limestone is added to the clinker and milled into the cement powder. It is also possible to add in 40–60 per cent slag or PFA with the clinker to make blended cement.

Of the 4,500 tonnes of cement manufactured every day at Ketton:

> 87 per cent is ordinary Portland cement
> 10 per cent is rapid hardening cement
> 2 per cent is masonry/ enhanced cement
> 1 per cent is special cement for the nuclear industry.

A total of 25 per cent of all cement sales is bagged cement in 25 kg lots, the rest is bulk cement supplied to customers in tankers. If there were sufficient demand for customised blended cement it could be produced without disrupting production of Portland cement.

Overnight cement truck deliveries and customer silo management can reduce bulk cement costs. With some major users of their cement, Castle organise deliveries overnight at times to suit the cement works, when there is little traffic on the road and during a quiet period at the works. Silo management enables Castle to fill customer silos when they fall below a preset level. Sensors in the silo send a signal to monitors at Ketton to ask for a delivery of 30 t of cement. Usually this is carried out between 1 and 2 a.m. The cement truck driver couples the outlet nozzle from the cement tank to the silo valve and then operates a switch to blow the cement into the silo.

GENERAL
CEMENT COLOUR

The simple explanation for the varying colours of cement, but by no means a satisfactory one, is that colour is determined by the chemistry of the clay and limestone, the raw materials. That is obvious, so what is in the constituents to cause the cement to be grey? First of all, the Ketton limestone is a light-brown colour and the clay a mid-grey colour. When they are heated to a clinker the colour is sucked out of them, just like the ash residues of coal or wood when burned. But why is it grey? The abundance of calcium and silicate compounds in the raw material after heating will cool to a whitish colour – calcium carbonate is white but tinted a greyish colour by the percentage of iron compounds and, to a lesser extent, the manganese compounds that are present in C4AF, which is black (that is tetracalcium aluminoferrite which is made up of four molecules of lime to one of alumina and one of iron oxide during calcining in the kiln). How the iron compounds combine with the lime compounds is directly related to the chemical composition of the limestone and the degree of oxidation in the kiln. The C4AF has a variable chemistry due to manganese and other trace elements present and this can result in different shades of dark grey to black coloured compounds. Ordinary Portland cements have high levels of C4AF which, when mixed with the whitish compounds of calcium silicate to form clinker, give rise to the grey colours.

The cement colour that is produced from a factory is pretty consistent year after year and is a good basis to design concrete mixes for visual concrete. The only word of caution is this – if the cement factory changes its kiln it may have a marginal effect on the cement colour produced.

In a crude sense we can say that the iron oxide found in the argillaceous constituents (the clays and shales) imparts the black colour to C4AF. As the properties of shale and clay are unique to each region, it explains the variation in grey from one cement works to another. What is intriguing is that the dark grey limestone and dark grey shale quarried in Padeswood in North Wales produce the palest grey cement in the UK. The percentages of iron oxide compounds in Padeswood cement and in Ketton cement are about the same, yet Padeswood cement is much lighter. That can only mean that the C4AF compounds produced at Padeswood are not as black as the C4AF compounds produced at Ketton!

What about the fineness of the cement? The finer the particles, the denser the surface packing, therefore the more light it will reflect and the lighter the resulting tone will be. Generally, ordinary Portland (OP) will have the same degree of fineness. Rapid hardening cement, however, is a much finer material and should appear slightly paler in tone or lighter in colour than OP made at the same cement works. Such small differences in tone due to finer particle size may not always be perceived by the eye and are a secondary influence on the colour.

Table 2
Details of the typical chemical compounds of Castle cements and white cement (most values will 'float' by 3–4 per cent over the production period)

	Ketton OP	Padeswood OP	Ribblesdale OP	Aalborg white	CBR white
C3S	52	51	49	62	56
C2S	19	21	22	24	21
C3A	7	9	9	5	11
C4AF	8	7	7	1.5	1
SO_3	3.2	3.2	2.5	2.0	2.7

Typical tri-stimulus Y values

26.7	33.1	25.1	86	81.5

Comments: Padeswood – lightest OP; Ketton – mid-coloured; Ribblesdale – darkest OP; Aalborg white – bright white in powder form; CBR white (Belgium) a slight green tinge in the powder form.

Tri-stimulus Y values are a numeric value of the reflectivity of a sample of a material when measured by a specialist light meter – the higher the number, the more reflective (lighter coloured) the material is.

The important point to remember is that cement supplied from the same source will produce the same colour. Therefore, on large projects where a regular supply of concrete is required, the cement must come from a single source. It is best to work with local materials and the source of cement used by the ready mixed supplier to keep costs down and to ensure consistency of supply (see Table 2).

IMPORTED WHITE PORTLAND CEMENT

If a customer requires white cement it is imported from Denmark, Spain, Belgium or Greece into the UK and distributed through one of the various cement manufacturers. There is a premium for white cement as it costs more to produce.

It is a consistent uniform colour with the advantage of having good chemical resistance, low alkalinity, high strength and excellent long-term durability. In precast production it is essential for high quality finishes to replicate natural stone or light-coloured surface finishes. White cement seldom varies in colour from the same single source, but grey cement can sometimes, which can be disastrous to standard colours in high volume production such as precast paving slab manufacture. That is why grey cement, unless it is specially colour controlled, is not specified in precast production.

BLENDED OR COMBINATION CEMENTS (SUPPLIED BY THE READY MIXED INDUSTRY)

The ready mixed industry is classified as manufacturers of insitu concretes mixed at the batching plant when supplying a blended cement using GGBS or PFA as a partial replacement for OP.

Thirty years ago, GGBS was virtually unheard of in the UK construction industry. It was only available from a single source at Scunthorpe, with production and sales reaching a meagre 25,000 t per year. Today, the situation is very different and GGBS is now readily available throughout the UK from five strategically situated works – in north of England, the Midlands, Wales and London.

15

Pulverised fuel ash, produced from coal-burning power stations, has been used in concrete for many years. It was first discovered some 2,000 years ago by the Totonacas who lived in Mexico and made a lightweight concrete with fly ash. The basic properties of PFA are pozzolanic – it is a material that reacts with lime to form a hardened paste. In modern times the use of PFA was first pioneered for use in concrete in the 1930s in the USA. With the advent of coal-derived, steam-powered plants in the UK in the 1940s, PFA was introduced for dam construction in Scotland following successful research work carried out by the University of Glasgow. The variability of fly ash in terms of its fineness, carbon content and lack of quality control proved problematic until the 1970s, so its uptake was spasmodic. PFA now accounts for about 20 per cent of ready mixed concrete consumption where it is used as binder at a 30 per cent replacement. Currently, around 500,000 t of PFA per annum is used in ready mixed and precast concrete production.

This widening use of both PFA and GGBS in concrete has been primarily driven by its lower cost compared to PC and, to a lesser extent, by the benefits of reduced early-age temperature rise, greater resistance to alkali–silica reaction and resistance to chloride ingress and sulphate attack.

GROUND GRANULATED BLASTFURNACE SLAG (GGBS)

Transportable, finely-ground dry GGBS was first produced in the UK at Scunthorpe in the early 1960s and was initially used for construction projects at the steel works. Around 1968 the ready mixed concrete industry started to take an interest in the concept of mixer blending with OP cements, and GGBS was successfully used for the construction of the Anchor Project Steel Works in Scunthorpe. During 1972 an Agrément Certificate was awarded for GGBS, which recognised its use as a separate component for the production of concrete. A British Standard for GGBS (BS 6699 'Specification for ground granulated blastfurnace slag for use with Portland cement') was first published in 1986. A revised version, taking into account the European test methods for cement (EN 196), was published in 1992.

Manufacture

Slag is a by-product produced in the manufacture of pig iron for steelmaking. It is a blastfurnace waste consisting of lime, silica and alumina with a similar chemical composition to Portland cement. The raw materials going into the blastfurnaces are iron ore, limestone and coke; the products emerging are iron and blastfurnace slag. These two products separate naturally, the heavier iron sinks to the bottom of the furnace and is taken off. The slag leaves the blastfurnace as a molten liquid at a temperature of approximately 1,500 °C.

For use as a cement replacement, the slag has to be rapidly quenched so that it solidifies as a glass to prevent it crystallising. Rapid cooling by water results in fragmentation of the slag into a granulated material, which is then finely ground to a powder and sold as a bulk material just like cement.

GGBS has the same fineness as OP but a much lower early strength gain. It will reach the same long-term strength as OP provided it does not replace more than 50 per cent of the OP in the final mix. Above 70 per cent, GGBS may take a considerable time to harden, and when it reaches 100 per cent replacement it is unlikely to set.

GGBS will not set by itself if mixed with water, it has to be activated by the hydration products of the OP to kick-start its own hydration. Once started, the further hydration of GGBS does not need the liberation of calcium hydroxide compounds from the OP to keep the reaction going.

Blended GBBS, with OP as a 50 per cent replacement, has similar properties to OP with respect to fineness and soundness. It will lower the heat gain during hydration compared to a pure OP and is useful in mass concrete pours or thick wall sections where temperature rise has to be restricted. In cold temperatures the low heat gain, coupled with its moderate rate of early strength gain, can lead to potential frost damage and therefore extended curing times will be necessary before formwork can be removed. Its high sulphate resistance and lower permeability makes GBBS blended cement ideal in sea wall construction.

GGBS – creamy colour

The iron content of GGBS is very low (generally less than 1 percent, measured as iron oxide) so virtually no iron is incorporated into the chemical structure and the granulated slag is a pale yellow colour. In common with other materials it becomes lighter after grinding due to the finer particle size. A finer powder creates a more closely packed surface area for more light to be reflected straight back to the eye, rather than allowing the light to pass through the particles, which will absorb certain colours in the spectrum, before being reflected.

Blastfurnace slag is very consistent in chemical composition because iron blastfurnaces are extremely sensitive and need to be fed with a consistent mix of raw material. The blastfurnace engineers analyse the slag rather than the iron to control the process, so its colour is as uniform as Portland cement.

fig 15
Colour comparison of Portland and blended GGBS and PFA cements
top left: 100% PC
top middle: 50% PC with 50% GGBS
top right: 70% GGBS with 30% PC
bottom left: 100% SRPC
bottom middle: 70% PC with 30% PFA
bottom right: 100% White PC

PC = Portland cement; PFA = pulverised fuel ash; SRPC = sulphate-resisting Portland cement; GGBS = ground granulated blastfurnace slag

Concrete character

The very pale colour of GGBS when blended with OP cement at 50 per cent produces a pale grey concrete at a very competitive price and is often specified for architectural concrete. It is unlikely that the eye will discern small differences in the proportion of GBBS due to batching tolerances of say + or − 3 per cent, where one load can have 47 per cent GBBS and another 53 per cent. To be absolutely sure, however, it is best to conduct site trials.

There may be a tendency for a bluish tinge to appear on the hardened concrete surface. This coloration is due to small quantities of polysulphide in the material, which can form blue/green compounds on the surface. It has no effect on the properties of the concrete. Several factors have been found to increase the likelihood and intensity of the initial blue/green colouration, namely a high percentage of GGBS relative to Portland cement in the mix, the use of resin-faced plywood, polished steel or GRP (glass reinforced plastic) formwork, curing under water or polythene, extended time between casting and removal of formwork, and washing the surface with acid.

The blue/green colouration fades as a result of exposure to oxygen in the atmosphere and it usually disappears completely in a week or so. Although rare, there have been instances where it has still been apparent after several weeks or even, in extreme situations, months. Generally, this has been attributed to the access of oxygen to the surface being hindered by constant immersion in water, curing membranes, coatings or sealants to the concrete surface and wrapping with plastic sheets.

Bleed-water and revibration

The setting time with a GGBS concrete is slower than an OP mix, and often excess bleed-water can rise to the surface after compaction which can hinder surface finishing, trowelling and floating. If left unchecked it may discolour the top level of the concrete permanently in walls and columns. To overcome this risk, the top layer of the concrete may require revibration 1 to 2 hours after completion.

Higher formwork pressures

The slower setting rate of GGBS mixes does increase the formwork pressure gradients, not significantly but noticeably and these must be considered at the temporary works design stage in line with the in recommendations CIRIA Report 108 (see the Formwork section for more details).

PULVERISED FUEL ASH (PFA)

PFA is a by-product that is produced in the burning of powdered coal at power stations. The combustion of pulverised coal at high temperatures in power stations produces different types of ash. The 'fine' ash fraction is carried upwards with the flue gases and captured before reaching the atmosphere by highly efficient electrostatic precipitators. This material is known as pulverised fuel ash (PFA) or 'fly ash'. It is composed mainly of extremely fine, glassy spheres and looks similar to cement. The 'coarse' ash fraction falls into the grates below the boilers, where it is mixed with water and pumped to lagoons. This material, known as furnace bottom ash (FBA), has a gritty, sand-like texture and is not suitable for concrete. It is used as a simple fill material.

Fly ash is a pozzolanic material high in silica, which reacts with the lime set free during the setting of OP cement in exactly the same way as GGBS. It will slow the hydration process and the rate of early strength gain, at the same time beneficially reducing heat gain of the concrete. Its inclusion in certain proportions reduces the permeability of the concrete, increases its sulphate resistance and is ideal for mass concrete and for thick wall sections and deep foundations to reduce heat gain.

The fineness and carbon content of fly ash has to be strictly controlled to make it a suitable cement replacement, and only those materials that conform to EN450-1, S and N to Category B carbon content are suitable. The S designation signifies that the PFA is finer and has water-reducing properties; the N indicates that it does not necessarily have water-reducing properties. Such a PFA can be used as a cement replacement for up to 55 per cent of the OP, provided that the loss on ignition value of the carbon content does not exceed 7 per cent and the sulphate content expressed as SO_3 is below 3.0 per cent.

The amount of unburned coal present in the ash can vary according to the type of coal being burned and the design of the furnace. The resident times in the furnace are only 2–4 seconds and some coals contain material that will not burn out so rapidly, the so-called 'inertites'. With the introduction of low nitrous oxide burners to reduce these toxic emissions, carbon contents of fly ash have increased in recent years.

Colour consistency

PFA is dark grey in colour, the shade varying depending on the source of the coal being burned and the process plant. The dark colour results from a combination of iron compounds present and carbon residues left after the coal is burned. As colour control is generally not an important criterion for sales of PFA, the product is only suitable for architectural concrete mixes when special measures are put in place to control fly ash colour. This may include instances where the requisite amount of product can be blended and stored to satisfy colour consistency.

fig 16
Cooling towers of a coal-burning power station

fig 17
PFA transporter lorry

ARCHITECTURAL INSITU CONCRETE - TECHNOLOGY

AGGREGATES

There is a wide variety of naturally occurring aggregates suitable for the production of concrete and a growing stockpile of secondary recycled crushed concrete and waste glass aggregates that is also suitable. Man-made synthetic materials such as sintered pulverised fuel, commercially known as Lytag, and expanded shales manufactured in Europe make excellent lightweight aggregates for concrete. It does not follow that all these materials will produce a concrete suitable for a fine finish.

The lightweight aggregates are unsuitable for exposed concrete finishes due to their porosity and lack of mechanical robustness. They also have a tendency to rise to the surface during compaction. Moreover, there is not a great deal of information about their use in fine finishes, except in the Art House where their use was specified for a single skin external wall designed to overcome the risk of cold bridging.

Recycled aggregates are defined as materials resulting from the crushing of demolition waste and reprocessing of construction waste or damaged materials on site. Recycled material is predominantly crushed concrete and brick masonry but may also contain significant quantities of deleterious materials, such as wood, metal reinforcement, plastic and trace elements, that are harmful to concrete. Crushers, separators and screens are used to remove the reinforcement, crush the concrete to size and grade the aggregates. The handling and processing costs of recycled

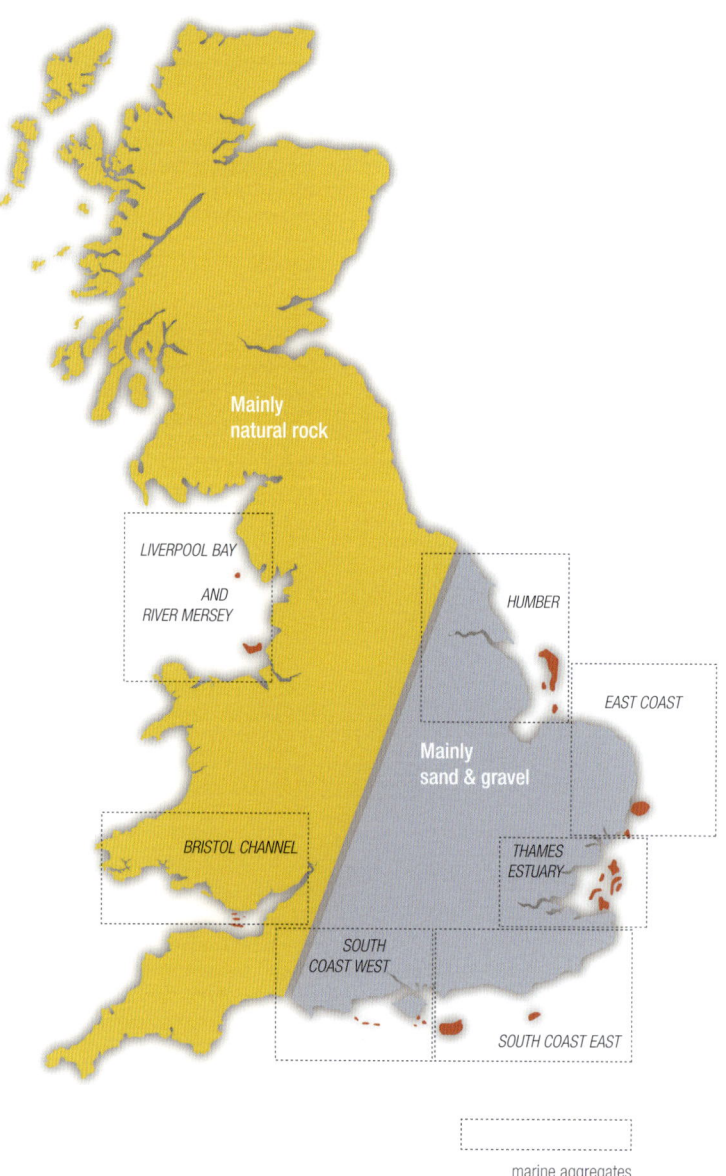

fig 18
Sources of land and marine based aggregates

Clee Hill 10mm gravel

Balidon 10mm gravel

Solent 10mm marine gravel

Criggean fines

Harden red sand

Clee Hill sand

Hoveringham sand

Horcroft sand

Solent marine sand

Longcliffe sand

fig 19
top: Marine-dredged aggregates operation, Portslade
bottom: Typical natural aggregates colours

fig 20
Classification of aggregates

aggregates will tend to make the material more expensive than locally available natural aggregates. The same can be said of waste glass aggregate, with regard to its long-term durability and the risk of alkali–silica reaction.

In this overview, therefore, only commonly available natural aggregates or crushed rock aggregates that occur as land-based materials or estuarine deposits will be described and their character and physical properties noted for colour, shape, texture and soundness. There will be a short note on lightweight aggregate material.

CLASSIFICATION

Natural aggregates can be divided into three main classifications:

> land-based natural sands and gravels

> marine aggregates

> crushed rock

Natural sand and gravels are found in gravel and sand pits close to rivers or as estuarine or coastal deposits in the sea, which have to be dredged. Sea dredged aggregates are referred to as marine aggregates. Crushed rock aggregates are sourced from inland stone quarries. It is a not a matter of choice between a crushed rock and a natural gravel for a site, but what aggregates are locally available because of the site's geological location.

Mainland Britain can be sliced into two geological halves by a diagonal line running from the Humber in the north, down to the Dorset coast in the south (see page 14). West of the line material is largely rock-based, east of the line it is natural sands and gravels. Estuarine deposits along the south and east coast are mainly flints and cherts, while those on the west and northern coastlines are derived from weathered rock which was transported there during the ice age.

The suitability of an aggregate for concrete is related to its geological history: how it was formed and, for a sand or gravel, how it was subsequently modified by weathering or subterranean changes. For a crushed material the decision process is similar and will depend on the condition of the rock and how it was formed.

LAND-BASED SAND AND GRAVELS

The most widely used and economical aggregates are land-based sands and gravels. Sand and gravel deposits occur as superficial drifts laid down by rivers or as glacial spread left behind when the ice sheet melted.

Sand and gravel were formed by the erosion of rocks and the particles were then transported and deposited by wind, water or ice. Sand quarries are usually shallow, sometimes only 5 or 6 m deep. Operations are likely to be shorter term than for a rock quarry with progressive restoration normally following closely behind extraction.

River deposits are generally the most satisfactory for concrete production as they have a consistent grading. As a result of scouring action by the river, the fragments are usually rounded and the erosion leads to the removal and elimination of weaker particles. River gravels and sands are fairly uniform and clean. The main deposits occur in the Thames and Trent valleys and their tributary rivers and valleys like the Stort and the Lea.

Glacial deposits, although widely spread across the country, are not always economically viable. Major extraction of this type of gravel and sand can be found in the Vale of St Albans, East Anglia, The Midlands and parts of the North-East of England. The Lancashire and Cheshire deposits are mainly sands, while in Clwyd they are mainly gravel.

MARINE AGGREGATES

Gravels and sands are dredged from river estuaries and offshore coastal regions adjacent to the mouths of the major rivers. Dredging operations are located in the Thames Estuary area, in the Solent, off the shingle coastline of East Anglia, the Humber and Bristol Channels, Liverpool Bay and the River Mersey estuary.

Marine aggregates are generally well rounded and smooth, just like land-based sand and gravel. They differ in their chloride content, which is high as they are in sea water, and in organic matter that may encrust the surface. The sands tend to be a bit on the coarse side and all marine aggregates have to be thoroughly washed and cleaned to ensure they are free of chlorides, dust, and clay and silt particles.

Extraction and production

Land-based sand and gravels are obtained by digging pits. Coastal gravel and sands are extracted by dredging the estuarine beds. Land-based pits are either wet or dry, according to the water level in the seam. The deposits are usually covered by overburden which has to be removed by mechanical shovels and scrapers. The sand or gravel seams are removed by mechanical shovels in stable-sided dry pits or by crane and grab in wet pits. The mixed gravel and sand is transported from the pit to the screening plant by conveyor belt or by trucks, and sometimes by pipeline when the gravel is raised by suction pump. Conveyor belts are used to raise the material from the receiving hoppers at ground level to the washing and screening plant.

A separating screen splits the sand from the gravel. The gravel is then fed to a refined screening plant which separates the gravel into single sized 20 mm, 10 mm and 5 mm stockpiles. Oversized aggregates are sometimes reduced at a crusher plant and then returned for screening into single sizes. The sand is washed and sieved using classifiers to meet grading standards for building sand and concreting sands.

CRUSHED ROCK AGGREGATES

Three main types of rock are used to produce crushed rock aggregates. Their classification, petrographic examination and prehistory will be critical in deciding whether or nor they are suitable for making concrete:

> *Igneous* – formed by molten lava flow. This group includes the granites, basalts, dolerites and gabbros. The granites and basalts are hard, dense materials and make excellent aggregates.

> *Sedimentary* – created by the settlement and cementing of particles that were deposited on the sea bed and lake bottoms millions of years ago. They include limestone, sandstone and gritstone. The harder and denser types of the carboniferous limestone found in the Mendips and Derbyshire are suitable for concrete. Ferruginous and siliceous sandstones are hard and dense and are suitable as aggregates.

> *Metamorphic* – formed by transformation of an existing rock material by heat and pressure into another type. This rock can be very variable in character. In this group marble and quartzite are usually dense and adequately tough for providing good aggregates.

The quarries supplying crushed rock aggregates grade and classify the materials as suitable or unsuitable for concrete aggregates in accordance with BS 882/BS EN 12620.

Production of crushed rock aggregates

The rock face in the stone quarry is blasted and the fallen stone lumps conveyed to the crushing plant by large tipper trucks which are loaded by a mechanical shovel.

The primary crushing plant reduces the lumps into pieces of 75 mm diameter or smaller. The pieces are fed into jaw crushers and disc crushers for further reduction and grading into 20 mm, 10 mm and 5 mm stockpiles as they are passed over a vibrating screen.

Sand is produced by rod mills or roll crushers where the crushed coarse aggregate material is fed between toothed and serrated rollers and crushed to a sand consistency and then graded by screening and separating into building and concrete sands.

LIGHTWEIGHT AGGREGATE MATERIALS

A lightweight aggregate concrete is a structural dream as a concept, but the practicality of its varied handling characteristics, its relatively low strength, elastic modulus and shear capacity have relegated it to use in concrete block making and to composite topping screeds.

Lightweight concrete is defined as a concrete with an oven-dry density of less than 2,000 kg/m^3. Any concrete with a density greater than that is assumed to be normal weight.

It is common knowledge that lightweight aggregate concrete using pelletised PFA, commercially known as Lytag, has achieved concrete strengths of 40 N and more. However, there is little to be gained by designing lightweight concrete suspended floors with high yield steel bars. Eurocode 2 imposes reduction factors on slenderness ratio, deflection and shear for lightweight concrete structures.

While lightweight concrete has superior fire resistance to normal concrete, it has a lower classification on durability so that the cover has to be increased.

The net result of all this is that the reduction in dead load is negated by the increased section depth necessary to overcome the code requirements on shear, deflection and cover. It is no wonder that so few structures are built using lightweight reinforced concrete.

Now, if we consider the use of lightweight concrete with prestressing for a suspended floor, the arguments are very persuasive. Prestressed lightweight concrete does not require a reduction for elastic modulus, nor for torsion. Nor is there a shear capacity reduction factor for increasing the slab depth.

For wall and column construction a normally reinforced solution will be adequate. The ideal mix for a fine finish with such a concrete can only be determined from trial mixes and by casting full-scale trial panels.

Lytag

This material is derived from pulverised fuel ash (PFA), a by-product from the generation of electricity at coal-burning power stations. The material is pelletised and sintered at 1,200 °C and then cooled to produce hard, spherical nodules with a 40 per cent void ratio and a dry density of around 800 kg/m^3. The main constituents of the material are silica and alumna and it is brown in colour. The Lytag is separated into three grades: fine 0.5–4 mm, medium 4–8 mm and coarse 4–12 mm.

It is usual to specify a Lytag coarse aggregate, 4–12 mm size, with a natural sand for a better surface finish and to make it easier for the concrete to be pumped. The Lytag is very absorbent, therefore additional water has to be added to the mix to compensate for this. Alternatively, the aggregates can be pre-soaked, which requires special storage facilities. Lytag requires about 12–15 per cent of its dry weight in water to satisfy the absorption demand. If this is not added then the available free water in the mix will be absorbed by the Lytag and the concrete will become harsh and unworkable.

It is also usual to blend an admixture with the concrete to prevent further high absorption into the pores, especially when pumping the concrete. The special admixture coats the Lytag particles, and is dispersed through the mix to improve workability and maintain a good consistency.

Loss of workability in transit can also be a problem. Should the workability be too low when discharged from the truck mixer, it is important to add water to the mix to restore the flow, provided that the poor workability is not due to the setting of the cement, which can occur in hot weather. It is therefore vital to agree an acceptance procedure for the controlled addition of water on site, including re-mixing and re-testing for flow.

For these reasons, a Lytag concrete will not give an even surface colour for every load using smooth, form-faced surfaces. A better approach, providing a more satisfactory result, is to detail a textured board-marked concrete. The porosity and voided nature of the aggregate make it unsuitable for an exposed aggregate finish.

The concrete should be vibrated as for normal concrete and if a smooth finish is required, to the top of walls, for example, a second pass will be necessary shortly after initial placement.

Expanded shale

Shale is a precious gift of nature, originating many millions of years ago as 'lias' in the organic sediments of the Jurassic seas. This compacted fossil rock is extracted by open pit mining and then transferred by tipper truck to the processing plant. It is crushed, dried and milled into a powder of less than 250 microns. The shale powder is then conditioned, mixed with water, pelletised into small, round granules and then coated with a fine limestone dust.

The pellets are graded into sizes and then conveyed to a three-stage rotary kiln. This technique enables the amount of expansion of the pellets to be controlled and the density and size of the granules to be engineered and graded precisely. The coating of lime powder increases the amount of surface vitrification, forming a dense, impermeable coating when the pellets are heated to 1,200 °C. The product is known as Liapor and is manufactured in Eastern Europe.

Liapor is like no other man-made aggregate. It is lightweight shale aggregate with an aerated core and a high compressive strength that is available from sand, 4–8 mm and 8–16 mm diameter aggregate. The density can vary from 325 to 800 kg/m^3 according to application and usage. It has an ideal spherical shape with a closed surface skin that has a slightly roughened texture. It is completely frost resistant and can be stored in the open under any weather conditions. Its impervious outer skin means that water absorption is not high and it does not need special admixtures or vacuum soaking before use. In fact, it can be treated just like a good quality natural aggregate.

Concrete using a Liapor grade 8 aggregate, comprising 4–12 mm aggregates with a bulk density of 800 kg/m^3 and combined with natural sand has achieved compressive strength of 80 N and a design density of 1,800 kg/m³. For lower concrete density, the natural sand is replaced with Liapor sand. A concrete density of 1,400 kg/m³ can be achieved with a Liapor grade 6 coarse aggregate and a Liapor sand, giving a 40 N compressive strength.

Clearly, high-strength, low-density concrete can be fine tuned to suit the requirements of a particular project. Liapor's scarcity in the UK means that it will not be readily available at local ready mixed plants and it will be expensive to import. However, the lower U values that can be engineered with such concrete make it possible to design single thickness external walls to meet Part L of the Building Regulations.

AGGREGATE SELECTION

The shape, surface smoothness and porosity of the coarse aggregates will affect the workability and the surface finish of the concrete. The colour of the fines content in the sand will influence the grey tint of cement as well as affecting the cohesiveness, stickiness and workability of the concrete.

The colour of sand can vary from a pale yellow to a dark brown, from mid-grey to pale grey, from a light pink to a mid-red, depending on its geological history and regional location. Coarse aggregates have an even wider variety of colours and contrast, for example with the granites there is white, pale grey, pink, red, dark grey; with basalts dark grey and grey-green; with limestone brown and grey; sandstone yellow, brown and pink and in gravels there can be brown, yellow and off-white, and the range goes on.

Ready mixed plants will have a limited range of aggregate materials. It is best, on economic grounds, to use the sand colours that are available in combination with a cement type to create a greyish or whitish sand-coloured tint to the surface. By abrasion of the surface, the coarse aggregate colour can be exposed to give a wider choice of tone and texture.

To create special concrete colours, for example blues, green and other unnatural earth colours, synthetic pigments are used in the mix. For further information refer to the section under the heading 'Pigments' (page 36).

MATERIAL QUALITY

All aggregates must have certain material properties apart from their size and grading consistency to meet standards for concrete.

Flakiness index

The flakiness index indicates whether or not there are too many flat, hollow aggregates in the material. This can lead to harsh, stony concrete mixes, poor finishes and lower concrete strength, none of which is desirable. The flakiness index of an aggregate is defined as the percentage by mass of particles in a sample of single sized aggregate whose least dimension (thickness) is less than 0.6 times their mean dimension. Slotted sieves are used to separate the flaky particles. The flakiness index of the aggregate will affect the stability and workability of mixtures and surface dressing treatments.

Mechanical properties

The mechanical properties of aggregates are determined by a test called the 10 per cent fines value, which shall not be less than 50 kN. The method consists of placing a quantity of the aggregate that has been graded between stated sieves in a cylinder, inserting a plunger and applying compression load to cause penetration of the plunger to set depth. The aggregate is then sieved and the 10 per cent value calculated from a formula given in the standard.

Harmful constituents

The aggregates must be free of any harmful impurities which may impair the appearance and integrity of the concrete. Certain sulphide minerals, such as pyrites found in marine aggregates, oxidise on the surface and cause unsightly rust stains and loss of strength if present in sufficient quantities. Coal and coal residues are unacceptable because of their black appearance on the surface and low mechanical strength.

The presence of lots of seashells and excessive flat and hollow coarse 'flaky' aggregates will reduce the workability of the mix and increase the harshness, which is not desirable in fine finishes.

Aggregates should have a low drying shrinkage, and those liable to suffer from the action of frost should not be used for concrete that may be exposed to freezing and thawing conditions.

Marine aggregates and some inland aggregates contain chlorides and require careful selection and efficient washing to achieve the 0.1 per cent chloride ion limit for concrete given in the BS. Whenever there is chloride in concrete there is an increased risk of corrosion of embedded metal. The higher the chloride content, the higher the curing temperature or, if subsequently exposed to warm moist conditions, the greater the risk of corrosion. Chloride may also adversely affect the sulphate resistance of concrete. All constituents may contain chlorides (for aggregates see BS 882) and concrete may be contaminated by chlorides from highway de-icing and by airborne salt spray either from vehicles or the sea. Calcium chloride and chloride-based admixtures should never be added to reinforced concrete, prestressed concrete and concrete containing embedded metal. It is recommended that the total chloride content of the concrete mix arising from the aggregate, together with that from any admixtures and any other source, should not exceed the limits expressed as a percentage relationship between chloride ion and mass of cement in the mix, in accordance with BS 882/ BS EN 12620.

Generally, the sodium chloride content of the fine and coarse aggregate shall not exceed, respectively, 0.1 and 0.03 per cent by weight for dry aggregates.

ALKALI–SILICA REACTION (ASR)

ASR may occur when certain forms of silica present in some sands and some coarse aggregates react with the alkalis present in the cement of a concrete mix to produce a gel. If sufficient gel is formed under certain conditions this can cause expansion and cracking of the hardened concrete. The expansion is only likely to occur if all three of the following conditions prevail:

> the equivalent sodium and oxide content in the cement exceeds 0.6 per cent

> the aggregate contains reactive silica

> there is continuous wetting and drying of the concrete face.

Ready mixed concrete producers are required to adopt procedures and use materials which will minimise the risk of ASR. The concrete mix design certificate will state these limits and how they comply. Generally, the risk is controlled by limiting the total alkali value of the concrete below levels set out in BRE Digest 330 or Concrete Society Technical Report No 30.

AGGREGATES FOR GOOD FINISHES

Consistency and uniformity of surface colour is important in designing a concrete mix. Generally, if the surface area of the aggregates remains the same for each batch of concrete that is mixed, whether it is wet batched or dry batched, the water demand for workability to overcome interparticle friction will be the same and so will the colour. When designing a concrete mix to give a high standard of finish, a different approach is required from designing to simply meet a specified compressive strength. The cement and sand together have to form a rich mortar paste which will coat and combine with the coarse aggregate into a cohesive mix. It is the richness and cohesiveness of the mortar that determines the quality of the concrete.

The selection of sand and coarse aggregates and the way in which they combine in the concrete are therefore critical in specifying a concrete to give a fine finish.

SELECTION OF SANDS OR FINE AGGREGATES

Sands are classified into three main types, based on the percentage passing the 600 micron sieve: coarse, medium and fine. The sand must not be too fine, as too much very fine material smaller than 150 microns will increase the water demand and create a sticky concrete, which increases the risk of blemishes and lowers concrete strength. Too coarse a sand will not result in a uniform appearance of the surface and can lead to greater surface voids. Well graded or medium sand generates a tight fit between the larger and smaller particles to ensure an even distribution in the matrix. This maximises the effectiveness of the cement binder and enables the particles to interlock, reducing the voids content, improving workability and giving a better finish. Sometimes a fine sand may be acceptable but the total sand content in the mix may have to be restricted. A coarse sand may be blended with a fine dust to create a well graded sand but the blended sand may not be uniform in colour. Local knowledge and experience of materials at the ready mixed plant will guard against the use of an unsuitable sand.

The dust content of the sand, anything smaller than, say, 63 microns, which is the same size as cement particles, will pigment the concrete. It is important that the percentage of this dust in the sand does not vary significantly. Variations in darker sand colours may affect the concrete tint much more than paler sand colours. Some trials with yellow-brown sand from the Thames Valley mixed with a mid-grey Northfleet cement suggest that variation of 6–9 per cent in the 63 micron content was not noticeable. However, site trials should be carried out to determine whether the variation in the dust content of a particular sand and cement combination is acceptable.

The sand content of the mix should not exceed twice the cement content of the mix to reduce the stickiness of the concrete and the adhesion of the wet concrete to the form face. This will allow air bubbles to escape more easily on the form face during compaction and reduce the risk of large blowholes forming.

COARSE AGGREGATE

Coarse aggregates vary between angular and rounded. The angular particles are the crushed rock aggregates, the rounded ones are natural gravels. Angular aggregates mechanically interlock better than rounded aggregates and result in higher strength concrete. However, the rough and variable surfaces of angular aggregates can cause increases and decreases in the water demand from one batch to the next and this can affect the surface colour. A high percentage of aggregates less than 10 mm will worsen this condition, reducing the aperture through which the sand and cement must pass and increasing the surface area of particles that the sand and cement have to coat. This leads to stony, dry patches appearing on the surface and the creation of surface voids, a phenomenon known as aggregate bridging. This can be avoided by reducing, or preferably eliminating, 5 and 10 mm aggregates in the mix, especially as the sand content has been reduced to reduce the stickiness of the mix.

Rounded gravels have a more constant surface area and are therefore less likely to be variable between batches, making them best suited for fine fishes. To reduce the risk of aggregate bridging it is advisable to specify a single sized 20 mm aggregate if this is available. Failing that, the next best approach is to limit the stone passing the 10 mm sieve to not more than 20 per cent.

SAND/COARSE AGGREGATE RATIO

When the sand and coarse aggregate are combined with the cement and water, the mix will start to flow and behave as one material. The concrete when mixed must remain homogenous and not segregate or separate when handled or placed into formwork. It should behave like a consistent porridge and have cohesiveness. To achieve this the total aggregate (sand plus stone) to cement ratio should not exceed 6:1. If it goes higher than 6 there is the risk of segregation in handling, with the water rising to the top and coarse aggregates settling to the bottom, resulting in a poor surface finish when compacted.

Water, admixture and workability are covered under separate headings.

fig 21
Separation of aggregates into stockpiles

CONCRETE PRODUCTION

Ready mixed concrete is the universal way in which concrete today is delivered to construction sites. It is a fast, efficient, same-day, doorstep delivery service that can blend a bespoke, made-to-measure or standardised mix of liquid rock to satisfy a customer's every requirement and match a 'Heinz' variety of material specifications. You can't say that of many man-made products in the world today. But ready mixed concrete has only been around for 60 years in the UK. Back in the 1930s, concrete was batched on site using simple mechanical mixers and material handling plant, with the aggregates stockpiled in the open and cement bags stored under cover. Sometimes the material would be mixed by hard labour and the shovels of a dozen navvies, and moved into position using wheelbarrows, hoppers and skips. A modern site batching plant today is a much more sophisticated process and is likely to cost over £500,000 to set up. It takes up valuable land space and that makes it impractical and far too expensive for construction sites in built-up areas.

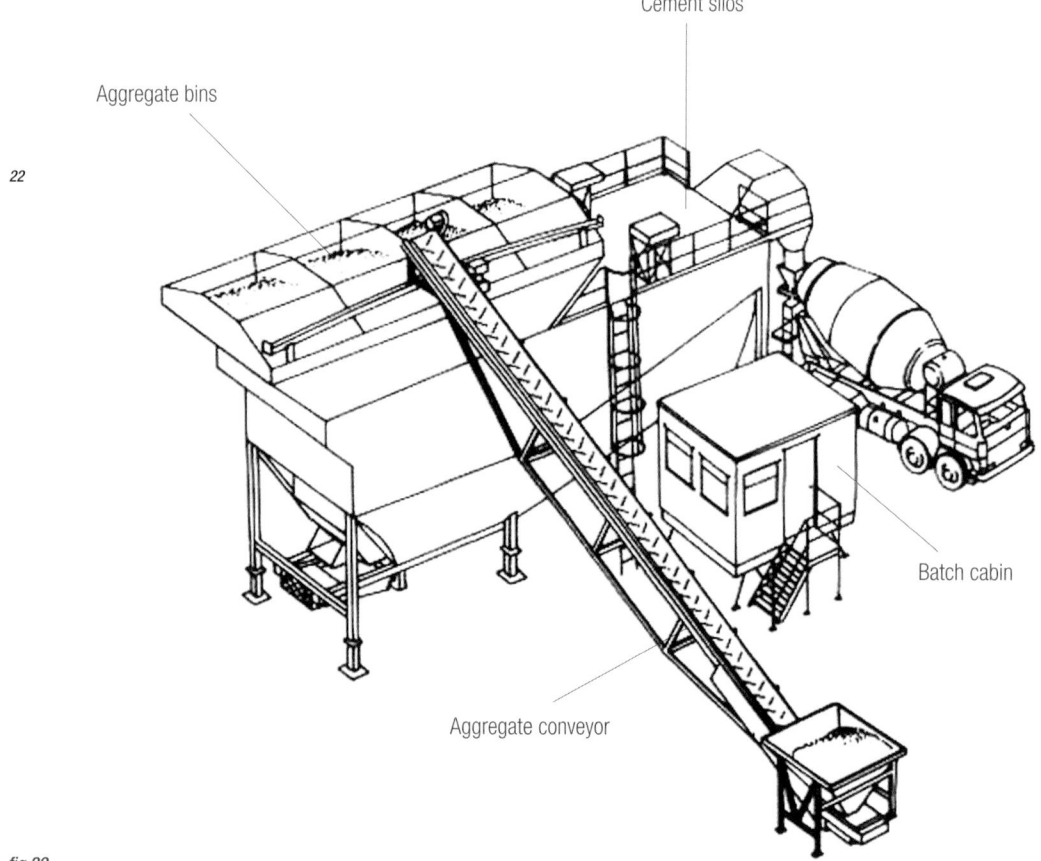

fig 22
Typical dry batch ready mixed plant

We are forever in the debt of Karl Ammentorp from Denmark who, in 1931, erected the first ready mixed plant in the UK at Bedfont near Heathrow Airport. Today the annual production figure for ready mixed concrete is approximately 25 million m^3, produced by around 1,200 ready mixed plants spread across the UK.

READY MIXED CONCRETE

Just thumbing through the telephone directory reveals the names of the local ready mixed companies. They are as common as builder's merchants and garden centres. The type of plant and equipment that is used to batch concrete will be dictated by the size of the local market, peak volume demand within the catchment it serves and when the plant was built. A concrete plant can supply sites within a radius of 5–8 miles, but there are exceptions. In a major city, for example, the radial distance could be as short as 3 miles due to traffic congestion and journey times. In urban areas, with lower populations and less congested roads, that could increase to 20 miles.

Most modern batching plants supplying large volumes of concrete in and around city centre sites will have sophisticated computer controlled, wet mix or wet batch production systems with enclosed cement silos and aggregate hoppers to minimise dust nuisance and noise levels. For smaller, and perhaps older, plants located in towns and rural areas a dry batch plant may be in operation, with open storage areas for the stockpiling of aggregates. It is important to understand the differences between these two operations, and the control and influence they have on the quality and consistency of the concrete supplied to a project. Wet mix systems premix the concrete before it is discharged into the truck mixer. In dry batching the materials are placed dry into the truck mixer drum, where they are mixed by rotation of the tilting drum as the required water and admixture dosage is added to the dry mix.

WET MIX OR WET BATCH PLANTS

At such plants the raw materials are stored in separate bins and silos. Cement and cement replacement products are kept in silos. These are the tall, cylindrical containers, usually visible from the roadside. There will be at least two silos, one carrying, for example ordinary Portland cement and the other carrying a cement replacement, for example ground granulated blastfurnace slag (GGBS) or pulverised fuel ash (PFA). The cement silos are filled by blowing the cement dust under pressure into the silo from a nozzle fitted to the cement truck, just like filling the tank of a car with fuel, except that this fuel is a very fine powder. The cement in the silo is fed to the wet mixing chamber when required in premeasured quantities using an internal screw pump and weighing flume.

The bins contain the sand and coarse aggregates. The sand may be a crushed rock that has been brought in by rail or a natural estuarine sand from a nearby sand pit which will have been washed and graded. The coarse aggregate may also be a crushed rock or gravel won from a nearby pit. It is more usual to find gravel and natural sand combinations or all crushed rock materials at a batching plant.

If a lightweight concrete is being supplied then a bin carrying pelletised PFA, commercially known as Lytag – a by-product of coal-burning power stations – will be available.

To fill the aggregate bins, supply lorries tip their loads directly into receiving hoppers installed near ground level or drop them onto open stockpiles. An enclosed conveyor belt transports the aggregates from the receiving hopper to the top of the aggregate storage bins. Loose material placed in stockpiles on the ground is shovelled into hoppers to be conveyed to storage bins later.

At a typical plant there will be a bin for the sand, for a graded 5–20 mm coarse aggregate and, perhaps, a lightweight aggregate. In London, for example, a modern wet mix plant may store the coarse aggregates in single sized bins – 20 mm, 10 mm and 5 mm – to facilitate better quality control and consistency of the concrete. The advantages of this practice are that it can help to reduce the cement content of a mix and the cost of the concrete by lowering the standard deviation of the concrete batched from the plant.

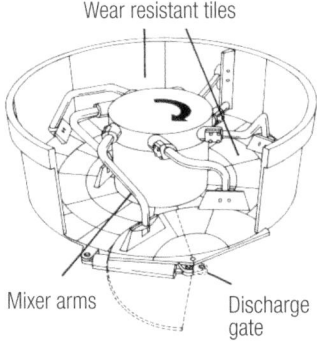
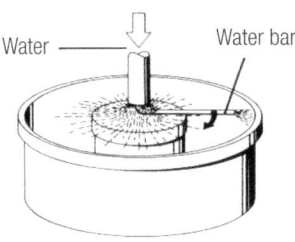

23

fig 23
Wet batch mixing hopper

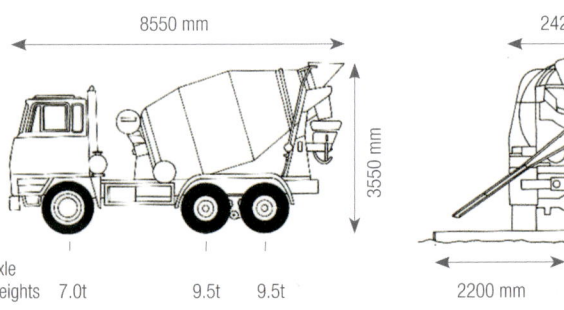
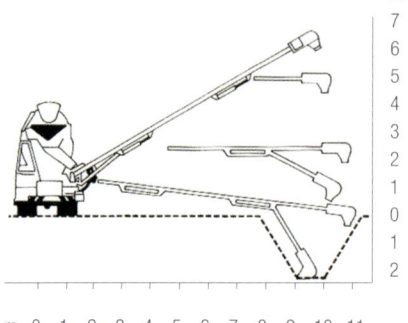

Put simply, better control by wet batching in mixing and proportioning the concrete lowers the design margin of the concrete to achieve the characteristic compressive strength of the concrete. The target mean strength of the mix is the specified characteristic strength of the concrete plus the design margin. The design margin ensures that the actual concrete's strength complies with the 95 per cent confidence limit that the concrete strength will be greater than the characteristic strength.

The design margin is the standard deviation of concrete batched from the plant multiplied by a constant 1.64. The smaller the standard deviation, the smaller the design margin and the closer the target mean strength is to the specified characteristic strength. Better control of the material properties and better accuracy of weigh batching reduces the standard deviation.

The usual mixer for wet batching is a pan mixer with either a 2 or 3 m³ capacity. The aggregates and cements are carefully weighed in their correct proportions and discharged dry into the pan mixer then mixed for a short time. The measured amount of water and admixture dosage is introduced and the pan rotated for a pre-set time of 30–60 seconds, depending on the character of the mix. When the concrete is thoroughly mixed, the gate at the base of the pan is opened and the concrete is fed into a waiting truck. Further batches are mixed until the required volume for the concrete truck has been reached.

DRY BATCH PLANT

The materials are stored in bins and silos as in a wet batch plant and they are discharged dry into the tilting drum of the truck mixer in a set sequence. The sand first, then half the cement and cement replacement, then all the coarse aggregates and then the remainder of the cement and cement replacement.

The concrete truck has a tilting drum mixer which can mix up to 8 m³ of concrete at a time. The water and admixture are added directly from the batching plant to the dry materials and the drum rotated at 80–100 revolutions a minute for 8–10 minutes to ensure thorough mixing in order to achieve a uniform concrete throughout the load. If the concrete is not thoroughly mixed there is the risk that the first batch discharged may be drier than the latter part of the load. This can have an impact on the final concrete colour as there may be marked changes in the water/cement ratio.

In the past, certain makes of concrete truck mixers were designed with mixing paddles in the tilting drums. However, this only agitated the concrete and did not fully mix the materials. They may be still in use today and are perfectly adequate if they are transporting wet mix concrete. It is difficult for the project architect, the site engineer, clerk of works and contractor to know the mixing efficiency of a truck mixer.

The consistency, flow and uniformity of the concrete throughout the tilting drum are crucially important in architectural

concrete. It is for these reasons that wet mixing is the preferred method of ready mixed production, but this type of plant is not always available

SELF-COMPACTING CONCRETE (SCC)

This is special concrete that has the advantage of being able to be placed into formwork using tremie pipe or flat hose without the need for compaction. It is not commonly available at many ready mixed plants in the country. It requires a wet batch plant, very careful control and monitoring of the mix ingredients, a concrete technician to test its fresh properties and a locally available sand that has a high percentage of very fine or powder material. It is also more expensive than ordinary concrete as the cement content can be as high as 400–500 kg/m³.

Structural concrete relies on full compaction after placing to achieve

fig 24
Typical concrete truck mixer vital statistics

fig 25
Truck mixer and concrete batching plant

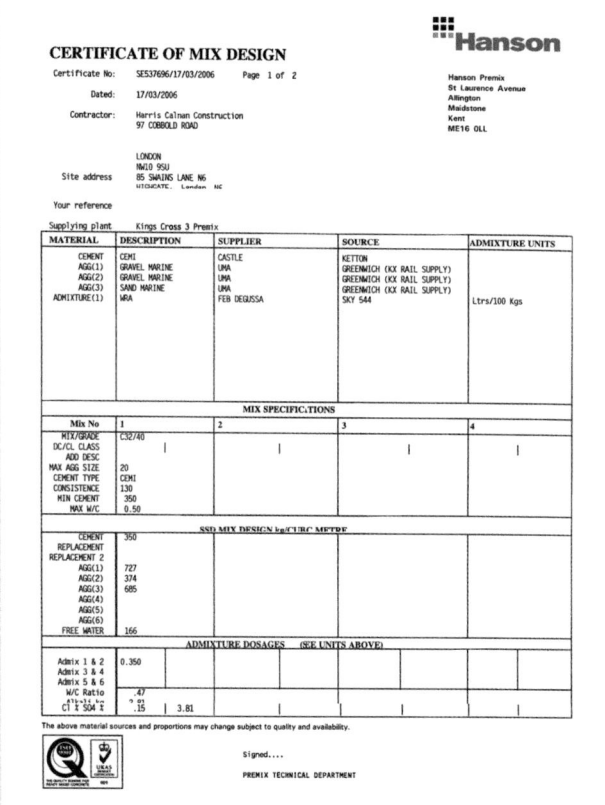

fig 26
Sand grading certificate

fig 27
Concrete mix design certificate

the required strength and durability. By compaction the wet concrete is consolidated, voids are removed to below a critical percentage and the finished concrete surface is left free from honeycombing and pronounced blowholes. Voids in the concrete are created by pockets of air that become entrapped between successive layers of placed concrete and by those formed during the drying of the excess water in the mix that has been added for workability. Self-compacting, high-performance concrete will reduce and perhaps eliminate air pockets if the concrete is tremied to the bottom of the formwork and not allowed to free fall more than 0.5–1 m during placement. It will minimise air voids induced during the mixing process and those formed because of the excess water required for workability.

VISCOSITY AGENTS

Fresh concrete is susceptible to segregation, because it is a composite material consisting of ingredients with different sizes and specific gravities. By introducing a viscosity agent into the mix, the viscosity of the paste can be increased effectively to inhibit segregation. For cohesive underwater concrete, segregation is inhibited by a high dosage of viscosity agent. However, in air a highly viscous concrete which is sticky may not release any air entrapped during placement and it may not easily pass through spaces congested with reinforcing bars. Consequently, SCC must have a reduced viscosity. The balance between the amount of workability admixture and viscosity agent is important for self-compaction, because the admixture and viscosity agent are trying to negate the effect of one another. By modifying the coarse aggregate volume and mix proportions, further improvement in self-compaction can be achieved without the need for viscosity agents. This has been successfully achieved by Lafarge in recent years.

LIMITING COARSE AGGREGATE VOLUME

In conventional concrete, the larger the proportion of coarse aggregates of large diameter, the better the mechanical properties of the hardened concrete and the better the surface finish, assuming the total amount of aggregate is the same. However, for fresh concrete to pass around obstacles and through reinforcement, it is better to increase the proportion of fine aggregate and to reduce the coarse aggregate size. For practical necessity, the proportion of fine to coarse aggregate in SCC is increased; as a result it does suffer from a drop in ultimate strength. Also, with SCC, if the coarse aggregates exceed a certain limit, there is greater contact between the larger particles which increases interlocking and the risk of blockages on passing through spaces between reinforcing bars. The possibility of interlocking is negligible if the coarse aggregate fraction is lower than 50 per cent of the total mix, provided that adequate mortar is used. Smooth, rounded river gravels are generally preferred because their use permits a larger coarse aggregate volume and improves workability in comparison with angular or rough textured gravels.

FINE AGGREGATE

Fine aggregate in SCC is defined as particles that are larger than 90 microns; anything smaller is defined as a powder. The amount of water confined by the fine aggregate is almost proportional to the volume of fine aggregate, so long as the fine aggregate proportion is around 20 per cent.

CEMENT AND POWDER MATERIALS

Selection of cement content and filler material, both defined as powder material, is critical because the properties affect self-compaction and govern the quality of the hardened concrete. SCC mixes contain higher than normal proportions of very fine material: the cement content can be as high as 500 kg/m^3 and in addition there is a PFA, GGBS or limestone powder filler and the very fine material in the sand. There is an optimum water/powder ratio for imparting a viscosity to the mortar paste that is suitable for self-compacting concrete.

When the volume of fine aggregate has been specified, then the volumetric water/powder ratio and dosage of superplasticiser can be determined.

HARDENED SCC

Properties of hardened SCC do not differ significantly from those of ordinary concrete of a similar basic composition. So, for compressive strength compliance, the standard concrete cube test will be adequate. However, results of verification testing of bond strength between concrete and reinforcement carried out in the UK suggest an improvement over ordinary concrete of the same strength. Research on drying shrinkage and creep suggests that there are no abnormal increases. It is highly unlikely that the Japanese civil engineering industry would be specifying SCC for major bridge structures if the properties of SCC were not predictable, consistent and within acceptable limits.

PROPERTIES OF FRESH SCC

The behaviour and characteristics of fresh SCC put it well outside the scope of current standard tests for workability and other properties of fresh concrete. In essence, SCC must have the following characteristics in its fresh state:

> flow and filling ability: it must be able to flow into all the spaces within the formwork under its own weight

> passing ability: it must flow through the small spaces between reinforcing bars under its own weight without blockages, and

> resistance to segregation: it must fulfil the requirements of (a) and (b) without segregating.

As with other types of new concrete, a number of non-standard tests have been developed for assessing the properties of fresh SCC. The most common method for assessing workability is the slump flow test, where the spread of a sample made in a slump cone is measured rather than the slump. The time taken for the concrete to spread 500 mm can also be taken with this method. Many other tests exist; some, like the L-box test which assesses passing ability, show potential to be adopted for standardisation. The long-term aim is to develop one or two simple and inexpensive tests which give consistent results when carried out rapidly on site.

TRUCK MIXING

It is not practical to transport full loads of SCC from the batching plant to the site. The concrete is so fluid that it behaves like a liquid and will spill out on steep gradients or when the vehicle has to brake or turn sharply. Usually SCC is batched with a slump of around 100 mm for transportation so it behaves like a normal concrete. The superplasticiser is added on site and thoroughly mixed and checked for flow before the SCC is discharged into the pump hopper. A flat hose or tremie with a receiving hopper should always be used to place the concrete. This will prevent air pockets being trapped and avoid segregation. Placing by skip is slow and tends to cause segregation and blockage during discharge. It is important that the properties of SCC are maintained for an adequate period of time – around 90 minutes or more – after completion of mixing.

SUMMARY

There is compelling evidence that self-compacting concrete can be produced satisfactorily and from many combinations of materials. The extra cost of the high cement content and special admixtures can be justified by the better finish, savings on the labour cost and time-consuming activity of vibrating concrete. Such value engineering exercises at present will probably discount the use of SCC on all but the largest projects. The elimination of compaction presents the opportunity for greater concrete automation into the concrete construction process. At present, the general adoption of SCC is hindered by the lack of user-friendly guidance, standard test methods and the need for specialists to interpret the results from site testing

ADMIXTURES

The commonly used admixtures in ready mixed concrete are water reducers, high range water reducers also known as superplasticisers, retarders, air entrainers and accelerators. In concrete mixes for fine finishes it will be the water reducing admixtures (WRA) and superplasticisers that are most often incorporated in the mix, so these two types will be described in more detail.

WATER REDUCING ADMIXTURES

The active components of admixtures are surface active agents. They are derived from either lignosulphonic acids, a by-product of the wood pulp industry, or hydroxylated carboxylic acids. These substances when mixed into the concrete are adsorbed on the cement particles, giving them a negative charge so that they repel each other and behave like tiny glass beads until the effect of the agent dissipates. Air bubbles are also repelled and cannot attach themselves to the cement particles, which become sheathed in water molecules, preventing close contact with other cement particles. By this action more water becomes available to lubricate the mix so that workability is increased. The extent to which workability can be improved depends on the mix characteristics but typically it makes a 50 mm slump (very stiff mix) with no admixture into a 75 mm slump concrete, and a 75 mm slump concrete with no admixture into 100–125 mm slump concrete. The admixture dosage is typically 0.3–1.0 per cent by weight of cement. It has to be reliably and accurately dispensed because it is such a small amount and is mixed in with the free water when it is added. The admixture is inert and does not impart colour or harm the concrete. On the contrary, it improves concrete durability, enhances performance, placement and cohesion and helps to produce a better surface finish.

SUPERPLASTICISERS

These are also water reducing admixtures but they have a much greater effect on workability of the concrete so are defined as high range water reducers. Chemically, they are sulphonated melamine formaldehyde condensates and sulphonated napthalene formaldehyde condensates.

The dosage rate can vary from 0.5 to 1.5 per cent by weight of cement and will be influenced by cement type and content, proportion of fine aggregate, temperature and mixing method. Trial mixes will be required to determine the optimum dosage.

The surface active agents have a short active life, typically 30–60 minutes, when mixed with the concrete, which is even shorter in warm weather. When the effect of the superplasticiser wears off, the concrete can return quite quickly to its original slump. For this reason the superplasticiser is usually dispensed into the truck mixer on site. For a given water/cement ratio, the admixture can change a 75 mm slump mix to a flowing concrete with a slump in excess of 200 mm. The improvement in workability is smaller at higher temperatures. At high dosage it will retard the concrete but does not affect the setting of concrete except when used with cement that has a very low C3A content, when there may be excessive retardation.

The only real disadvantage of a superplasticiser is its relatively high cost compared to a WRA for use in normal concrete mixes. They are essential for high-strength, self-compacting and flowing, self-levelling concrete.

fig 28
Admixtures improve concrete handling and performance

PROFILE OF TYPICAL READY MIXED CONCRETE OPERATIONS

An in-depth review of batching plant operations and truck mixer efficiency from a leading ready mixed supplier gives a greater insight and understanding of the day-to-day matters in dry batch and wet batch operations and how concrete consistency is managed.

John Anderson, Area Operations Manager, and Derek Ballard, Plant Manager, Hanson Premix Greenwich

The South-East operation division of Hanson Premix covers an area outside the M25 that includes the counties of Essex, Kent, Hampshire and Sussex. You can run a diagonal line from Bournemouth to Harwich to get the expanse of the catchment area.

There are 35 concrete batching plants in the division with six or seven plants located in each county. Of these, ten are wet batch or central mixing plants, the rest are dry batch plants. The rural or town batching plants supply around 30,000 m^3 of concrete per annum. They will have between three and six truck mixers at each plant carrying predominantly 6 m^3 of concrete in the drum and a few 8 m^3 truck mixers. In city centre batching plants, particularly in London, it's the other way round as there are more 8 m^3 than 6 m^3 truck mixers.

29

fig 29
Material storage silos at Hanson Premix, Greenwich

The preference for the smaller truck mixer in country locations is related to access problems and weight restrictions when reversing down driveways and unmade farm roads. In the year 2000, the fleet in the South-East had 170 6 m^3 trucks and just four 8 m^3 trucks. Today, there are 30 8 m^3 trucks representing 25 per cent of the fleet. It makes economic sense because the bigger the load that's carried, the lower the cost of the concrete.

The batching plant is located close to where the population is centred and near to the industrial and business heartland. The road network, town expansion plans, growth projections and industrial development zones are studied before positioning a new plant in the area. Journey time for truck mixers should not be more than 40 minutes' travel each way. On average, it is a 6-mile journey each way and that means reaching out to customers within an 8-mile radius. It is important to locate the plant close to a railway siding because of the cost savings in the supply of material and the lower rental and land use cost. If there is a sand or stone quarry nearby that would be the best location as no haulage of materials is required. These would be the most important factors when selecting the plant location.

It would be difficult to obtain permission to site a batching plant in a built-up, inner city area or town centre due to the noise and environmental impact they can make. No one wishes to have heavy concrete truck mixers trundling down their High Street all day. The plant will be located on the outskirts of the town in the old-fashioned industrial estate if there are no railway sidings or quarry nearby.

DRY BATCH PRODUCTION

Dry batching is when the bulk aggregates and cement are transferred dry into the drum of the truck mixer. There is no mechanical intervention to pre-mix the constituents before they enter the drum. The water is added to the drum before the dry materials. The sand and coarse aggregates are semi pre-blended on the ribbon conveyor as it enters the weigh hopper before being discharged to the drum. Generally, most plants will put 60 per cent of the water in the drum before the solid constituents because this method blends it together much better. If you put more of the water before the mixing at the back of the drum, you get a lack of uniformity of the concrete as it discharges. Basically, drum mixing is not as efficient as pre-mixed or wet batching.

The aggregates are fed into the mixer with the cement in lots of 3 m^3. The coarse aggregate and sand go in first, followed by the feeding in of the cement. This is the best approach as the water already in the drum flows through the stones and then evenly saturates the fine particles of the sand and the cement. In drawing down material from the aggregate bins the stone is drawn off first and the sand is placed on top of it, followed by the cement. Balling is caused when neat cement or an excess of sand falls into the water and produces large lumps of material which are wet on the outside and dry on the inside; the dry mixing procedure avoids this. There are usually four bins in a country plant, with the sand at the end. It is better from a performance and strength characteristic point of view to have single sized coarse aggregates, for example one bin of 20–10 mm stone and another of 5 mm stone.

We endeavour to buy all our raw material from within the Hanson Group, when we can usually obtain a single-sized material. We also buy from other sources, which tend to be 20–5 mm blends. When we are commissioning or designing a batching plant we think carefully about the position of the aggregate bins and the ribbon conveyor. We usually have five storage bins, one of them for recycled materials which may contain crushed concrete, crushed terracotta pots or china clay. We use recycled materials in non-structural applications such as lean mix or oversite concrete. Each storage bin contains 50 t of material; they are open topped, square in plan and measure 3.5 × 3.5 × 5 m high. They are made of mild steel 6–8 mm thick with galvanised chutes, sills and access wear plates. A storage bin will last about 5–10 years before we refurbish it. We have tried using sacrificial lining surfaces such as polycarbonate sheeting, but this has tended to accelerate the corrosion due to the high moisture content of the aggregates. It is better to fit large wear plates and install new plates just on the wearing areas as this will extend the life of the whole unit.

We would usually have two bins of aggregates, one with 20 mm stone and the other with 10–5 mm stone and two bins of sands. There could be two types of sand, for example a coarse sand, usually marine dredged, and a silty sand with more fines, which we blend together to produce the best concrete. The finer sands are better for concrete finishing, the harsher sands for superior strength. For example, at our Croydon plant one sand is marine dredged and lacks a percentage of material from 2–0 mm so we blend it with 25 per cent unscreened silty sand from Hastings.

We buy our cement from Lafarge in the South-East – it either comes from Northfleet in Kent or Westbury in Wiltshire, the quality and the colour of both is very consistent. All our cement blends are purchased from Civil & Marine Ltd who supply us with our GGBS from Purfleet near London and Newport in South Wales.

CASE STUDY: HANSON PREMIX DRY BATCH PLANT AT TUNBRIDGE WELLS

The plant uses marine dredged gravel which is a blend of 20–5 mm material and marine sand. The gravel comes from Erith near Faversham, a 20-mile journey. We did try to supply the materials from our quarries in Maidstone and Rochester, but even though Erith is further away, it is a shorter journey time to Tunbridge Wells. The materials are dredged from the Thames estuary and screened, sieved, washed and stored and regularly tested for grading and mechanical integrity. The testing is carried out every 2 days and if there are any noticeable variations in the concrete strength made from these materials the batch is investigated and precautionary measures taken to return it to a consistent quality. This can arise if a particular bed has become contaminated with silts or other materials.

We have two 30 t silos with OP cement from Lafarge Northfleet and one 50 t silo of GGBS from Civil & Marine Ltd which comes from Purfleet. We generally sell blended cements to our customers, which are usually a 50/50 blend of OP/GGBS. This is the most cost-effective concrete we can produce for strength and workability. When you consider the consistency of concrete colour, blended cement will have a batching tolerance of +/− 2 per cent but we make adjustments as the lots are batched to ensure that there is no tolerance on the cement content. For example, if in the first batch we have 52/48 OP/GGBS, in the next batch we would compensate by reducing the OP and increasing the GGBS to even out the mix.

This levelling process uses a simulated object orientated technique (SOOT). If we exceed the batch weights on the first part of the load, we can compensate in the second batch. The plant has a chart which will tell the batcher exactly how much cement material to compensate after it has been automatically weighed. The compensation is calculated automatically so that the full load of concrete is batched to the exact mix specification. This is the control system we put in place for Canary Wharf. The benefits are that the concrete is better controlled, we reduced our batching tolerances and the standard deviation, and as a result save money.

All the storage bins are open topped and all the aggregates and sand are stored in them, material is fed to the weigh hoppers using a radial transfer conveyor. To avoid a condition called 'shot hole', where very fine materials like sand and granite dust stick to the sides of the bins and are difficult to dislodge, the storage bins are fitted with vibrators to dislodge them. If all the bins are full, we will have 3 hours of production for an average day. We will have made orders by phone for material deliveries for the following day by midday on the previous day. Five loads of coarse aggregates will be delivered using one loading truck on turnaround and one truck for the sand. The trucks will be standing waiting at the gates first thing in the morning. Bigger plants in city centres will have stock bays with material stockpiles and a loading machine. Smaller plants cannot justify such facilities as the loader costs £60,000 and the driver's wages are £25,000–£30,000 per year. That would add £2–£3 per m^3 to the cost of concrete that we sell. On average, at Tunbridge Wells we can handle 48 m^3 production per hour.

If there is a bigger pour, we can combine with another nearby plant, such as Maidstone or Hastings. If Tunbridge runs short of materials during the day, we can call on the Crawley and Maidstone plants to supply concrete, as they are only 25 miles away. This is one of the benefits of having standard material supplied from the same source at our batching plants throughout the South-East.

TRUCK MIXERS AND DRIVERS

When the tilting drum of the truck is used in dry batching, it is rotating at 10 revs per minute for 10 minutes to complete the mixing cycle. If the concrete is not mixed well due to the mixing blades being worn out or some other fault, it will take longer to mix. This would increase standing time and make it take longer to discharge the concrete, which would affect our efficiency and the price of our concrete. So it is not allowed to happen, but you have to take my word for it!

The plant supervisor is responsible for the quality of the concrete, along with the truck driver. The truck drivers are the best technicians for assessing the quality of the concrete. They see the material going into the drum; they can check the mixing efficiency by a workability rev counter on their machine, which measures the drag of the concrete on the drum. It works on a system of hydraulics; a needle gauge will quiver due to the changing drag of the concrete material on the drum surface dependent on changing dryness and wetness of the material. As it becomes more uniform, the changes in the drag become minimal and the needle doesn't quiver. At that point we can say the target slump has been achieved and the mix is uniform. The driver then takes the ticket and drives off to the customer.

If the needle gauge is still wavering after the set 2–3 minutes at 10 revs per minute of the drum, then there is a problem with the mix. Drivers are penalised if they take concrete to a customer that isn't properly mixed or is out of specification and rejected.

If the driver fails to tell the batcher that there is surplus water in the drum that could lead to inconsistency then the driver will not be paid for the concrete load and is deemed responsible. In this way we have the driver and the batcher working as a team and we aim to supply a quality product with no rejections.

The drivers are all self-employed and they collect and distribute our concrete to our customers. We supply the truck and the drum and they lease it back from us. The owner/driver has a range of trucks and drums to choose from. The drivers can mix and match their preferred options, rather like a company car, some will prefer 500 hp trucks, others 200 hp trucks. They can choose from a list of manufacturers, for example they can select a DAF truck with a McPhee drum or similar combinations of truck and drum. They advise us of their choice and we look into the performance history and efficiencies and any weight problems and provide feedback with our advice. For example, a truck and drum combination may be too heavy for the truck, but if they prefer that combination we will cap the carrying capacity, so instead of taking 6 m^3 loads that truck is restricted to 5.5 m^3 loads.

ADMIXTURES

At out of town plants we keep plasticisers, water reducing admixtures and air entraining admixtures, which are mainly for roadways. About 60 per cent of all our production uses water reducing admixtures. We can offer other admixtures on request – accelerators, retarders, superplasticisers. We work closely with our admixture suppliers, who will design the concrete mixtures with these additives. We don't keep pigments to make coloured concretes; we need a specification and plenty of notice if we are required to produce them. Admixtures are generally dispensed with the water but there are some that are placed at the end of the mixing period.

A typical London plant will use 10 different admixtures and have 16 dispensers.

fig 30
Aggregate bins and conveyor at Hanson Premix, Tunbridge Wells

fig 31
Truck mixer wash out area

CASE STUDY: CENTRAL WET BATCH PLANT, GREENWICH

This is a relatively new plant. It replaced the Blackwall plant, which has now been closed down. It is a wet batch tower plant that was purchased second-hand from Germany. There has been a recession in Germany that has hit their construction market, with one out of every three concrete plants either being closed or scrapped. We purchased a 10-year-old plant, dismantled it, shipped it to the UK and erected it at Victoria Deep near Greenwich. This plant will cater for the needs of the city area of London and Canary Wharf. To buy and erect a new batching plant of this capability the investment is around £1.5 million. We bought the second-hand plant for £150,000 and it cost us £400,000 to erect and make operational – that is less than half the cost of a new one.

This is a wet batch or central pan mixer plant which batches concrete in 3 m^3 lots before discharging it into the drum of the truck mixer. The plant is able to mix concrete in three different ways:

> dry batching

> semi-dry batching – where the cement and sand are mixed as slurry and poured into the drum of the truck mixer which contains the coarse aggregates, then the drum of the truck mixer mixes the concrete, which reduces the wear and tear on the central mixer

> wet batching – where the aggregates (sand, stone, cement and water) are blended together and mixed before discharging into the drum of the truck mixer. While this is the most efficient

fig 32
Wet batch plant at Hanson Premix, Greenwich

way of mixing concrete it causes high wear and tear on the mixing plant which is repaired every 6 months at a cost of £6,000 a time.

We have devised some practical guidelines for mixing. When the concrete strength is 40 N and above, we always wet batch. If the concrete is less than 40 N, say for non-specific oversite concrete, we do semi-dry batching. For lean mixes we will dry batch.

We produce about 100,000 m^3 of concrete per year from the plant, predominantly mixes with high workability and slumps between 150–200 mm. We carry PFA from Drax and West Burton power stations for our pump and high workability mixes. We prefer it to GGBS as it does not bleed as badly and gives better consistency, but the PFA is darker in colour. We stock PC and limestone dust filler, which we are using for a current project to produce a pale concrete colour.

We have single sized limestone aggregates, marine dredged aggregates, two types of sand and Lytag aggregates. We stock five different cements and cement combinations and can produce every combination of concrete our customers may require. We have three cement silos of 100 t each and two of 150 t. We hold eight plasticisers, two types of air entrainers, plus workability aids for slip forming and superplasticisers for high early strength concrete.

We have eight bins for storing aggregates, each of 100 t capacity. All of our sand and aggregates are transported by river to our distribution wharf; the cement always comes by lorry from the cement works. We have a fleet of 11 truck mixers; we work 24 hours around the clock and produce on average 600 m^3 of concrete per day. The plant has capacity to supply 2,000 m^3 per day if needed.

We cover the south side of the river, the Isle of Dogs and East London as far as Beckton and west as far as the City. The output per hour from the plant is 50–60 m^3 and it is very consistent. We are prevented from delivering higher outputs per hour due to having only one central batching mixer. If we had two it would double our hourly output. We have a large loading shovel and open stockpiles of aggregates from which the shovel loads material into the storage bins.

CENTRAL TESTING

We have central facilities to test all the concrete cubes we make every day, the cubes are stored on our sites in curing tanks and after 3 days are collected and taken to Bristol where they are crushed and analysed.

ENVIRONMENTAL PRECAUTIONS

We recycle a lot of the wash-out water from the truck mixer, which we put back into the newly batched concrete. The residues from the truck mixers are discharged into a slurry pond which is kept agitated. In this way we recycle 40 per cent of our waste water into the fresh concrete mix and adjust the fines content of the mix with the slurry from the slurry pond. Dust and airborne pollution and noise impact are controlled and kept in accordance with the Local Air Pollution Prevention and Control (LAPPC) regulatory regime.

CONCRETE RECIPE BOOK

Every day we produce, on average, 15–20 different concrete mixes. Our computer stores about 2,000 different concrete mixes, it uses a polynomal with our batch data so that for a given cement content you can produce a range of different workabilities and performance standards. For example, for a C20 concrete we will have a lean mix, a 50–75 mm workability mix, a 100 mm slump mix and a 150 mm slump mix for pumping. A C20 concrete will have a number of variable mix types, the same goes for a C30 and C40 concrete and so on. We probably carry too many mixes and we should aim to standardise and rationalise them.

fig 33
Aggregate bins and cement silos, enclosed to reduce dust pollution

fig 34
Truck mixer being loaded

CONCRETE COLOUR

The mechanism that gives colour to concrete is the light absorption qualities of the finest particles in the mix. For ordinary concrete mixes, the colour of the cement particles – PC and PC/GGBS blends – and percentage of particles in the sand that pass the 63 micron sieve will dominate the surface colour. The shade or tone of the colour will lighten with a higher water/cement ratio and darken with a lower water/cement ratio. The final colour of the concrete may take months or years to settle, depending on the rate of drying out and surface carbonation. In drying out and carbonating, the concrete will lighten up by several tones.

The surface appearance of insitu concrete will also be patinated due to subtle variations in the water/cement ratio intermixing on the surface. Many micro-pores will be visible on the surface, created by the capillaries through which water vapour has been expelled as the cement hydrates. Cement requires about 22 per cent of its own weight in water to fully hydrate, the remaining 25–28 per cent of the water that was added for workability is a surplus and this excess water is driven out as water vapour. This is a characteristic of normal concrete. The micro-pores will absorb surface moisture and darken the concrete when it rains and will attract dirt stains. For a high quality, dirt-free surface that stays clean and repels water, a silane or siloxane coating should be applied.

In the construction phase there will be a noticeable difference in the tone of the concrete colour when casting concrete in winter and in summer. In cold winter months the formwork has to remain in place for longer periods as the concrete requires more time to gain strength. As the concrete hardens and the cement hydrates, the excess water vapour is driven out but cannot escape and so condenses on the form face and saturates the concrete outer skin. On removal of the formwork the concrete is darker due to the saturated surface, but this will dry out and the colour will lighten as the water evaporates and the concrete carbonates. In summer, when vertical formwork is removed after a day, there is not the same build-up of water vapour so it appears much lighter in tone.

For the surface to remain a stable colour, especially if a dark shade is required, it is best to expose the coarse aggregates on the surface and reduce the cement mortar concentration. For a wider range of concrete colours that are not based on the natural colour of cement and sand, synthetic pigments may be used but can be expensive and tend to give an uneven surface tint. Dry shake pigments are very economical when used to colour a floor slab which is to be power trowelled.

PLAIN, SMOOTH, NATURAL COLORATIONS

These are based on the cement and sand colour combinations and will be grey tinted shades when grey cement is used or lighter colours if PC is blended with 50 per cent GGBS. The concrete produced is very economical and the colour will remain stable, although it will lighten by several tones as it carbonates. The cement and sand combinations available at the nearest concrete batching plant will offer a limited but very economical choice of concrete colours.

EXPOSED AGGREGATES

For textured, stable coloured finishes without pigmentation that will not carbonate or lighten in shade with time, an exposed aggregate finish can be very successful. The limits on colour will be imposed by stock availability at the local concrete batching plant. For vertical faces it is usually best to have 10–5 mm stones for such finishes as this will give good surface saturation and reduce 'hungry' patches (with no aggregate showing). As the concrete is placed from a height and free falls by gravity, the randomness of the stone concentration on the surface may cause some areas to have less saturation than others. The effect overall is like the random distribution of pebbles washed up on a beach and the wall or column surface should be seen in this light.

For a floor, it is possible to scatter additional single-sized aggregates on the top of the wet concrete and tamp them down to ensure good surface saturation with minimum variation.

fig 35
Pigmented concrete finishes
top left: Colour wax coat, *top right:* As struck pigmented finish
bottom left: Acid wash finish, *bottom right:* Grit blast finish

APPLICATIONS

SURFACE RETARDERS

A surface retarder is applied as a coating on the vertical form face instead of the release agent. It prevents the surface skin of cement from setting. When the concrete has hardened the surface is water jetted to expose the aggregates to a depth of 2–4 mm, depending on the strength of retarding agent applied to the formwork. The surface is then cleaned with dilute hydrochloric acid to remove traces of lime that can smear the aggregate surface.

For flat surfaces the retarder is applied by spray to the fresh concrete once the water sheen has evaporated. It is water jetted the following day and cleaned with dilute hydrochloric acid. The waste water must be properly discharged away from the work area and protective clothing must be worn.

ACID WASHING

The hardened concrete can be acid washed with dilute hydrochloric acid to remove the surface laitance without exposing the coarse aggregates. The surface has to be thoroughly washed afterwards to remove all traces of acid residues to avoid subsequent staining. Operatives will need to wear protective clothing to guard against accidental spillages. The acid is brushed over the surface to etch it and then washed over with water. Acid will attack some aggregates, for example limestone and marbles, altering their surface texture, which may impair their surface quality depending on the depth of the etch.

GRIT BLASTING

Depending on the pressure and grit size used, it is possible to achieve a variety of different surface finishes, from a light abrasion which removes just the surface laitance down to a deep etch to reveal the coarse aggregates. The grit used in blasting is made from processes mineral slag or metal fibres. The use of natural silica sand is prohibited for health reasons as it is harmful if inhaled, unless the operative is fully protected and the work area is sealed off. The abrasive grit for concrete is chosen according to the particle size – fine for removing the surface laitance, medium for light blasting and coarse for heavy blasting. Water is often introduced into the air jets as a means of reducing the dust created.

CONCRETE TERRAZZO

On flat surfaces the aggregates are exposed by disc abrasion and polishing to achieve a smooth, hard-wearing surface. In succession, coarse and medium carborundum and diamond-studded abrasion discs cut into the hardened concrete surface to a depth of about 2–4 mm to reveal the aggregate matrix. The surface is then polished with fine abrasion discs until smooth. It is more efficient when water is used with the cutting discs – known as wet grinding. For large floor areas, ride-on machines are used for the cutting and polishing process. Handwork is used for small areas and for finishing the corners and edges of panels.

fig 36
Grit blasting
top right: Medium grit blast
bottom right: Heavy grit blast

ARCHITECTURAL INSITU CONCRETE - TECHNOLOGY

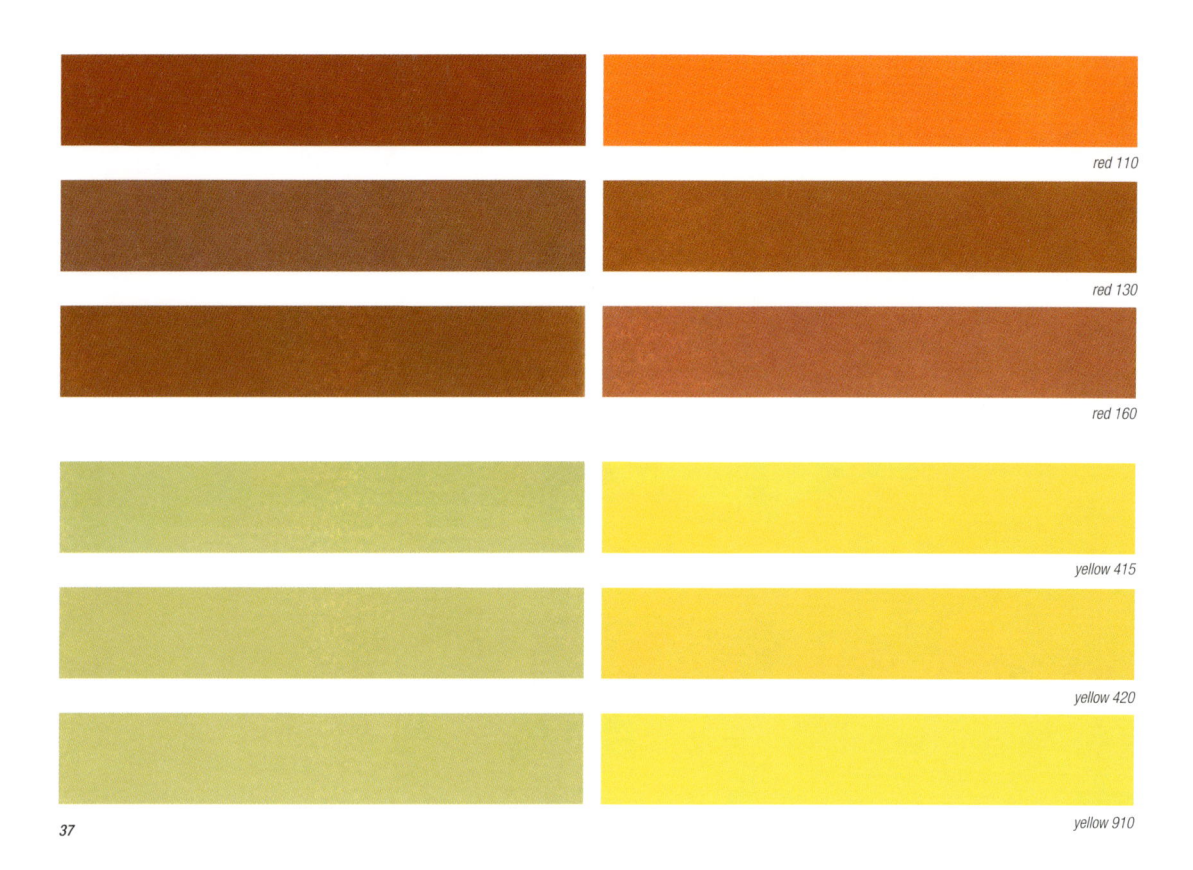

red 110

red 130

red 160

yellow 415

yellow 420

yellow 910

37

PIGMENTS

INTEGRAL THROUGH COLOUR

When a pigment is introduced into the mix, the pigment colour will dominate the final colour because it is much finer than the cement. The amount of pigment needed to colour a concrete will vary according to the cement content, the pigment type and the method of incorporating it into the mix.

Naturally occurring pigments are inert oxides and hydroxides of iron and titanium and copper complexes of phthalocyanine found in mineral rocks. They range in colour from red oxide to brown oxide to yellow oxide. The full description of pigments specified for use in concrete or mortar is given in BS 1014. A pigment may be supplied as a fine dry powder or an aqueous suspension or slurry and is virtually inert in combination with the ingredients of concrete. It is intended to impart a specific colour to the finished product. The mineral rocks containing raw pigment deposits are quarried, heat treated, crushed and then ground to a flour-like consistency to create industrial pigments. And, like sands and coarse aggregates, pigments do have unique characteristics – some are needle-shaped, some are spherical, some are much smaller than others, while some, like the phthalocyanines, are hydrophobic.

Red oxide pigments from different sources, for example, may have the same particle size but their bulk density and water absorption may differ significantly. Some have a bulk density of 1,500 kg/m^3 and

fig 37
Pigmented colour tones with grey and with white cement
Left hand column: 3% Bayferrox pigment using grey cement
Right hand column: 3% Bayferrox pigment using white cement

water absorption of 20 ml/100 g and others a bulk density of 900 kg/m³ and water absorption of 35 ml/100 g. Yellow iron oxide pigments have needle-shaped particles that can vary in bulk density from 500 to 800 kg/m³ and in water absorption from 50 to 90 ml/100 g. Green or blue phthalocyanines are hydrophobic and have a particle size ten times finer than a red oxide pigment, and a bulk density of 500 kg/m³.

The various pigment mixes of red, yellow and brown oxides are blended to create intermediate colours, and have to be carefully batched so that the bulk density and water absorption are known and can be adjusted in the mix.

It is for this reason that synthetic oxide pigments were developed by Bayer and others to create a more homogeneous pigment particle, with more uniform bulk density and water absorption. Synthetic pigments are more intense in colour than organic pigments and have excellent long-term colour stability. The pigments are produced in primary colours – red, black and yellow – by aniline or Penniman–Zoph process. In the aniline process, nitrobenzene is reduced to aniline in acid solution using fine iron filings as the reducing agent. During this process the iron filings are oxidised to produce an iron oxide, which eventually turns a blackish grey in colour. By controlling the oxidation it is possible to produce black and yellow slurries with a high tinting strength. After washing and filtration, the slurry is dried out to produce black and yellow pigments or heat treated and calcined to produce red oxide pigments. In the Penniman–Zoph process, iron filings captured from scrap sheet metal are dissolved in acid solution in a hydrolysis process, involving the oxidation and hydrolysis of iron sulphate in the presence of metallic iron to produce an iron oxide yellow pigment with needle-shaped particles. A range of brown pigments is blended from these three primary colours. Green and blue pigments are processed from copper oxides and cobalt deposits and are very expensive. Synthetic pigments are preferable for all architectural insitu concrete work.

COLOUR SHADES FOR CONCRETE

The use of synthetic oxide pigments with a high tinting strength means that colour saturation is achieved at a lower concentration than with organic pigment. Any higher dosage will not necessarily increase the intensity of the concrete colour. If a light-coloured cement such as white cement or blended GGBS is used it would need less pigment to reach colour saturation. Light-coloured cement is important where vibrant, strong colour shades are required.

Every pigment has a colour saturation point, after which further increases in the dosage rate will fail to make an appreciable difference to the colour intensity. For example, black iron oxides have the highest tinting strength, generally achieving saturation at 6 per cent dosage by weight of cement. Brown iron oxides have a slightly lower tinting strength, levelling off at approximately 7 per cent dosage, followed by 8 per cent for red oxides and 9 per cent for yellows. However, these figures will vary significantly for different shades of each pigment type. The current European standard for pigments in cementitious products (BS EN 12878) requires that pigment dosage rates be limited to a maximum of 10 per cent by weight of cement because the strength of the finished product could be reduced owing to the displacement of cement.

Colour intensity of the pigment is also influenced by the method of incorporating pigments into the mix. Using blending or ball mills to grind cement and pigments together, only half the concentration of pigment is needed, compared to adding it to the mix separately. However, the high cost of installing ball mills to produce coloured cements and silos to store it, coupled with the lack of demand, has made this option uneconomic in the UK to date.

Thus, for most coloured concrete production, pigments are introduced into the mix by dispensing them with the mixing water, as a powder in water-soluble bags or through a plasticising admixture suspension or as freeze-dried granules that behave like a sand.

One of the big headaches of coloured concrete and mortar production has been the problem of lime bloom or efflorescence. This is caused by the carbonation of calcium hydroxides that migrate to the surface where they form white deposits. For pigmented and architectural concrete this condition could spoil the finish, and methods of minimising and controlling lime bloom should be seriously considered.

In addition to a good concrete mix design with a low water/cement ratio, prevention of rapid drying out of concrete in the early days is the best way to eliminate secondary efflorescence, which can occur throughout the life of the concrete, until it has fully carbonated. A surface coating of either a transparent vapour-permeable membrane that is non-yellowing and will not break down under UV light, such as a silane, would be beneficial.

Primary efflorescence problems are more difficult to control and usually occur on the first day or so after the formwork has been removed. The solution is a light acid etch or water jetting to remove any efflorescence.

For ready mixed production of coloured concrete the guidelines set out in the next section ('The right mix') should be followed. Trial mixes should be carried out to check colour and surface finish.

It goes without saying that if there is good control of mix proportions and, principally, the cement content, then good coloured concrete will result. Large variations in water content and water/cement ratio between batches of concrete will vary the pigment concentration and affect the finished colour. The higher the water content, the lighter the shade; the lower the water content, the darker the shade.

Accurate weigh batching, control of aggregate moisture content and water/cement ratio are essential for good colour production. Whether the concrete is wet batched or truck mixed, in the absence of better data the following steps should be followed.

(1) Batch a minimum of 3 m^3 in a 6 m mixer drum.

(2) Add 50 per cent of the water and ribbon feed aggregates, cement and pigment.

(3) Mix thoroughly before adding the remaining water.

(4) Truck mix for at least 15 minutes. For a wet batched concrete the time will be less as the truck mixer is an agitator.

(5) Colour test the mix before discharge, by sampling from the front and back of the mixer using a slump test.

It is important to clean and wet concrete skips and concrete handling plant prior to use.

The selection of the formwork release agent in architectural work is as important as the selection of the forming system of facing material. Trial panels should be cast early in the contract to check compatibility. This point is discussed in more detail in later sections.

It should be borne in mind that pigmented concrete, especially dark shades, will fade and lighten as the concrete carbonates and may appear patchy and variable from one elevation to another. Detail the panel design to accommodate this change and so heighten the impact of such natural variation.

DRY SHAKE SURFACE COLOUR

This material can be applied directly to a wet concrete surface that has been tamped and floated to level. It will give the top few millimetres of a grey concrete surface coloration by pigmentation.

The dry shake powder can be distributed by hand or by automatic spreader machine. It is broadcast over the fresh concrete when no excess water can be seen on the surface. If applying by hand, the first shake consumes about two-thirds of the prescribed materials, one-third being held back for the second shake to cover any lean patches. After the first shake the concrete is floated but not trowelled. After the second shake the concrete is floated *and* trowelled. Excessive hard trowelling should be minimised to ensure uniformity of appearance.

The application rate will vary according to the requirements of the specification but is usually between 3 and 5 kg/m^2. The higher dosage creates more intense, lighter colours. The dry shake, which contains pigment, cement, quartz aggregates, dispersing agents and other ingredients, is spread from a height of about 1 m. It can be cured using clear acrylic urethane coating or similar products.

THE RIGHT MIX

A concrete mix for a good finish and consistent colour has to follow procedures which will ensure that the constituent material is the same throughout and batched in the same proportions.

The NBS Concrete Specification and other guidance documents on the design of concrete mixes are based on performance standards, which means that the concrete has to meet a minimum compressive strength or durability requirement. This allows the supplier to adjust and vary the raw ingredients during the contract to meet these performance standards. This is unacceptable for concrete where colour control is of paramount importance and where the raw ingredients have to be carefully selected and monitored and kept to exactly the same proportions for each and every truckload that is batched.

The yardstick for acceptability for such a concrete mix is consistency and uniformity of materials, with minimum variation, therefore a prescriptive mix should be specified. These mixes are quite easy to produce and are well within the capability of most ready mixed suppliers. The cost of concrete is largely governed by the cement content, such mixes are about the same average price as a 40 N structural concrete and are not therefore expensive.

The choice of suitable aggregates and cement types has been described in the previous sections. It is important that the availability of materials is discussed with the local ready mixed supplier before drafting the specification.

38

PRESCRIBED CONCRETE MIX

The concrete mix shall be a prescribed mix whose constituents shall be weigh batched and truck mixed generally in accordance with current BS standards.

The proposed concrete mix shall comply with the requirements for fair face concrete work and shall be of uniform colour.

The proposed concrete mix shall comply with the following conditions to satisfy uniformity of colour and surface finish.

(1) The concrete workability shall be sufficiently cohesive for internal vibrator compaction, handling by skip and concrete pump and to free fall 1 m without segregation or causing excessive bleed-water to rise to the surface. The target slump for concrete shall be between 125 mm and 150 mm.

(2) The concrete must have cement content not less than 325 kg/m^3. The cement shall be a PC or PC/GGBS or PC/PFA blended cement taken from the same source to eliminate changes in cement colour. For a blended cement the ratios must be stated.

(3) The cement content and water/cement ratio will be fixed for all concrete supplied to the contract and must not be adjusted at any time during the contract. The workability can be adjusted by increasing or decreasing the admixture dosage.

(4) The water/cement ratio shall not exceed 0.5. Once the ratio

fig 38
Concrete ingredients: sand, cement, stone and water

fig 39
Mix consistency

fig 40
Slump test

has been agreed by the architect, it must not be adjusted at any time during the contract, as any variation in the water/cement ratio and the cement content will affect the concrete surface colour.

(5) The concrete mix must meet the minimum compressive strength and durability conditions stated in the structural concrete specification.

(6) The total aggregate/cement ratio shall not exceed 6.

(7) The sand/cement ratio by weight shall not exceed 2. This will reduce the risk of blowholes forming. The sand should be well graded medium sand without excessive fine dust and be of the same colour, since the fines act as a pigment. The quantity of fines passing the 150 micron sieve must be declared and the widest variation reported, as any significant variation can affect the final concrete tonal colour.

(8) The coarse aggregate shall preferably be a single sized, rounded aggregate of nominal 20 mm diameter (19–15 mm). A blend of 20 mm and 10 mm single sized aggregate may be acceptable, provided not more than 20 per cent passes the 10 mm sieve. Single sized crushed aggregate and blended crushed aggregates will be considered if rounded gravel is not locally available, but will be subject to batching trials.

(9) Any plasticiser, water reducing admixtures or additive used in the mix must be stated.

(10) The concrete supplier shall submit the mix design details to the architect for review and comment. Trial mixes may be required to check variation in surface colour due to non-uniformity in the dust content of the sand.

(11) The slump test shall be used to monitor the consistency of the concrete mix supplied to site. If the concrete is not within the target slump parameters specified and agreed (e.g. nominally 125–150 mm) there is a risk that the water/cement ratio may have changed, which may affect the finished concrete colour.

QUALITY CONTROL

Accurate weigh batching and control of aggregate moisture content is essential in the production of good visual concrete. Aggregates should be stored under cover in silos or bins to prevent rain wetting stockpiles. The moisture content of aggregates should be monitored regularly and free water content adjusted to maintain the correct total water/cement ratio.

The use of wet batched or pre-batched concrete is preferable in visual concrete production. Truck mixers should only be used to agitate the mix.

If no wet batch plant is available near to the site, dry batching will be acceptable subject to assessment of truck mixer efficiency and workability consistency. This will entail slump testing concrete at the beginning, middle and end of a full load delivered to site for non-exposed work. Truck mixers should be thoroughly cleaned at the batching plant, particularly the rim of the entry hopper of the drum and delivery chute, prior to the loading of dry batched materials. There should be no traces of previous concrete or marks on the truck mixer that might discolour the mix.

Truck delivery tickets must show clearly the batch weights of all mix constituents, including the free water content and water/cement ratio. The mix constituents, especially the cement content and water content, must not vary between batches as this will change the surface colour of the compacted concrete.

At the start of operations the concrete should be sampled from the truck mixer before discharge and slump tested to check the uniformity of the mix. Reliance is placed on the concrete supplier to carry out quality control checks on each concrete load before dispatching it to the site. The concrete supplier must state what controls and procedures will be in place to ensure consistency of supply.

The contractor should keep a slump record for every day of concreting and a note of where each load was placed in the permanent work.

FORMWORK AND PRACTICE

INTRODUCTION

The character and quality of a concrete finish will be defined by the character of the formwork in contact with the concrete. Concrete by itself has no shape or form; it is a liquid rock which will set into some amorphous shape if unconfined, like solidifying lava. By placing it in moulds and confining it by formwork it can be shaped into walls, floors, columns; imparting structural rigidity and architectural definition. The surface in contact with the wet concrete imprints the texture, smoothness or roughness of the formed face and gives the concrete its surface quality. The primary purpose of formwork is to contain the concrete, but very often formwork materials that perform this function quite adequately cannot be relied on to give a satisfactory surface finish.

The selection of the formwork face is therefore critical in the visual appearance of concrete and is the most important element in the whole process of fine finishes. If the concrete mix, its consistency and colour, is good but the formwork is poorly constructed, the panel joints are badly fitted, the support system not rigid enough to adequately resist the pressure from the liquid concrete, then the surface appearance will be poor. In this section, the commonly available construction timbers, plywood, steel and synthetic liners are reviewed.

The timber materials are imported into the UK from all over the world and it is likely that in a few years' time some products will not be available and new materials will come onto the market. It is important that the designer understands the different material types and the different finishes they will give, how many reuses they may offer, the cost comparisons and whether or not the materials originate from renewable, sustainable forests and timber yards or are felled in virgin forests.

Untreated timber, metal formwork and other options will also be discussed, including board marking, no nail technology, support systems, tie bolt holes, panel layouts, good workmanship, cutting and drilling, reducing wastage, release agents and formwork aftercare, and the section's conclusions are collated under the heading 'Concrete workmanship'.

Designers must make the key decisions on formwork choice, such as whether or not to use Forest Stewardship Council (FSC) grade timber, give consideration to recycling options and provide detail of the panels to reduce wastage. It is unreasonable to leave all these decisions to the contractor, whose prime objectives are to reduce costs and simplify the construction and assembly. In my experience, more often than not this will result in substandard workmanship and poorer finishes and lead to contractual disputes.

The designer should have a clear understanding of the finish required, the material that will achieve it and should provide a statement giving guidance on the products to be used and procedures to be followed. This reduces the burden of risk on the contractor, who will not have to second-guess what the architect really requires, and avoids the need for casting numerous trial panels to achieve a finish that satisfies the subjective judgement of the architect at the expense of the contractor.

The contractor, on the other hand, has to construct the support system so that it can adequately resist the hydrostatic pressure from the concrete, with minimum movement and deflection. On smaller projects this aspect of workmanship tends to be overlooked and can lead to major problems with grout leakage and honeycombing. The guidance notes given in this section will instruct the designer on what details to expect the contractor to submit in terms of formwork and temporary works drawings, how to assess their fitness for purpose and the care needed in handling materials for greater reuse.

ONE CLASS OF FINISH

There is only one class of finish for exposed, visual concrete and that is the best that can be achieved given the choice of the form face, the consistency of the concrete mix and the care taken in workmanship. It can be confusing to try to categorise concrete with different classes of finishes, viz. F1, F2, F3, F4 and F5 as published in various technical handbooks and guidance literature. For a standard or a basic concrete finish which will not be seen, blemishes, abrupt changes (not more than 3 mm), blowholes and staining from resin and tannins in the ply are acceptable but not grout loss, honeycombing or alignment deformation. Where only one face of a wall is fair face, it is important that the unseen face is also well constructed.

The classification notes in the National Building Specification are very practical and could be used with some further explanatory details:

> basic smooth finish
> fine smooth finish
> fine, board-marked finish
> other types of finish.

As it stands, these headings are rather bald. They require an explanation and description of exactly what form face material is to be used and how it is to be cut, sealed, laid out and fixed to achieve the desired finish.

UNTREATED TIMBER

This material will only keep its shape for a limited time, not swelling and warping and splitting, unless it is waterproofed and edge sealed. Into this category fall the board-marked solid timbers such as larch, Douglas fir, spruce, hemlock, oak and cedar and the like, as well as the high-quality birch ply panels for plane surface finishes. The softwoods, used for general shuttering ply with a 'good one side' face, can be successfully adapted for a durable finish, provided that the surface is waterproofed and sealed to prevent the tannins and resins being drawn out of the wood during the concrete hydration. This will minimise stains and blemishes on the concrete surface.

Generally, for site work it is best to use a quick-drying acrylic urethane coating that is water based, which is preferable for indoor working. Two coats will be adequate to give the timber end grain and the surface some protection against moisture ingress. Occasionally, there can be an adverse reaction with varnish coatings that are solvent based, which can break down the glue bond between plywood layers and cause brown staining and furring of the ply. This has occurred once in the author's experience when a yacht varnish was used on birch ply. It is advisable, therefore, to test the varnish on a sample piece of ply before using it and to check the product suitability with the manufacturer.

Water is one of the worst enemies of wood. A large proportion of the damage done to formwork is the direct result of moisture changes in the wood and subsequent dimensional instability. Water generally enters through open cracks, unprotected end grain, defects in surface treatments, nail holes, drill holes and the like. The water-repellent coating to the surface and end grain will inhibit water from penetrating the wood by capillary action and reduce fluctuations in moisture content, enhance dimensional stability and encourage a number of reuses of the timber before it is finally discarded.

SAWN BOARD TIMBER

A wide range of finishes can be obtained from sawn board timber, varying from smooth to deeply grained timber. The deeper the grain, the more likely it is that some of the grain will be lost after each use. In all cases it is best to seal the face with a water repellent that has a durable hard film, which will minimise grain loss after casting.

Rough sawn timber with a hairy or splintered surface should be avoided, due to difficulty in striking, loss of surface definition with each casting and patchy appearance.

Where individual boards are of different thicknesses, to express the joint lines between them it is economic to use standard stock widths available at the timber merchant. The merchant will have a cutting service to reduce the timber width and thickness but this will create waste, unless all the offcuts are used. Thickness can vary from 16 to 50 mm, widths from 50 to 150 mm plus. The length of the board will depend on the length of the trunk it was cut from and will range from 2.4 m up to 4.8 m. These lengths can be cut down to size by the stockist to suit site requirements.

Seasoned, kiln-dried softwoods and hardwoods are suitable for board marking. They should be supplied at a sensible moisture content, not too dry as this will cause swelling in contact with moisture and make removal difficult. Variation in surface absorbency, natural resin and tannins, salts or wood

fig 41
Board-marked concrete trial panels

fig 42
Material tested

fig 43
Comparison of finish

sugars may affect the colour and surface durability of the concrete. It is important to seal the timber and make it water repellent before applying the release agent.

Some timbers, such as keruing and merante, are known to cause serious problems with retardation of the concrete face and discoloration. The advice of an independent timber specialist such as the Timber Research and Development Association (TRADA), APA – The Engineered Wood Association, the Forest Products Research Centre or the Building Research Establishment (BRE) should be sought if there is any doubt about a species.

PLYWOOD

Plywood consists of cross-laminated veneers of wood, bonded with synthetic resin adhesives. Coniferous softwood species are used extensively for formwork plywood. Hardwood species, especially birch and tropical hardwoods, will generally provide a finer surface finish and give increased strength to the ply and a potentially greater number of reuses in repetitive work. Plywood is also manufactured with a combination of hardwood and softwood veneers, called 'combi' ply for short. Although the appearance and quality of the top layer of the ply is of paramount concern, combining different softwood or hardwood veneers or having them all of the same type will yield different results with regard to strength, reuse factor, dimensional stability and cost of the sheet material. Plywood is made by gluing together a number of thin veneers, or plies, of softwood or hardwood. There are always an odd number of veneers and each ply is at a right angle to the one below, as this gives the material its strength. The more the type of veneers used the stronger the plywood becomes. Both the type of glue and type of veneer determine the suitability of a sheet for a particular application. The finish quality of plywood varies considerably, some plywood has attractive grains while others can contain knots.

Plywood is graded for exterior or interior use depending on the water resistance of the glue used to stick the plies together. Code letters show this grading on each sheet.

Exterior grade plywood with weather and boil proof (WBP) glue can be used for concrete work.

The softwood veneers are not ideal for fine finishes as they tend to release tannins which stain the concrete, are prone to formation of wood sugars if exposed for long periods in hot sun and will only give a very few reuses before the surface deteriorates. They are very useful as a backing ply and for non-exposed concrete finishes and are competitively priced.

Untreated timber exposed to ultraviolet light can suffer degradation and form wood sugars which will retard the concrete.

The majority of timber imported from Canada is Douglas fir and redwood, while plywood from Finland, Sweden and Latvia, for example, is mainly spruce, whitewood, redwood and birch in combinations of hardwood and softwood veneers.

Formwork plywood should always have statistical data giving strength, bending and sheer criteria, preferably with allowable pressure span tables. The strength of panel and how it is supported, will affect deflection and ultimately the number of reuses. Many manufacturers use independent test criteria from the APA or TECO (Timber Engineering Company).

NATURAL BIRCH

Birch ply sheets may be through birch (birch throughout) or combi

44

sheets with birch as the face veneer. The long-established timber yards of Finland have FSC-grade material or the equivalent, accredited by the Forest Stewardship Council which promotes responsible management of the world's forests. It is very likely that other countries such as Latvia, Sweden, Poland and Russia operate a similar system and can provide appropriate proof of certification. Local timber merchants may have stocks of both 'combi' and through birch ply panels that are suitable for small projects. It is best to inspect the quality before making a choice. For the larger projects the material will be available through a timber importer or their agents and supplied through a distributor. Samples that are the size of an envelope can be sent in the post for examination, supported by timber specification sheets. But how can a sheet's fitness for purpose be established?

Ply comes in different face quality grades: B, S, BB to WG, with B being the best quality with no defect, only pin-hole repair, no discoloration and WG being a basic finish which allows for 40 mm size knots, surface discoloration, surface repairs, glue penetration, scribe marks and splits from sawing up to 5 mm from the panel edge.

In summary:

> B is for high quality painting, staining and lacquering.
> S is for good quality painting, staining and lacquering.

fig 44
Natural birch ply

> BB is a standard grade that has plugged repairs, for interior paint finish and coating with transparent and non-transparent overlays and films, and also veneering.
> WG is for use where surface appearance is not important (and is the reverse side of a sheet of another grade).

For fine finishes it would be best to try the B, S or possibly BB grade. When the birch ply is to be recycled into the permanent architecture, then a B or S grade should be selected and no face nails or screws should be used to fix the panels to the formwork.

The glue that bonds the veneers should be weather and boil proof to be suitable for external use and so that it meets the requirements of BS 6566, Parts 3 to 8 as appropriate.

Panel thickness for construction should be between 18 and 25 mm for site handling and rigidity and there should be about seven to eleven veneers in the ply. For forming curved walls and similar constructions, the individual ply sheet thickness should be thinner, at around 5 mm, to allow for bending and then built up to make the 20 mm thickness.

Clear, fast-drying acrylic urethane, polyurethane and wax emulsions that are recommended for waterproofing untreated timber should be applied in controlled conditions, under cover where possible. For site application, the timber should always have been stored in the dry before treating, as the moisture content can be critical to the performance of the coating.

FILM FACED PLYWOOD

Formwork surface coatings and treatments are many and varied. It is best to ensure that the coating product has been proven and that it is compatible with the release agent.

Special surfaces have been developed to enhance the appearance of the concrete finish, and also to maximise the potential number of reuses and reduce the cost of the formwork by so doing. In addition, these coating films have the benefit of masking the wood grain or defects as well as preventing wood sugars and resin bleed from the plywood. Many panels are coated on two sides to balance the panel and prevent warping. However, some plywood manufacturers have developed technology that produces a stable panel coated on one side only, thereby reducing costs.

45

PHENOLIC FILM FACED (PFF)
PFF is the generic term for high resin content (64 per cent), high performance surface coating with a film density from 120 to 800 g/m². Colours available are black or brown.

46

PFF is a fully cured and impervious coating made from mixing phenolic resin with paper. The paper is only a means of carrying the resin to the board in a rigid format.

Most commonly used PFFs have film density of 120 g/m² and 240 g/m² although they are available up to 800 g/m². The additional amount of film improves the wear and tear of the coating, and increases the cost, but does not alter the surface finish. The high density coatings are used in slip-forming and climbing formwork applications.

If the PFF sheets are fabricated into a panel system, with good handling and site practice at least four to six uses are achievable without deterioration in the finish of the concrete. It is important to bear in mind that the amount of resin in the film will dramatically affect the number of reuses. A good quality 120 g/m² film will have 80 g/m² of resin and 40 g/m² of paper. Many budget panels on the market which offer 140 g/m² or 160 g/m² films, may only contain 40–60 g/m² of resin, so the resin density should be checked carefully.

This type of impervious overlay is common in Scandinavian, Latvian and Russian plywood, which are mainly derived from birch and birch 'combi' veneers. Birch veneers are thinner than softwood veneers and have a dense, close-grained texture. As a result of this there are more veneers to make up the ply thickness. To keep the panel thickness to a minimum, and to overcome the imperfections of using cheaper, lower grade birch veneers, a dense phenolic resin overlay is bonded to the contact face. The resin thickness is usually 0.5 mm or 500 microns. It is dark mahogany in colour, shiny in appearance and smooth to touch.

Such film coatings will give a smooth, slightly shiny finish to the concrete surface and enhance the patina, which is due to the migration of water and mortar paste on the surface as it sets. This can leave a subtly flecked finish with

fig 45
Phenolic film faced ply - WISA Birch

fig 46
Phenolic film faced ply - WISA Beto film

dark marbling, which is extremely smooth to touch. This will fade as the concrete begins to dry out and carbonate.

As this is a non-absorbent coating, a release agent that has sufficient adhesion to create an adequate barrier and not run off the face is required. Any excess release agent must be wiped away as it will mark the surface. The correct release agent will ensure that the formwork will strike cleanly and leave a clean, blemish-free surface.

Phenolic film faced panels may undergo a degree of localised swelling and rippling to the surface film due to moisture ingress beneath the veneer or in prolonged exposure to hot sun when they lie flat. The panel edges, if they are not factory sealed, should be coated with a fast-drying acrylic on site to prevent water ingress at the edges softening and swelling the ply and causing ripples on the film. To minimise the risk of this happening, on first use a weak solution of cement grout should also be applied over the surface, which should then be washed down immediately and wiped clean. The release agent should be applied for first use.

47

MEDIUM DENSITY OVERLAY (MDO)

To overcome the problems of veneer imperfections, absorbency, grade quality and release of tannins, the top ply is commonly faced with a film of paper that is impregnated with resin – the reverse of the phenolic film face. For MDO film the resin saturation levels vary between 30 and 60 per cent. With some brands the ply can be seen behind the overlay, especially when the resin impregnation is as low as 30 per cent saturation. If the backing ply is visible behind the MDO then the film is quite porous and will allow moisture migration and the possibility of tannins being drawn out of the backing ply as the concrete hydrates, especially if it remains in contact with the overlay for long periods. Some MDO products can cause light-brown staining to the concrete surface when casting suspended floors and beams. This does not happen for walls and columns using the same product, provided that the forms are struck after a day or two.

Absorbent MDO films were pioneered by the Canadians, as they suited their softwood veneers. In recent times Chile, Brazil, Indonesia, China and other countries have entered the market and produce MDO ply that outperforms the original North American MDO, by using a higher grade of paper, greater resin impregnation and improving the gluing methods.

An MDO ply will give a matt, non-shiny surface finish to the concrete. The various MDO products on the market will each give subtly different surface texture, smoothness and tone. A more absorbent and porous surface will allow some air pockets formed on the concrete surface to escape and moisture to migrate from the liquid concrete into the backing ply, and vice versa when the concrete has set. This will tend to give a mildly darker tonal finish than a dense impervious film face due to the slight reduction in water/cement ratio.

The ply is not prone to a temporary surface ripple that may occur with a phenolic film face. The MDO paper overlay is absorbent and allows moisture to migrate through it to balance the pressure on both sides when it comes into contact with the wet concrete.

It is only by inspection of the timber, conducting an absorbency test, seeing finished work, checking the paper density and resin saturation that the finish quality can be accurately ascertained. The manufacturers of some MDO plywood invariably fail to provide the data needed to make such judgements. The maxim is – if in doubt, don't specify it.

The most difficult item to try to anticipate is the number of reuses that can be achieved. This is completely dependent on the care, attitude and approach taken by the contractor. Manufacturers' claims about the number for reuses of their products tend to be more appropriate for precast production under factory conditions.

HIGH DENSITY OVERLAY (HDO)

HDO will give much better performance than an MDO as there is higher resin content in the paper film. It is more expensive and is close to the surface finish quality of a phenolic film face. The density of the overlay, the resin saturation and price give an indication of whether the product is an MDO or HDO ply. Generally, a film density of $120\ g/m^2$ indicates an MDO and anything above $300g/m^2$ will be an HDO. The surface will appear much harder and shiny and will resemble a phenolic film surface, although not quite as shiny. HDO ply would be used instead of an MDO ply for the exposed soffits of floors and beams to avoid the risk of brownish staining from the tannins through the porous film of the MDO ply.

Suitable release agents will be those that are appropriate for phenolic film faced sheets.

SUMMARY

Formwork plywood is produced with a resin paper or phenolic film

fig 47
Medium density overlay ply

face in order to create an engineered panel which can be reused many times for casting and forming concrete.

Typical phenolic film faced panels
WISA – Birch, WISA – Betofilm, Gethalfilm, Garudaform, Fepcoplex, Kronoply FF Spezial, Syktyvkar, Riga Form, Tulsa Film Phenolic.

Typical paper faced panels
MDO: Tulsaform MDO 323, Ultraform, Coastform, Thomasi Plastform MDO 333, WISA – Form Duo, WISA MDO, Ainsworth Pourform 107.
HDO: Ainsworth Pourform HDO, Thomasi HDO 252.

Sizes available
2440 × 1220 mm (8 ft × 4 ft); thickness available 12–25 mm plus. Other sheet sizes are available but they need to be ordered in advance in units of 50 plus. It is necessary to check with suppliers.

Applicable standards
BS EN 313/314 – Plywood classification, terminology and bonding quality.
BS EN 635/636 – Plywood classification by surface appearance and specification.
BS 5268 – Structural specification.

ORIENTED STRAND BOARD

Oriented strand board (OSB) is an engineered panel product made of strands, flakes or wafers sliced from small diameter, roundwood logs and bonded with an exterior type binder under heat and pressure.

It is manufactured in wide mats from cross-oriented layers of thin, rectangular wooden strips compressed and bonded together with wax and resin adhesives (95 per cent wood, 5 per cent wax and resin). The layers are created by shredding the wood into strips. These are sifted and then oriented on a belt. The mat is made in a forming bed, the layers are built up with the external layers aligned in the panel direction and internal layers randomly positioned. The number of layers placed is determined by the required thickness of the finished panel, typically a 150 mm layer will produce a 15 mm pressed panel thickness when the 150 mm mat is placed in a thermal press. Individual panels are then cut from the mats in standard sizes.

Strand dimensions are predetermined and have a uniform thickness. The majority of Structural Board Association (SBA) member mills in the USA use a combination of strands up to 6" (150 mm) long and 1" (25 mm) wide.

The strength of OSB comes mainly from the uninterrupted wood fibre, interweaving of the long strands or wafers, and degree of orientation of strands in the surface layers. Waterproof and boil proof resin binders are combined with the strands to provide internal strength, rigidity and moisture resistance.

OSB cuts and handles just like ordinary wood.

FINISH
The mat of surface strands leaves a bamboo lattice finish on the concrete. It is best to seal the face with water repellent to minimise moisture movement and dimensional change during construction.

PERFORMANCE
What is the difference between plywood and OSB? Plywood is made from thin sheets of veneer that are cross-laminated and glued together with a hot-press. Imagine the raw log as a pencil being sharpened in a big pencil sharpener. The wood veneer is literally peeled from the log as it is spun. Resulting veneers have pure tangential grain orientation, since the slicing follows the growth rings of the log. Throughout the thickness of the panel, the grain of each layer is positioned in a perpendicular direction to the adjacent layer. There is always an odd number of layers in plywood panels so that the panel is balanced around its central axis. This strategy makes plywood stable and less likely to shrink, swell, cup or warp.

Logs are ground into thin wood strands to produce oriented strand board. Dried strands are mixed with wax and adhesive, formed into thick mats, and then hot-pressed into panels. The strands are aligned and the 'strand plies' are positioned as alternating layers that run perpendicular to each other and mimic plywood. OSB is engineered to have strength and stiffness equivalent to plywood.

Performance is similar to that of plywood in many ways, but there are differences. All wood products expand when they get wet. When OSB is exposed to wet conditions, it expands faster around the perimeter of the panel than it does in the middle. Swollen edges of OSB panels can 'telegraph' through thin coverings such as asphalt roof shingles.

Different qualities in terms of thickness, panel size, strength and rigidity can be given to the OSB by changes in the manufacturing process. OSB panels have no internal gaps or voids and are water-resistant, although they do require additional membranes to achieve impermeability to water. OSB has begun to replace natural plywood in many environments. The most common use is as cladding in walls, floors and roofs.

FIXING
OSB does not have an annual ring

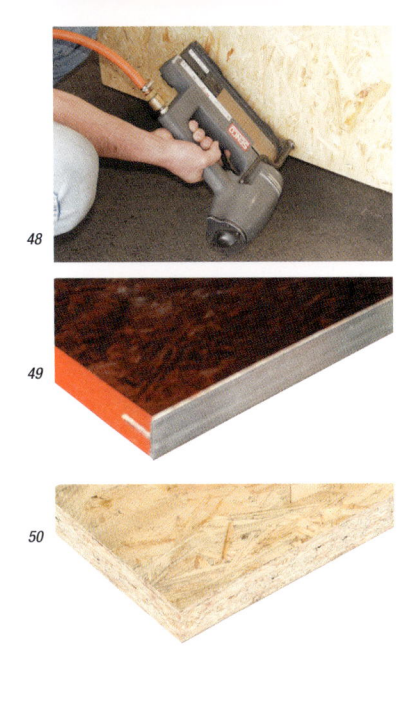

structure, which considerably reduces the strength of ordinary solid wood, and therefore does not tend to split. Furthermore, OSB does not dry out, i.e. there are no cracks caused by drying which could also have a negative effect.

It is advisable to follow the fixing notes for chipboard to err on the side of caution.

PHENOLIC FILM FACED OSB

The OSB panel has a phenolic resin coating on both sides. This coating is an assurance that, apart from an application of release agent, the panel requires no further processing. The OSB will behave anisotropically. As a consequence of this the direction of the span must be followed strictly when laying panels on the support system.

Phenolic film faced OSB that is edge sealed absorbs moisture much more slowly, which will reduce edge swelling and time-consuming reworking, repair and replacement of panels.

The surface treatment, release agent and concrete finish will be similar to a tyipical phenolic faced ply.

CHIPBOARD (FOR SINGLE USAGE)

Chipboard is made by bonding together wood particles with an adhesive under heat and pressure to form a rigid board with a relatively smooth surface. Chipboard is available in a number of densities; normal, medium and high density. Normal density is fairly soft and 'flaky', high density is very solid and hard (often used for worktops and fire doors), medium density is somewhere in between.

Exterior grades of chipboard are available, but most are only suitable for internal use as all but high density tend to soak up water like a sponge. Once waterlogged, chipboard tends to swell and break down.

High density chipboard, which is hardwearing, rigid and heavy, is often used for the carcasses of kitchen furniture, worktops and

fig 48
Shot-fired fixing pins

fig 49
Phenolic film faced OSB

fig 50
Oriented strand board (OSB/Kronoply)

fig 51
Surface finish

47

flooring. None of the best grades of moisture-resistant chipboard (C3 to C5) are suitable for use in wet conditions when the moisture content is likely to exceed 18 per cent.

Chipboard is normally available in 2,440 × 1,220 mm sheets (or subdivisions). Finished veneered sheets are available in smaller sheets so that the four decorated edges do not need to be cut. Thickness range from 12 mm to 25 mm.

In high density chipboard, the finish after cutting is generally good. As with plywood, veneered chipboard should be cut with the saw blade going into the finish face to prevent the finish veneer chipping. To reduce the amount of damage when cutting chipboard, a strip of masking tape should be applied along the line of the cut and the cut made through the tape.

FIXING
With high density chipboard, nails, screws and fixing can be used. The screw-holding power is improved if double-threaded or chipboard screws are used. The manufacturer's instructions should be followed for special purpose board fixings.

FINISHING
Chipboard should be coated with clear water repellent such as a wax emulsion. The chipboard is prone to swelling when it comes into contact with moisture and wet concrete. If it is face sealed and fixed to a backing ply it should perform adequately for one casting.

METAL

Steel faced formwork is expensive for one-off uses; it only becomes economic for at least 20 or more reuses. Unless it is new and unmarked it will lead to imperfect surfaces as every dent, twist or scratch on the surface will be imprinted on the concrete.

To obtain a high-class finish with steel, pre-ageing is recommended by lightly shot-blasting the surface. This roughing of the surface improves the surface appearance and reduces the risk of dark discoloration appearing on the concrete when the forms have been left in place for prolonged periods.

Any marks detectable on the form face, such as those caused by strong backs, plate attachments, plate joints, welds and changes in surface texture will be reproduced on the concrete surface.

Provided that the forms are well constructed, cleaned of any rust, are oiled and minor damage repaired, steel faced formwork can be used successfully for as many as 50–100 times. Chemical release agents developed for steel forms should be used to achieve best results.

Repair of a steel form face is a highly specialised operation and is best carried out by the manufacturer. Any repairs to the surface are likely to show up on the concrete surface.

OTHERS

FORMLINERS
These include plasticised PVC, polyurethane, fibreglass and materials that are flexible and easily mouldable, giving good definition of finished concrete with a high reuse factor. They can be imprinted with a wide variety of surface texture, patterning and even replicate board marking. They become economical when they are used at least 10–20 times insitu.

Usually the liner is bonded to a 20 mm ply base for insitu work. Standard formliners are available in sizes up to 1.5 m wide and 6 m long. It is best to discuss the project requirements with the manufacturer.

Concrete or metal spacers should not be used in formliners, only plastic covered spacers. It may be prudent to have them colour matched to blend with the integral concrete colour. Where possible, the use of cover spacers should be avoided by hanging the reinforcement in position.

The combination of formliner and backing ply is used as the primary forming system, which is then attached to waling and strong back supports to complete the formwork assembly. PVC liners can be joined by fusing liner panels together to make larger panel sizes. Polyurethane liners cannot be heat fused but joints can be sealed using adhesive recommended by the manufacturer.

While there is not sufficient scope to review all types of formliners, a note on the most common types will

fig 52
Chipboard

53

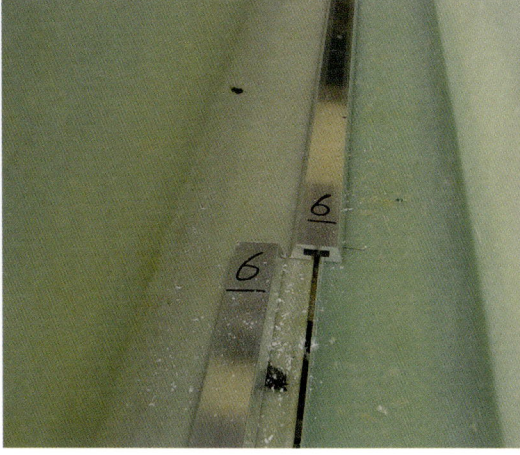

54

assist in appraising options for a particular project.

Rigid formliners

These are usually made from one of three materials: fibreglass, ABS or a hard PVC plastic. They are made cheaply for single use or with a medium gauge thickness for up to ten uses. This material is not suitable for intricate patterns, particularly sharp corners or undercuts. Care must be taken to limit the formwork pressure on the liner to 1,000 psf. Liner expansion and contraction due to temperature fluctuations must be guarded against. They should be covered with black tarpaulin to limit UV damage if they are exposed to the sun for an extended period.

A quality release agent should be used, one that is non staining and that does not cause deterioration of the formliner.

Elastomeric formliners

These include plasticised PVC, polyurethane and materials that are flexible and easily mouldable, giving good definition to finished concrete, with a potential reuse of up to 100 times. Usually the liner is bonded to 20 mm ply for insitu work or supplied without backing for precast production. Standard formliners are available in sizes up to 1.5 m wide and 6 m long. Elastomeric formliners are most economical for large projects where there is a lot of repetition as they are expensive. The total cost will depend on the texture, weight of material per m^2 and whether the liner is bonded to ply or supplied loose.

Release agents for PVC formliners are as for rigid formliners and should be non staining and stable against PVC. However, release agents for polyurethane formliners must be a wax emulsion, either water based or solvent based.

To clean formliners a household detergent and a bristle brush should be used.

Cracks or breaks in formliners should be repaired with recommended fillers or plastic tape. Repairs will show up on the concrete. For large, irregular repairs, breaks or cracks, panel replacement is advisable.

Glass reinforced plastic (GRP)

GRP formliners are used for making bespoke coffered, ribbed or curved soffit moulds. They must have at least ten repeat uses to make the cost economic. The manufacturing process involves the making of a master mould in timber or plaster to the required cast shape. The layers of resin and glass fibre are then built up on the master mould, including stiffening members, until the required thickness is achieved. When it has cured and fully hardened it is removed and another shape layered on the master mould. It is a labour-intensive, skilled and expensive exercise. The mould will need to be supported over the profile to maintain its shape under the weight of concrete and foot traffic, which adds to the cost of construction.

GRP is well suited to forming complex shapes but the brittleness of the material at corners means that it is best to avoid sharp arisses to ease removal from the concrete. It is best always to maintain corners at radii of not less than 10 mm. Minor damage to GRP forms can be repaired on site using polyester resin repair kits recommended by the manufacturer. The manufacturer will be able to recommend the best release agent to use for a good finish.

GEOTEXTILES

Several woven fabric materials have been developed since the mid-1970s for use as geotextiles to act as a separation membrane between subsoil materials and as earth reinforcement. One particular fabric, branded Zemdrain, has been used with great success to enhance the durability of concrete by significantly reducing the permeability of concrete and eliminating blowholes and blemishes on the surface.

Zemdrain is made from polypropylene fibres and is pinned to the contact face of the formwork. The liner drains the excess water and releases air pockets that migrate to the form face. The result is a dense, durable surface with no blowholes. The reduction of the water content on the surface means that the concrete is a much deeper tone. The Zemdrain fabric leaves a cloth-like imprint on the surfaces, which is matt and slightly textured. The supplier will provide guidance on application and construction procedures.

fig 53
GRP cast finish

fig 54
GRP soffit rebate

PLASTIC SHEETING

Thick-gauge polythene, visqueen and similar plastic sheeting can be draped or stretched tautly across the formwork to create a marbling effect on the concrete face. By draping the sheet like a loose curtain hung from the top of the pour and pinned at the edges and attaching it to the form face at wide centres, the sheeting is stretched and pulled by the concrete as it fills the form to create circular crease lines on the face.

If the plastic sheeting is stretched tight across the formwork without being able to crease then a smooth surface is revealed. Plastic sheeting will not require release agent and may become damaged when being removed so it should only be considered for single use.

DISPOSABLE COLUMN FORMERS

A number of manufacturers now offer disposable circular column formwork made from spirally wound rigid paper tubing that is impermeable and resistant to water absorption. The tube is internally lined with a smooth-faced plastic release sheet for good surface finish and appearance so there is no down time on site applying release agent to the inside of the tube. The tubes must, however, be protected from rain during storage and when in use. A good sized tarpaulin, which totally covers the tubes in storage, would usually be sufficient short-term protection from the weather, provided that the tubes are stacked off the ground on pallets. The disposable tubes are supplied to the required length and diameter. If tubes need to be shortened on site, they should be cut from the bottom of the tube using a fine toothed blade. This will ensure that the tear-off strip action is not impaired. The cut rim of the liner must be resealed with a suitable formwork tape.

Tubes can be supplied in diameters ranging from 150 mm to 1,200 mm and in lengths to suit the project. To complement the standard circular form, the inside of the tube can be fitted with a moulded liner that can give square, hexagonal, octagonal, fluted, oval, L-shaped or rectangular columns, plus more shapes and the option of patterned surface finishes.

Installation

(1) The rebar cage must be reasonably plumb before the tube is sleeved into position. Tubes should always be erected with the arrows pointing upwards, which ensures correct operation of the tear-off strip.

(2) The base of the tube can sit on the concrete without a kicker, provided that the concrete is level. A square timber template is fixed to the concrete to maintain the tube in position.

(3) The top of the tube is restrained with a square timber frame to enable props to be used to plumb the column.

(4) As the tube is quite lightweight, under concrete pressure head it is liable to lift. The tubular column should be held down by wedging the top of the tube to the rebar cage or tying it back to the concrete slab.

Concreting

For columns up to 4 m high a skip or bucket should be used to place the concrete. For columns higher than 4 m it is better to pump or tremie the concrete into place. The concrete should be placed slowly to minimise trapped air and only vibrated from inside the tube.

Removal

It is necessary to wait until 48 hr after casting before stripping the disposable form. The tear-off strip should be pulled down the full length of the tube. Flaps open the tube to ease it off the concrete. Then the liner can be removed from the column and the surface finish checked. It may be best to leave the casing in place to protect the concrete surface from accidental damage.

FORMWORK PRACTICE

STORAGE AND HANDLING

Panels are usually delivered to site in a relatively dry condition and they should be protected from direct exposure to the elements, either by storing in a building or by covering with a secure waterproof tarpaulin.

The pack of sheet panels must be stored on level ground, well clear of mud or standing water (puddles) and away from any risk of contact with vehicles or machinery. If the panels are to be stored on site for more than a few days, the outer packing and strapping should be removed.

fig 55
Disposable column former

All formwork should be protected from rainwater damage and standing water. Prefabricated panels must not be left flat on the ground after use, but stored upright. The top surfaces must be covered to prevent them absorbing moisture and swelling.

When handling formwork and craning it into position, no metal chains, metal carriers or abrasive tools should be allowed to scrape or come in contact with the fair face.

CLEANING

Shutters should be cleaned immediately after they have been struck. A brush or plastic-headed tool should be used to remove any concrete residue.

When panels are dry and clean, release agent can be reapplied and the panels then stored with the fair face away from the sun. If panels are allowed to be directly exposed to the sun they will dry out and the veneer can be damaged, creating problems with the overlay.

REPAIRING

When concrete is being poured and compactors are used the panel faces can be damaged. Once the panel has been cleaned, minor repairs can be carried out using a proprietary patching system, typically a two-part epoxy filler. The manufacturer of the product should be consulted to ensure correct usage. Any repair to the face may be noticeable on the next use, no matter how carefully it is carried out, due to the difference in permeability between the overlay of the plywood and the filler used.

FORMWORK SHUTTER ASSEMBLY

Due to the normal difficulties encountered on any construction site in terms of lack of space and absence of a controlled environment, it is preferable for forms to be assembled off site or under cover where possible. However, it is recognised that this is often impracticable.

For fine finishes, the formwork must be rigidly assembled, with panels well fitted and closely jointed to allow no grout loss, no discoloration or blemishes due to panel movement, deflection or distortion. The sheets of ply or timber must be supported by a propriety formwork support system fabricated from metal, aluminium beams, steel props, wood trussed sections and the like and be designed by a specialist formwork company to resist the forces, loads and stresses acting on the form face.

Such temporary works design forms part of the contract, with fully detailed drawings and supporting calculations submitted to the architect and engineer for comment before any work can be started.

The key elements of typical support systems for wall, column and floor slabs are shown in the illustrations. These elements are:

> *soldier* – upright vertical beam that supports the walings. On occasion the walings are designed to support the soldiers

> *waling* – a horizontal beam which supports the timber sheet material that is in contact with the concrete

> *tie bolt* – a rod of high tensile steel or mild steel that passes through both sides of vertical wall panels and ties the soldiers to prevent them moving under concrete pressure

> *clamps* – horizontal flat metal sections used as waling in column construction

> *brace* – a stiffening element in a support system which prevents swaying and provides stability

> *strut* – a compression element that does not carry load, set at an angle or rake, which acts as a brace to prevent overturning or twisting

> *steel prop* (adjustable or non-adjustable) – for supporting vertical loads. Props should be braced with scaffolding tubes or a bracing framework for lateral stability.

SHUTTER DESIGN

The shutters must be designed using appropriate information from the load tables published on the timber used and the formwork pressures published in CIRIA Report 108. A temporary works specialist should be engaged to engineer the support system. Allowances must be made for the increased moisture content of plywood panels after several uses, which can lead to greater flexing between supports. Adequate support must be in place so that finished concrete will remain within specified tolerances for flatness.

FIXING

Wherever possible, the need for nail and screw fixing in film faced or smooth natural ply panels should be eliminated as this will cause surface fractures and splits and lead to swelling and localised damage. Back fixing with screws that do not penetrate through the contact face and the use of an inexpensive backing ply to fix to are recommended practice. The backing ply is nailed to the support system. The combination of backing ply and contact ply can increase the rigidity of the formwork assembly, and can reduce the thickness required for facing ply.

All fixture points, whether screw or nail holes, should be countersunk or punched below the surface and the exposed plywood filled and sealed with a water-impervious product. A two-part epoxy resin can be used as a filler, but the filled area must also be completely sealed.

CUTTING TOOLS

Formwork can be worked using normal woodworking tools, provided that they are sharp and in good condition. A fine-toothed saw is recommended to prevent chipping to the reverse face. Cuts should always be made from the contact face and holes drilled with a high-speed head to ensure the cleanest cut for bolt holes. The drilling of pilot holes from both sides is strongly recommended to reduce breakout on the reverse face.

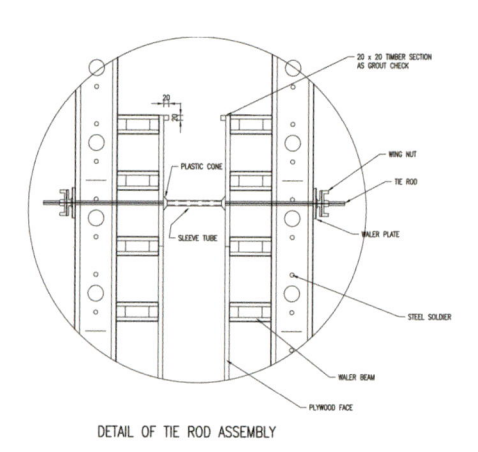

EDGES

All cut edges and face holes must be resealed using a water-impervious coating, such as such a fast-drying acrylic urethane. It is recommend that two coats be applied to limit the ingress of moisture into the core. Restricting the ingress of moisture to the edges or around cut areas will reduce swelling and help the appearance of the concrete.

Due to the nature of timber-based products, the edges may not be completely straight. To ensure no loss of material through the joints a layer of clear sealant should be applied between the panel edges when the form is assembled.

TIE BOLTS

Metal rods or tie bolts connect the opposite sides of vertical formwork and resist the applied concrete pressure acting on the vertical soldiers and horizontal walings, depending on how the support system has been designed. They act as tension ties and become fixed points of support for the vertical and horizontal beams to span.

Incorporating tie bolts reduces the amount of supporting formwork, reduces the working space required and speeds the assembly. Generally, high tensile DyWidag rods are used to withstand the large forces generated by the liquid head of concrete. It is important that if the tie is a recoverable type and is to be reused this is stated on the formwork design drawings.

Recoverable ties that are made from high tensile steel must have a minimum factor of safety of 2.0 on the minimum ultimate strength of the tie and, moreover, the design working load acting on the tie must be below yield. If the working load exceeds the safe working stress on the tie and is above yield, then the tie cannot be reused.

The tie rod is the most critical component in the design of vertical formwork because failure of one can lead to failure of the whole system. In designing the system it is the performance of the tie assembly, the waler plates, bearing plate and washers that should receive the fullest consideration.

High tensile ties are more highly stressed that mild steel ties and will tend to elongate under load. The designer should make allowance for the elastic elongation when ties are very long. The elongation of all tie rods is the actual increase in length of the tie in tension from the underside of the washer.

With fine finishes, the ties in the lower half of the section must not extend at all, otherwise there is a serious risk of grout loss ensuing from the base of the shutter and of a dark colour band forming on the contact face due to moisture movement as the vertical soldier deflects.

The tie rod either has screw threads at its end or ribs along its entire length to enable the wing nuts to screw the tie rod to the waler or washer plate. The tie rod is sleeved inside the formwork using a plastic tube and removable plastic cones, which allow recovery of the tie. The plastic sleeve and cones act as a strut and spacer for the tie when the wing nuts are just tight. The cones press against the form face to prevent

fig 56
Detail of tie rod assembly

fig 57
PERI Trio wall panel system

fig 58
PERI Vario wall panel system

fig 59
Adjustable column formwork (Outinord)

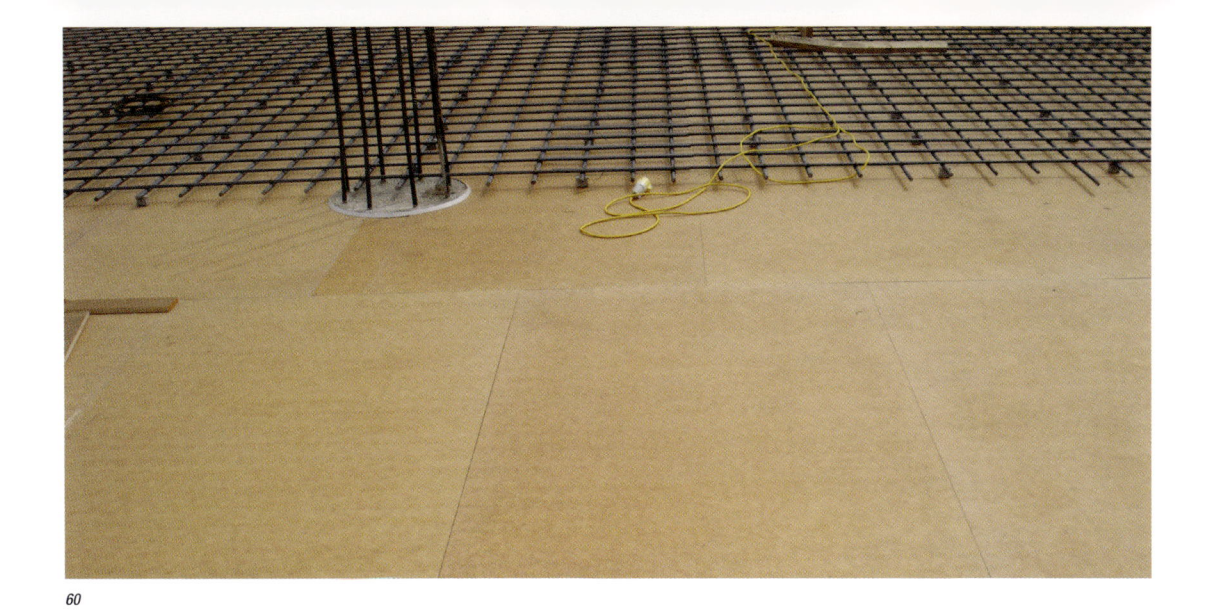

grout entering the plastic tube and bonding the tie.

When tightening formwork against preformed kickers, there is a tendency for site operatives to overtighten the lowest row of ties and thus pretension them. This can lead to excessive elongation of the tie under concrete pressure, causing gross loss and even tie failure. This practice must be avoided.

This can also occur for any tie if wing nuts are overtightened and can cause buckling of the plastic cone and possible grout loss and spalling of the hardened concrete of the tie hole when the tie rod is removed.

CONCRETE PRESSURE ASSESSMENT

The pressure on the formwork created by the liquid concrete as it is poured shows a gradual increase with depth of placement until a plateau is reached where the pressure does not increase with rising height of pour. This is due to the arching effect of the concrete in the formwork and the partial setting of the concrete in the lower section.

Recent test results for rates of concrete rise of 3 m/hr monitored by Dundee University on self-compacting concrete indicated that pressures near the base of the formwork increased with each lift of concrete and then began to decay. This continued for a period of slightly more than an hour after the concrete first covered the load cell in the base of the pour. Then the pressure steadily reduced, despite continuation of the concrete pour. For very fast rates of rise, the maximum pressure was reached when the placing of concrete in the column form was complete or nearing completion, and soon after the pressure started to drop. The results illustrate the significance of the rate of rise of concrete, the influence of the concrete setting characteristics and the ambient temperature on the concrete pressure acting on the formwork.

The maximum lateral pressure of concrete on formwork work depends on six main factors and can be determined from tables published in CIRA Report 108.

(1) *The vertical height of the formwork H in metres* – this is not necessarily the pour height.

(2) *The rate of rise of the concrete vertically up the form R in metres per hour* – the faster the rate of rise, the greater the pressure. For example, for a wall of height 4 m and a rate rise of 1.5 m/hr a pressure of 65 kN/m^2 will be generated, while at 5 m/hr it is 85 kN/m^2 at 5 °C concrete temperature.

(3) *The ambient temperature of the concrete at the time of placing* – the higher the temperature, the quicker the concrete will set and lower the maximum design pressure. In cold winter weather the concrete will set slowly,

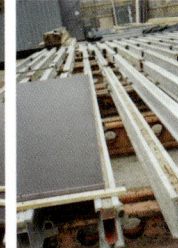

fig 60
Flab slab – MDO panel layout with butt joints

fig 61
Locating and placing facing panels to supporting waling for a wall

fig 62
Fixing board-marked panels with a nail gun

fig 63
Panel of column formwork that has been screw fixed – screw heads are sealed with plastic filler

64

consequently the formwork pressures will be higher.

(4) *The plan dimensions of the pour* – is it a column or a wall? The formwork pressure acting on a column for the same rate rise and height will be greater, due to confinement of the concrete, than for a wall pour. See CIRIA Report 108 for definitions of 'column' and 'wall' for assessing design pressures.

(5) *The density of the concrete* – a value of 2,500 kg/m^3 has been assumed in all the pressure table calculations. When a lightweight concrete is being considered then the CIRIA Report 108 suggests an adjustment in proportion to the density.

(6) *The type of concrete* – is it a pure PC mix or blended mix with GGBS or PFA, with or without admixtures, or with or without retarders? The faster setting pure PC mixes have lower design pressures than slower setting blended cement mixes. There are five groups classified in the report: Groups 1 and 2 are considered to behave in the same manner and are all PC mixes without retarders; Groups 3, 4 and 5 are blended cement mixes with GGBS and PFA, or PC mixes with a retarder.

> Group 1: PC without admixtures.
> Group 2: OPC, RHPC and SRPC with admixtures, excluding a retarder.
> Group 3: OPC, RHPC and SRPC with retarders.
> Group 4: GGBS and PFA blended mix with OPC containing less than 70 per cent GGBS or less than 40 per cent PFA, without admixtures.
> Group 5: GGBS and PFA blended mixes with OPC including admixtures, excluding a retarder.

(RHPC – rapid hardening Portland cement; SRPC – sulphate resistant Portland cement.)

RIGID ASSEMBLY AND MINIMUM DEFLECTION

Prefabricated sheets of ply should be made into a rigid assembly that does not deflect, that does not twist or warp when lifted by crane or cause individual sheets to bow or move out of alignment during construction. Significant movement of formwork during and after concreting may cause permanent changes in surface colour, grout loss and poor surface appearance.

CONSTRUCTION JOINTS AND PANEL LAYOUT

Rebates are formed at agreed construction joints, to give a neat grout-tight edge. The preferred position and detail of all construction joints including butt joints must be shown on the architect drawings and referenced clearly on the contractor's formwork drawing, together with a layout of panels for approval by the architect. Advice from the contractor on the economic use of panels to minimise waste and to improve construction efficiency should be taken into consideration.

REUSE OF FORMWORK

Plastic-faced cleaning tools should be used and the formwork surface wiped with a cloth to avoid scratching or indenting the surface with metal faced tools or abrasive papers. After casting the concrete the ply should be cleaned with a cloth and recoated with release agent.

With proper care and attention and protective measures, the formwork should give up to four uses. All plywood should be stored carefully for later reuse in the permanent work if required.

FORMWORK STRIKING TIMES

Walls and columns

Formwork should be stripped 24–36 hr after casting in summer months. This will give a consistent colour to the concrete and avoid exposing the forms to the prolonged heat of hydration and the chemical action of the release agent. It is important to strip formwork to give the same equivalent maturity throughout the project as even small variations in maturity time can cause initial colour variation. Maturity time will be extended in cold weather, and these times can be evaluated from cement content and actual curing temperature.

Floor and beams

The minimum curing period given in the structural specification must be complied with for control of deflection and for adequate development of bond and shear strength of the concrete. Cubes

fig 64
Tie bolt, plastic cone and plastic sleeve

FORMWORK AND PRACTICE

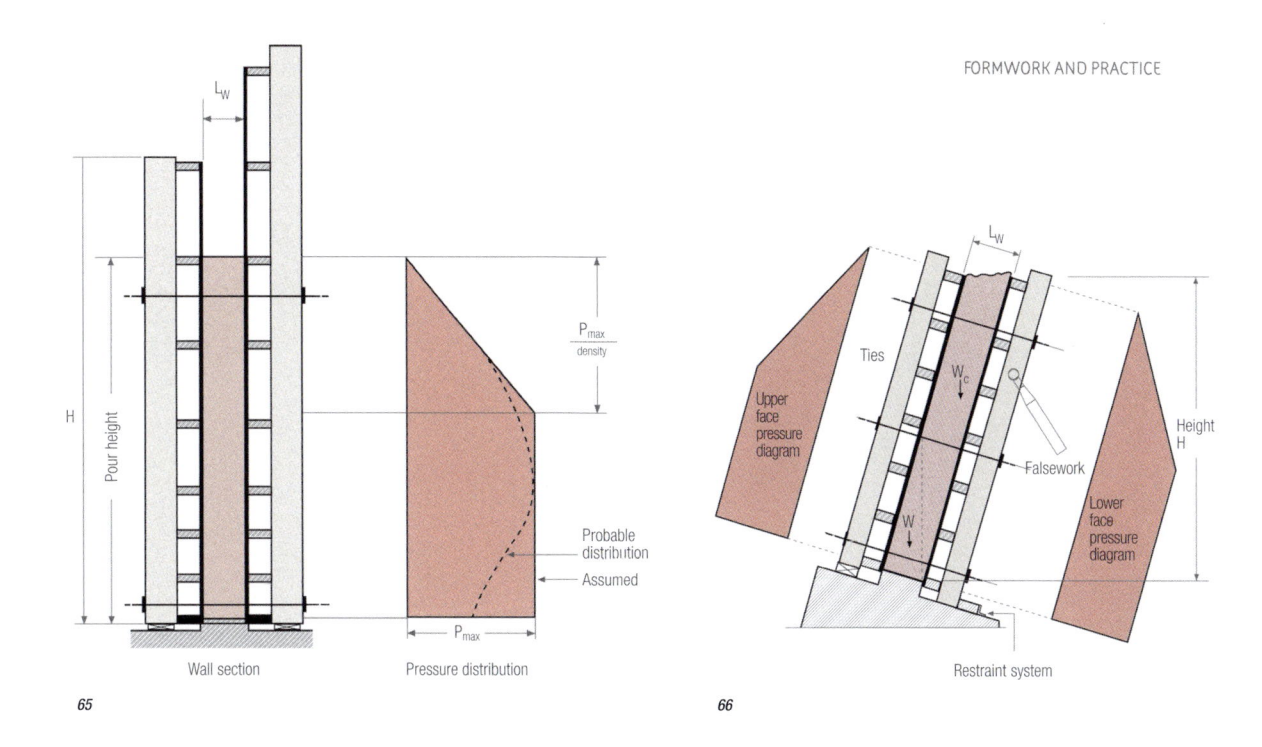

fig 65
Vertical formwork pressure distribution

fig 66
Pressure distribution on an inclined face

cured alongside the slab under the same conditions may be considered to show compliance with minimum strength for removal of formwork, subject to the approval of the structural engineer.

RELEASE AGENTS

Release agents are essential in the production of fine concrete finishes Their primary function is to prevent the cement hardening at the contact face and the concrete bonding to the formwork. Some release agents will produce very good concrete surface finishes, helping to eliminate the risk of large blowholes, surface staining or dusting of the concrete surface while maintaining a clean, cement-free surface to the formwork.

The three generic types of release agent that have been commonly used for the fine finishes shown in the case studies are:

> chemical release agents
> solid release wax coating, and
> neat oils with surfactants.

The author has tended to specify chemical release agents and occasionally solid wax coatings since they dry as a film on the form face, remain active for prolonged periods and will not wash off in the rain. Neat oils with surfactants tend to wash off in the rain and have a shorter life once applied. They may be best in factory-based production methods under cover.

Operatives prefer to use a spray to apply the release agent, as it is faster and less laborious than rubbing a wax coating on the form face. The wax will soften when the temperature reaches 35 °C, which is becoming more common on sites in the summer. The wax is neutral and not toxic so it is ideal for use in confined areas. Spray-applied solvent-based chemical release agent can be formulated to be odour-free for indoor use.

A good release agent should provide a clean and easy release or strike of the formwork without damage to either the concrete or the form face and have no adverse effects on either the form or the concrete surface such as staining, air bubbles or blistering. It will assist in obtaining the maximum reuse of the form face, provided that the contractor takes proper care in handling the formwork. It must be suitable for use in the temperature regime and weather conditions expected on site. Also, it should be easy to apply and ready to use without site mixing and not harmful to the operative if used in accordance with the manufacturer's instructions and guidance notes.

The selection of the most suitable type of release agent will usually be confirmed from successfully completing site trials.

ARCHITECTURAL DETAIL DRAWINGS

The designer should indicate the arrangement and orientation of ply sheets, board marking and form liners, etc., and the preferred position of construction joints, tie bolt holes and any sheet cutting plans. This should be shown on plan and elevation drawings and fully annotated with material type, thickness, tie bolt centres, butt joints, construction joints, vertical pour heights, shadow gaps and the like. These drawings should be part

67

68

of the tender package so that the contractor has a clear understanding of the design intent and what has to be priced. This will give plenty of opportunity for the contractor to review these points with the designer before they submit their tender. It is unreasonable to expect the contractor to carry out this detailing work or for them to be given this information after the contract is let. This will invariably lead to extra work, extra cost and design compromise.

For plywood, always try to specify sheets that are whole units, which do not require site cutting and so reduces the risk of making poorly formed butt joints. Sheet sizes are currently 1,220 mm × 2,440 mm (4 ft × 8 ft). By minimising the need to trim and saw cut, wastage is eliminated and the risk of damage and frayed edges is removed. The actual plywood or timber that is specified should always be inspected and samples and certificates of manufacture should be obtained to confirm identity in case there is a mistake in the supply. The thickness tolerances of sheets must be checked as some manufacturers allow variation in a given thickness, which may be unacceptable for achieving a smooth, flush floor soffit or wall line. Usually, the maximum thickness tolerances are + or – 0.3 mm, which is acceptable for a flush, smooth form face.

Sheets should be laid out to the pattern dictated by the design and built up to form a wall length, floor area or column box. Work should be carried out to modules that are between grid lines and storey height. The contractor should be encouraged to make up wall lengths that can be prefabricated and moved by crane into position. This will reduce labour cost and time in having to dismantle and reassemble the sheets, speed up the construction cycle and increase the number of reuses of the sheets.

The preferred position of tie bolt holes should be shown. For most columns they are not necessary. As a general rule, for walls the first tie bolt is typically 500 mm above the floor. Vertically, they are then spaced between 1 m and 1.5 m apart and horizontally they are at the same centres. This gives an indication of the arrangement, and the contractor will price for their inclusion.

It is important to indicate where shadow gaps and rebates are to be formed. The examples in the case studies give shape and detailing suggestions. These features will be tested in constructing the trial panel.

fig 67
Positioning a 12-m long external formwork panel (PERI Vario)

fig 68
Positioning a 5-m high internal wall panel (PERI Vario)

FORMWORK AND PRACTICE

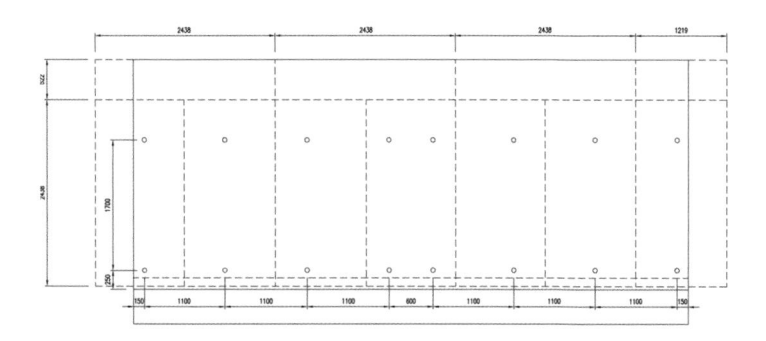

ELEVATION SHOWING ASSUMED PLYWOOD LAYOUT AND TIE PATTERN FOR TWO TIES IN HEIGHT

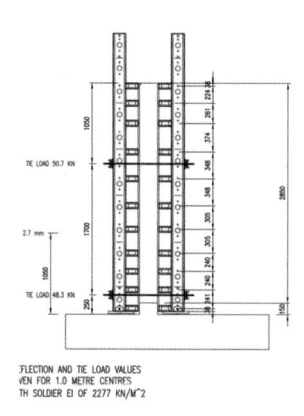

TYPICAL SECTION THREE TIES

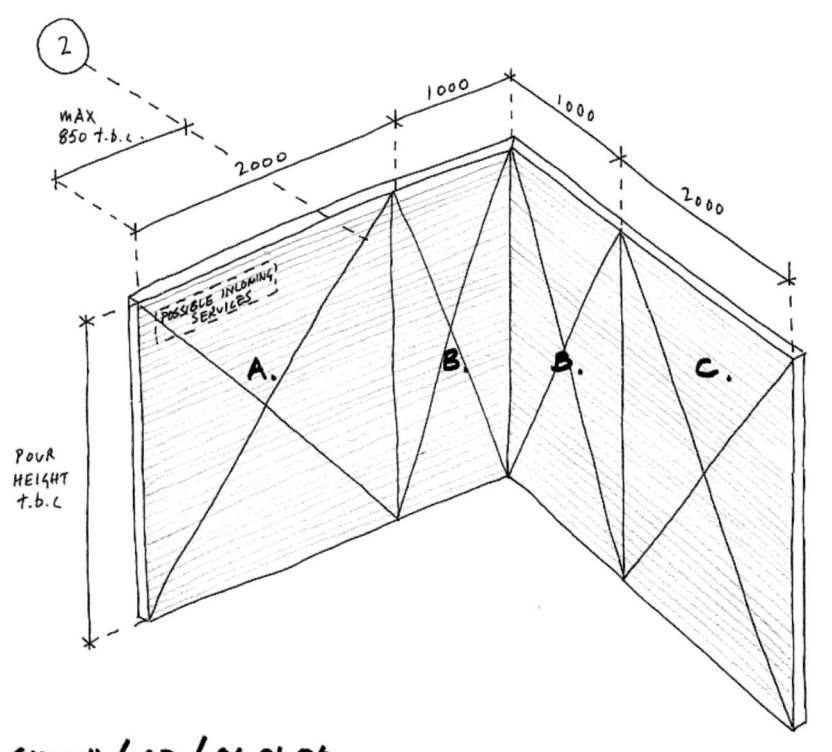

69

SK004 / GD / 06.01.06

SKETCH OF TIMBER BOARD SAMPLE PANELS IN BASEMENT INDICATING DETAILS OF JOINTS

1. CORNER DETAIL — CREATE TIGHT JOINT TO MINIMISE GROUT LOSS

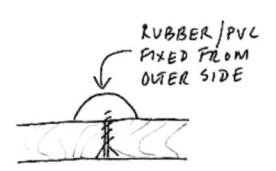

2. PVC/RUBBER REBATE — RUBBER/PVC FIXED FROM OUTER SIDE

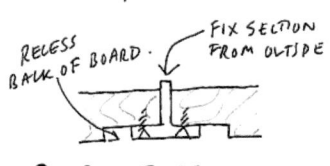

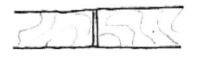

3. PVC T-SECTION — RECESS BACK OF BOARD, FIX SECTION FROM OUTSIDE

4. BUTT JOINT - NO REBATE

70

Architectural detailing examples
fig 69
Typical single storey high wall panel layout and section

fig 70
Sketch of timber board sample panels in basement indicating details of joints

ARCHITECTURAL INSITU CONCRETE - TECHNOLOGY

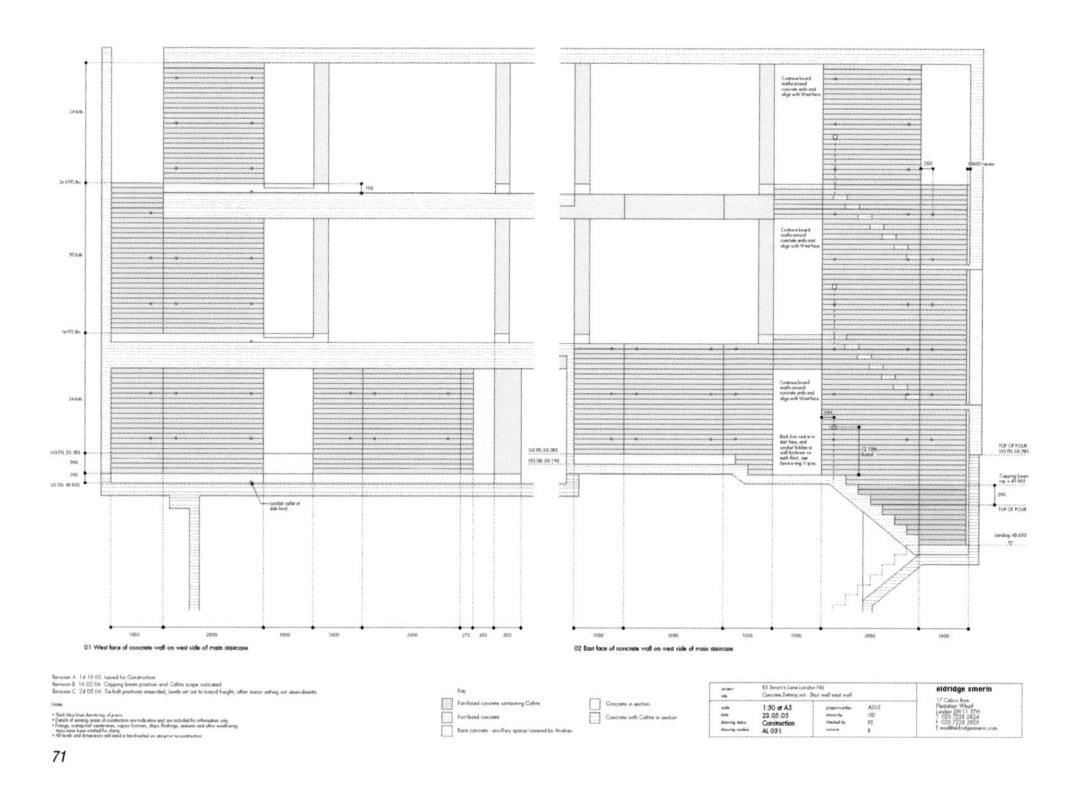

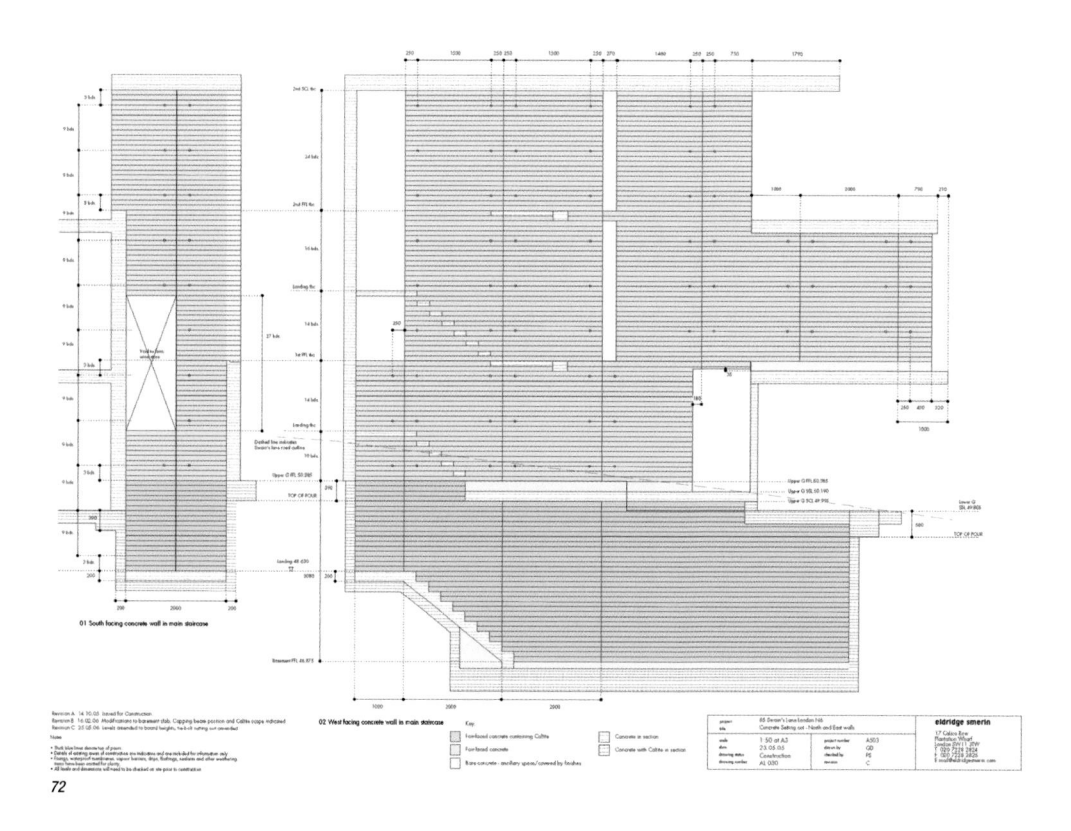

figs 71, 72
Elevation drawings – board-marked walls
and preferred tie bolt locations
(Swains Lane/Eldridge Smerin)

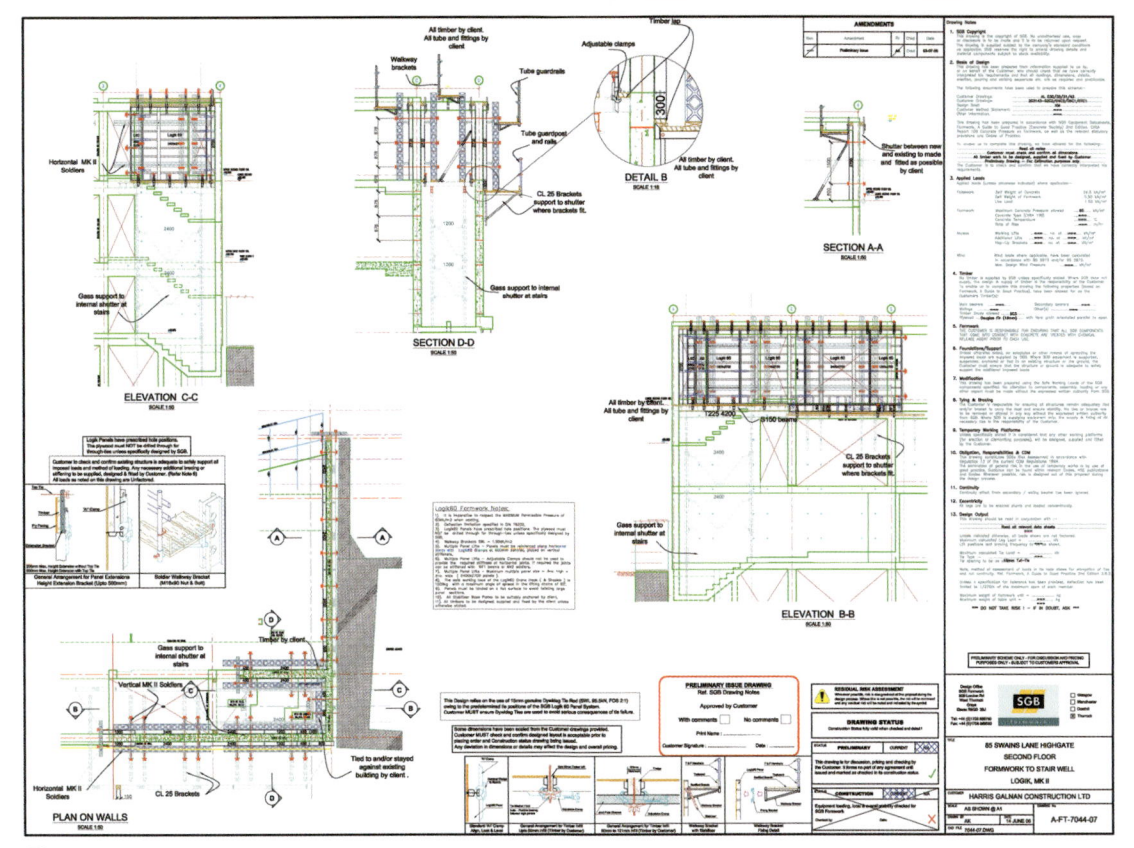

fig 73
Construction drawing: formwork support system and assembly details
(Swains Lane/SGB Systems)

fig 74
The finished board-marked wall
(Swains Lane/Harris Calnan)

CONCRETE WORKMANSHIP

INTRODUCTION

The handling, transportation, placing, compaction and consolidation of concrete are of critical importance in achieving the best finish possible. The concrete will be delivered to site by truck mixer with a uniform and consistent colour and at the correct workability and cohesiveness for placing by skip, by pump or by tremie pipe.

During transportation the concrete must not be allowed to segregate or stand for any length of time or allowed to lose workability. In all cases the concrete must be taken to the point of placement as quickly as possible.

Placement and consolidation of the concrete are closely integrated and should be considered as one operation when planning a pour. The rate of concrete delivery should match the rate at which the concrete can be placed and compacted in the vertical formwork or the slab decking. Compaction or consolidation of the concrete by internal vibration densifies the concrete, removes the air voids and can produce a blemish-free finish.

Sometimes, no matter how carefully planned and well executed the operation has been, minor blemishes and imperfections are still left on the surface when the formwork is removed. An approach to carrying out a skilful repair is described below but it should be considered a last resort. It is far better to leave a minor blemish and do nothing than to repair it, as there will always be a marked contrast in the surface appearance, unless you have a very skilled person doing the work.

HANDLING AND PLACING

On large building sites concrete will be discharged into the formwork by a skip or by a pump. The choice of placing method will depend on the type of structure to be poured, the site access and cost constraints.

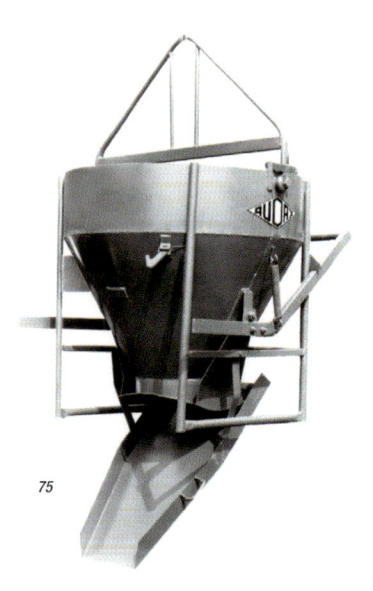

SKIPS

Columns and walls are usually poured using a concreting skip that is moved by crane. The skip will have a capacity of 0.5, 0.75 or 1 m^3; some have a bottom-opening chute with a wheel controlling the opening and others have fixed, sideways-opening chute operated by lever that discharges concrete at an angle. The bottom-opening chutes, which can have a flat hose or tremie pipe attachments, are the best as the concrete is directed over the point of placement with greater control. The capacity of the skip and what it can carry will be governed by the reach and lifting capacity of the crane. Placing concrete by crane and skip is a slow operation as each load has to be taken to the point of placement, the skip held in place while the concrete is discharged, then returned to the truck mixer to be reloaded. It is the best method of placing controlled amounts of concrete in columns and walls which do not take more than 5 m^3 or 6 m^3 to fill. The truck mixer should be discharged well within the hour as, in hot weather, the cement may start to set and the mix begin to lose workability. That can cause a blockage of flow in the skip and lead to segregation of the concrete in trying to unblock it by internal poker vibration.

Concrete mixes with slumps of 100 mm or less and a maximum aggregate size of 20 mm may be difficult to discharge quickly from skips with narrow or tapering openings as arching and blocking of the flow may arise, particularly in skips with sideways-opening chutes. It is important that the chutes are clean, the internal faces are wetted down before being used and the lever mechanism and wheel controls that open the chute or discharge ports are working well.

The build-up of hardened concrete on the outside of the skip may be prevented by spraying the surface with chemical release agent.

fig 75
Skip with sideways opening discharge chute

fig 76
Bottom opening skip with flat hose discharge tube

77

PUMPING CONCRETE

Concrete pumping is the usual method of placing concrete on large and small building projects, both for small single pours and for large pours when a number of pumps are required. The main reason for its popularity is due to the cost benefits of a faster placing rate, fewer labour requirements and competitive hire costs – whole-day or part-day charge. It also releases the crane to handle other material for the project.

The most popular pumps are the lorry-mounted mobile pumps fitted with telescopic booms which can be on site at short notice on the day of the pour. They take up little site space and are usually parked on the road or hard standing. The boom can reach to the far corner of the site, vertically to the upper floors, extend over an obstruction, and even feed through a window opening on the third floor for casting internal walls in a refurbished building.

When the demand for concrete is on a daily basis or it is for a high rise building it is more efficient to have a static pump line installed on the site with a placing boom.

Pumps are capable of moving up to 100 m^3 of concrete per hour, depending on the pump type, the horizontal and vertical length of the pipeline, the number of bends and the concrete mix. In practice, the pump output is around 20–30 m^3 per hour due to supply intervals awaiting the next truck mixer and operational movements to alter the position of the boom or pipeline. Most pumps can transport concrete 60 m vertically and 300 m horizontally, but not at the same time. Distance is compromised by increasing height.

The concrete mix should have a workability of 125 mm or more to ensure that the concrete will pump easily without blocking. Often, pump operators insist on adding water to increase workability, without regard for the requirements of the project or concern for colour consistency. This practice must not be allowed to prevail. The concrete mix should be reviewed by the ready mixed supplier, who will design a mix suitable for pumping that also conforms to the requirements of consistent colour and minimum blemishes. It need not be over-sanded nor have a pumping admixture included. These requirements are necessary for porous, absorbent aggregates like Lytag – a sintered PFA aggregate – but not for a dense natural or crushed rock aggregate.

The new valve designs, hardier pipeline systems and improved pumping technology available today make it possible to pump large aggregates and harsher and drier concrete mixes without encountering the clogging and abrasion problems that plagued pumping in the early years.

In planning a pour, work should be arranged to start from the furthest point back towards the pump so that the static line sections of the pipes are removed rather than added as the pour proceeds. Before commencing pumping operations, the pipeline is primed by passing a grout mix through it. The first 200 kg or 300 kg of mixed grout and concrete are used for lubricating the pipeline and must be discarded as they will cause surface discoloration.

On large pours, a stacking area for trucks is necessary to minimise disruption to road traffic and queuing. It is useful to have a pump

78

fig 77
Mobile concrete pump

fig 78
Extent of reach of mobile pump

61

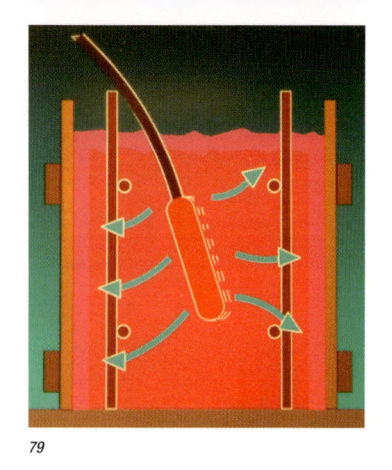

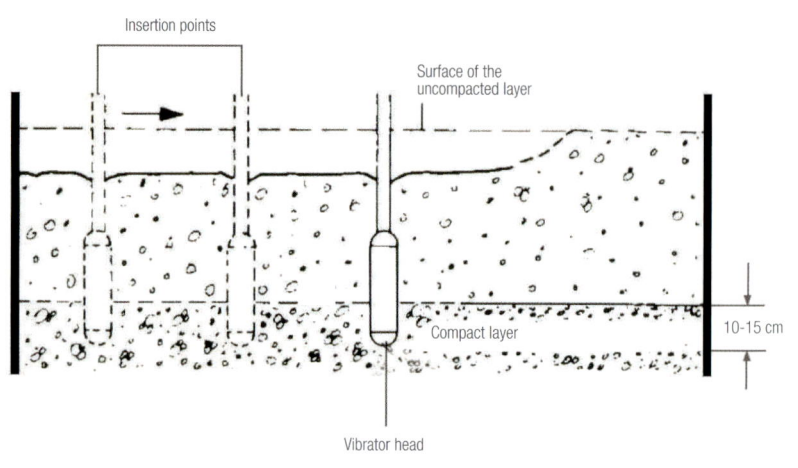

with a hopper, which allows two truck mixers to discharge concrete at the same time.

TREMIE PIPE AND LAY-FLAT HOSE

A tremie pipe with a hopper is commonly associated with underwater concrete operations, where the pipeline is usually 250 mm in diameter. The tube has an end plug to prevent water rushing into it when the concrete discharge begins. The bottom of the pipeline is kept in the body of the placed concrete to seal it against water ingress. The pressure head of the concrete is maintained well above the water-line to maintain the concrete flow. This technique has been adapted for concreting in dry conditions to place concrete at the bottom of shuttering to minimise free fall and overcome the risk of segregation.

When concrete is to be placed in walls or columns that are 4–8 m high, a plastic tremie pipe of, nominally, 100–150 mm and sometimes 75 mm diameter (depending on the workability and working space) is inserted into the middle of the formwork between the rebar to within 1 m of the bottom. The concrete is discharged through the pipe to avoid the mix getting caught up with rebar and tie bolts, trapped by cover spacers or becoming segregated when the coarse aggregates strike the rebar as the concrete free falls to the bottom of the shutters. Free falls of more than 2 m is not good practice and should not be permitted, especially when a fine finish is required. The pipe is removed and the length shortened as the concrete fills the shutters. A series of tremie pipes is inserted into the wall at set interval for casting a long length of wall. The concrete is placed by pump or by skip into a receiving hopper, which has a discharge nozzle that slides into the tremie or fits over it.

Farmers use a lay-flat hose for irrigation because the hose is flat when not in use so can be rolled up and is easy to transport. A heavy-duty lay-flat hose with a coupling and an end closure tie can be used instead of a plastic tremie pipe. It is often attached to a bottom-opening skip or hopper. These hoses are frequently used for placing self-compacting concrete as they can be sleeved between the rebar easily when flat. Care has to be taken to avoid air pockets becoming trapped in the hose as the concrete is poured and to ensure that the hose does not become snagged as it expands to release the concrete.

COMPACTION AND CONSOLIDATION

When the concrete is placed in the formwork, as it drops to the bottom to meet the rising level of the wet concrete within the formwork it traps air. That air has to be removed by compaction using internal vibration, otherwise the surface will be riddled with unsightly large and small blowholes. Internal vibration also consolidates the mix to release air pockets and voids contained within it and to meld the concrete into a closely packed, uniform material.

fig 79
Correct compaction technique for walls and columns

fig 80
Thorough underfilling of inserts through one-sided pouring and vibration followed by revibration

81

A fully compacted and consolidated concrete will be dense, have an even colour, will be impermeable and durable.

Concrete placed in walls and columns is compacted on site using internal poker vibrators. Ground-supported floors and suspended slabs will use vibrating screed rails combined with internal poker vibration. With floor slabs the air pockets rise to the surface to leave the soffit free of blowholes, and so the compaction technique is not as critical. The main issue here is to avoid cold joints and pour lines appearing on the soffit. That can be overcome by a continuous and steady flow of concrete to the slab and systematic working methods.

INTERNAL POKER VIBRATION

Poker heads for compaction vary in size from 25 to 75 mm diameter. The section width, pattern of reinforcement and character of the concrete will determine the correct poker diameter to use.

The compaction and consolidation of the concrete is imparted by the vibration of the poker head. The type of drive and the efficiency of vibration are critical in the production of a fine finish. The constant amplitude, high-frequency electrical compactors developed by Wacker Ltd are the most efficient hand-held vibrating pokers currently available and are ideally suited for producing a fine finish. The vibrating mechanism is contained in the poker head, which is isolated from the electrical cable that is attached to it so no vibration is transferred to the hand. This eliminates the risk of developing hand–arm vibration syndrome (HAVS), which has become a major health issue on building sites. HAVS is caused by repeated and frequent use of hand-held vibrating tools, for example power drills, pneumatic and motor-driven poker vibrators and concrete breakers. It may also be caused by holding or working with machinery that vibrates. It has been estimated that up to 1 in 10 people who work regularly with vibrating tools may develop HAVS.

The radius effect of the vibrator head determines the optimum diameter of the poker. The Irfun and similar designs vibrate at constant amplitude of between 12,000 rpm and 10,000 rpm, and never less than that. A microchip in the circuitry informs the head to draw more current to maintain rotational speed at or above 10,000 rpm as it is immersed in a rich, dense concrete mix, when the resistance to shear and the force required to liquidise the mix increase. As a guide, the radius effect of the poker can be assumed to be ten times the poker diameter, i.e. for a 35 mm poker it is 350 mm, for a 50 mm poker it is 500 mm. It is advisable to ask the manufacturer to provide this information. But what about the insertion intervals and maximum horizontal depth of compaction to ensure that there are no blowholes and that the concrete is an even colour for the full height of the wall?

CORRECT COMPACTION

The concrete is vibrated upwards from the lowest tip of the poker. It creates a cone or cylinder of compaction or fluidity by transferring the agitation to the coarse aggregates and the mortar. The angle of the cone is steeper for low workability and wider for highly workable concrete. With an Irfun compactor, the whole of the casing is vibrated to create a cylinder of fluidity. The radius effect of the poker indicates the zone of compaction radiating from the poker head and gives the maximum interval between insertion points. It is good practice to have the zones of compaction overlapping each other. For example, for a 35 mm diameter, the insertion intervals would be 500 mm with a 200 mm overlap. If the same energy in compaction is given to the concrete, the same force pushes the fine particles of cement, sand and water to the form face and surrounds the coarse aggregates. This results in an even colour of the compacted layer and eliminates the risk of vertical colour banding due to a drop in vibration effort, such as is associated with pneumatic and mechanically driven compactors.

To ensure that the air pockets are removed effectively, the poker head has to be immersed quickly into the concrete layer and drawn up slowly. This avoids the upper layer being compacted first, as this can make it difficult for the air bubbles and water to migrate from the lower sections. The poker should be withdrawn at 40–80 mm per second, depending on the mobility of the concrete. Withdrawing the poker quickly is one of the main causes of poor compaction. The effectiveness of the compaction can be recognised

fig 81
Constant amplitude electrical internal vibrator (Wacker Ltd)

63

by the air rising to the surface at the beginning of the withdrawal cycle and then reducing and stopping when full compaction is achieved.

The dangers and problems arising from under-vibration are far greater than those associated with over-vibration, since it is virtually impossible to over-vibrate a properly designed and proportioned concrete mix.

The horizontal layers of compacted concrete should not be greater than 500 mm as this will increase the risk of large air pockets of 10 mm or so being trapped on the surface.

As each new layer is placed over the compacted layer, the poker should be inserted about 150 mm into the lower layer to knit them together and to unify the concrete.

To avoid the risk of seeing visible pour planes of slightly different tones, particularly in the summer when the cement sets more quickly, the pouring rate of the concrete must be faster than 2 m/hr vertically. It is important not to use the vibrator to move concrete into place as this may cause mortar to separate, resulting in streaking and segregation which may appear on the surface.

Touching the reinforcement above the pour with the poker when the lower section of the concrete is setting is not good practice. Besides loosening the bars, it may impair the bond of those that are embedded. Also, a dark ghosting of the rebar pattern, especially a mesh fabric, may appear on the hardened face, which can be unsightly.

The poker head must not be allowed to touch the form face as this will cause burn marks and abrasion, which will show up on the concrete. The poker for walls or columns should always be inserted between the layers of reinforcement and within the centre or middle part of the structure. It must be kept 75 mm to 100 mm away from the form face.

A rubber shield should be used over the poker head for working in very congested reinforcement to minimise accidental disturbance of the bars and reduce any marking on the form face.

When compacting small columns and thin walls it is vital to avoid the temptation to insert the vibrating poker once more through the compacted layers for good measure as this will increase the hydrostatic pressure acting on the formwork by as much as 30 per cent. This is often the cause of grout loss and shutter movement, especially at the corners of columns and at the bottoms of walls.

REVIBRATION

The top 75–100 mm of the concrete in deep section, such as deep beams and sometimes columns and walls where GGBS cement is used, has an increased tendency to display blowholes and for the fine material and bleed-water to rise to the top. If left, this can create a tidemark on the finished face and may lead to plastic settlement and plastic cracking. Revibration of the top layer within an hour or so after first compaction and before the surface has set will reduce the risk of plastic cracking, blowholes or tidemarks forming.

The poker head should be running before it is inserted into the concrete and should sink under it own weight into the concrete. When it is withdrawn, the concrete should flow back into the space occupied by the poker head.

FLAT SLABS

Vibration beams, in combination with internal poker vibration, are used to compact concrete in floor slabs. Details on vibrating beams can be found in literature on ground-supported slabs and is outside the scope of this book.

When using the internal poker vibrator to compact the slab, the poker head should be allowed to fall at an angle into the concrete almost parallel to the soffit, then pulled slowly along, fully immersed below the surface, to effectively vibrate the slab concrete. Vertical insertion points may only be necessary over deep sections, such as rib and coffer beams, and along the edge of the slab.

REINFORCEMENT AND COVER

Before use, reinforcement should be stacked off the ground and be well supported so that it does not become covered with mud and grease. Cut and bent bar, mesh and fabric reinforcement should be delivered to meet the construction schedule and bundled neatly, with bar-mark labels visible so that they can be easily identified.

The rebar needs to be free from loose rust, mud, grease and any substance that will affect the bond with concrete or can cause staining on the shutter soffit. This is critical for floor slabs and beams with exposed concrete soffits. Where possible, use clean blue steel and, before placing the rebar on the deck, brush away any scale or rust. Place the rebar as fast as possible in position when assembling a large area of flat slab construction, use prefabricated single-directional or two-directional rebar panels where possible to speed

fig 82
Compacting concrete for a flat slab

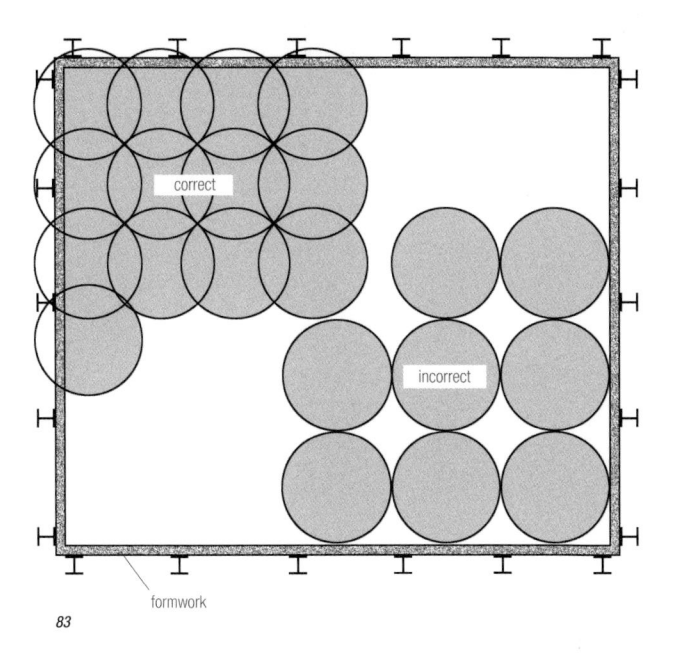

rebar installation and concrete the area as soon as possible.

For vertical elements such as columns and walls there is no harm in rust scale being on the bar as this will not affect the finish. The critical concern here is that the starter bars for the next lift should be coated in cement mortar to prevent rust-saturated rainwater running down the newly cast concrete, staining the fair face surface. It is advisable to place polythene over the top of the wall or column, securing it by pushing it through or wrapping it around the rebar to act as an umbrella, directing rainwater away from the top of the concrete. This prevents excess free lime, which accumulates on the top of the wall, from being washed down the concrete face to cause lime and efflorescence stains.

Reinforcement must be fixed in the correct position and held rigidly in place and have the correct cover. Single bars are secured by using tying wire, which creates debris on the soffits of slab when the ends of the wires are snipped. These loose ends must be swept up and removed as they will show as rust marks when they oxidise.

The top layer of reinforcement in slabs is secured in position using reinforcing chairs. The shape of the chairs and where they are placed must ensure that the reinforcement does not sag or become displaced when walked on. The bottom reinforcement is held away from the form face by cover spacers and the position of spacers and arrangement of the rebar must be rigid when walked on.

The rigidity and design of the cover spacers is important as they may show on the concrete surface, particularly on the soffits of an exposed slab and the vertical face of columns and walls.

SPACERS

For vertical reinforcement, rigid, plastic 'open wheel' type spacers are the best choice as they have few points of contact on the form face. As they are not of solid construction they allow mortar and aggregate to pass around and through them and so are not seen on the surface. They should always be fixed in the perpendicular position to avoid creating a trap for aggregates. If they are positioned horizontally they can create traps and cause voids on the surface due to aggregate bridging. For the bottom reinforcement in a slab or beam, a dense, heavy-duty concrete spacer is essential. The preferred types for fair face soffit work are the double and triple 'figure of eight' or 'jelly baby' shaped concrete spacers. They should be positioned upright so that there is minimum contact with the form face. The concrete spacers should be comparable in strength, durability, porosity and colour to the surrounding concrete. For grey concrete mixes, the standard grey extruded concrete spacers are fine. For white concrete and pigmented concrete it is best to ask the manufacturer to colour match the concrete spacers. Site-made concrete cover spacers should not be used.

Mesh block spacers, continuous plastic chairs, grade plate spacers and wire hoop spacers are suitable for ground-supported slabs.

Spacers are usually spaced about 1 m apart, depending on the type and the reinforcement.

SURFACE PROTECTION OF REBAR AND NEWLY CAST WALLS

Protruding reinforcement starter bars should be coated with cement mortar to avoid rust stains running down the wall face. The tops of cast walls should be covered with polythene sheeting after removing formwork to prevent lime streaks or lime stains caused by rainwater running down the wall face. If there are any minor discolorations due to efflorescence or lime streaks, the surface may be lightly sanded to remove them.

fig 83
Compaction of a flat slab
Maintain correct spacing of poker insertions to ensure an overlap

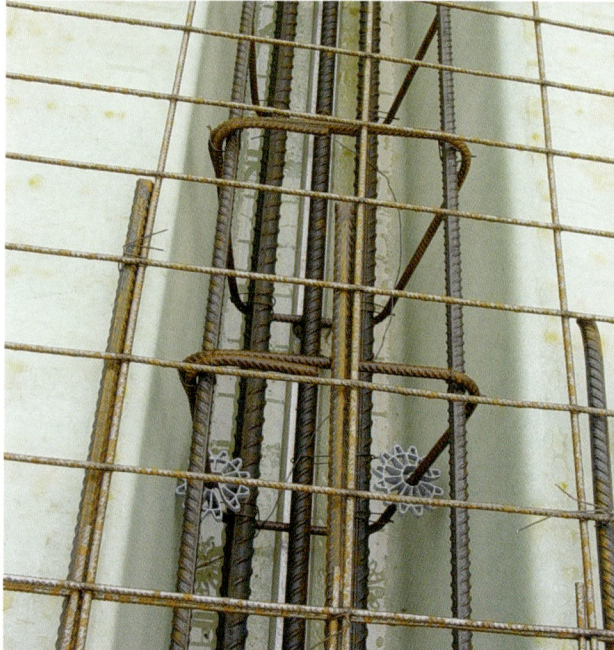

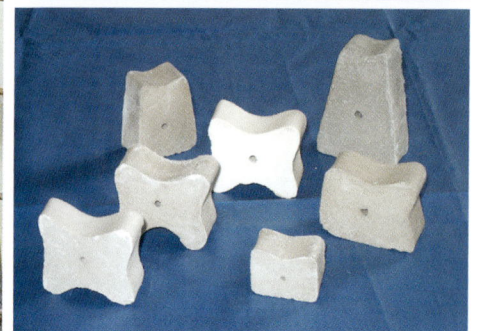

figs 84, 85
Spacers
Plastic wheel spacers suitable for lightweight vertical rebar

fig 86
Selection of plastic spacers (Creteco Ltd)

fig 87
Selection of concrete spacers for slabs, beams and columns (Creteco Ltd)

CURING

WALLS AND COLUMNS

For fair face concrete finishes the curing of concrete by the formwork is usually quite adequate, provided that the formwork is removed only when the surface of the concrete has reached a strength of, say, 10 N. In hot summer months this will be achieved after 24–36 hr. In cold winter months the period will be much longer and can be determined either by pull-out tests on a trial slab or wall area or by using maturity tables published in the codes of practice. Further curing when the formwork is removed is of little benefit and can impair the finish. The exception is for water-tight basement construction, when a curing membrane should be spray-applied to the surface as soon as the formwork is removed. Most curing membranes will stain the surface and darken the background colour or add sheen. An acrylic urethane spray coat which is also a waterproof coating will not mark the surface very much and will add a light sheen, but it should be tested beforehand. A dilute silane or siloxane coating applied by brush is the best option as it leaves no mark and is non-staining.

FLOORS

For exposed power-trowelled floor finishes and textured and abraded concrete a number of propriety surface sealers and curing membranes can be applied. For post-textured and abraded surfaces, when the surface laitance is removed, there is no harm in applying a spray-applied curing membrane as it will be removed. Otherwise, an acrylic urethane spray applied after power trowelling when the surface sheen has gone or a covering of polythene in close contact with the concrete will be effective. It is important not to allow the polythene to come loose at the edges or for air to get under the sheets in local spots, as this will cause dark and light coloration in the floor which may not fade.

TRIAL PANELS

The trial panels on a project are there to assess the quality of the surface finish, the accuracy and rigidity of the formwork system proposed by the contractor and the concrete workmanship. They are not built to judge the final colour of the concrete as that can take 6 months to 1 year before the concrete carbonates. The trial panel, whenever possible, should be incorporated in the permanent work in a non-critical area so the panel is built under the same working conditions.

The truck mixer should carry a full load of concrete which should be checked for uniformity of the concrete mix, by taking slump tests at the beginning, middle and end of the load. The concrete should be placed by the method that will be used for the actual work, so the required plant and equipment must be on site for the pour.

The effectiveness of the release agent can be assessed and waterproof coatings checked for any surface staining and discoloration.

If the joinery and formwork produce true, accurate and well-defined edges and corners and the concrete workmanship results in no blemishes, no grout loss and very minor blowholes, then the concrete trial panel has met the requirements of a fine finish.

CLEAN, DIRT-FREE SURFACE

The concrete surface, once it has hardened, has lots of minute air holes on the surface formed by water vapour capillaries. This is how the concrete dries out; the excess water trapped in the matrix is expelled as water vapour passes through these micro-pores.

In addition, there may be blowholes around 2 mm or 3 mm in diameter that are visible on the surface, caused when air was trapped when the wet concrete was placed in the forms.

The blowholes are conduits for dirt and moisture and the capillaries allow moisture ingress into the concrete and encourage microbes and dust particles to encrust the surface. If left unchecked, the concrete will soon start to look dark and grimy.

Its important that any water repellent that is used to coat the surface does not prevent water vapour escaping from the concrete so it can dry out. It must not prevent CO_2 from entering the capillaries to carbonate the outer skin of the concrete and lighten the colour as the CO_2 combines with the available free lime present to form calcium carbonate.

The water-repellent coating must not be visible when it dries on the surface and must not change the concrete colour or yellow or degrade under UV light.

What is this magic potion? It is a dilute silane or siloxane coating that forms an invisible barrier on the surface to prevent rainwater ingress yet allows water vapour to escape and CO_2 to penetrate the surface. It forms chemical bonds with the substrate. The visual appearance of the treated surfaces will show no change resulting from its application. The benefits include enhanced resistance to efflorescence and freeze–thaw damage. The cured system is resistant to UV light. Rainwater runs off it like off glass, while the concrete remains clean and unblemished.

It is used to protect and maintain bridges, car park structures, airport pavements, industrial plants, precast cladding panels and other concrete structures. It penetrates and chemically bonds with the concrete substrate to provide long-lasting protection against water-related staining or deterioration. It will not produce a surface film or impair the natural breathing characteristics of the treated surface.

The manufacturer should always be consulted regarding its suitability and method of application. It is best brush-applied externally to reduce wastage when working in slightly windy conditions. Internally, spray and brush-applied coats are suitable.

STAINS AND BLEMISHES

It is inevitable that, no matter how carefully and thoroughly the formwork may have been constructed and the concrete placed and compacted, some stains and blemishes will arise. We need to understand the cause of such blemishes and how they can be avoided and remedied.

It has been assumed that the concrete mix design has followed the guidelines set out in this book, that it is not over-sanded, that it has a workability of between 125 mm and 150 mm slump and it has a minimum cement content of 325 kg/m^3. Likewise, it is assumed that the concrete placing techniques have followed the guidelines given here for placing and compaction.

If not, then for the next pour or trial casting all these points must be addressed and the recommended procedures carried out.

GROUT LOSS

The most usual surface imperfections are due to grout loss. This can be seen at the corners of columns which are nibbled and devoid of grout; at horizontal construction joints at the base of the wall where a skin of concrete mortar has hardened over the joint or the concrete is bare and honeycombed between panels. This is caused by movement of the formwork at those locations due to the support system and/or tie bolts not being robust and rigid enough to resist the hydrostatic pressure of the wet concrete.

Solution

> *Columns*: the formwork system must be adequately braced and secured at corners to prevent movement. The formwork should be designed to resist the hydrostatic pressure generated by concrete with a placing rate of 6 m/hr. If there are any doubts about security, horizontal clamps should be doubled up over the bottom half. Clear sealer or thin draught-strip should be placed between the panels' butt joints at corners, to create a grout-tight seal when the shutter is closed. Above all, the formwork system must be checked and designed by a specialist formwork supplier to resist the full hydrostatic pressure without movement. The support system recommended by the formwork supplier should be hired and used. Ignoring this advice and relying on the contractor's experience to build the formwork system without calculation is a common mistake.

> *Walls*: the support system must be designed by a specialist formwork supplier to ensure minimum deflection of soldiers and waling and no extension of the tie bolts. The pressure should be taken for a placing rate of 3 m/hr with concrete temperature of, say, 5 °C in winter when the risk is greatest. Tie bolts must be designed to be within their safe working limit of load. If they extend by as much as 1 mm at the bottom, where the pressure is greatest, there will be some grout leakage at the base. The tell-tale sign will be a skin of grout forming a hard crust over the joint and/or dry honeycomb patches at the bottom of the face, and the concrete standing slightly proud of the lower wall line.

EFFLORESCENCE

Surface discoloration is caused by white efflorescence and lime stains over the concrete surface after the formwork is removed.

This is caused by early removal of the formwork, when the concrete has not fully hardened, and occurs in cold, damp conditions. The excess free lime on the surface of the immature concrete absorbs moisture and CO_2 rapidly to crystallise calcium carbonate or efflorescence deposits on the surface.

Solution

The formwork should be maintained in place until the concrete has achieved 10 N. Pull-out tests or maturity tables can be used to give guidance for minimum striking times.

DARKER COLOUR TONE

The darker concrete colour tone of slab soffits and walls using the same mix is caused when the formwork is left in place for extended periods.

This is caused by the excess water in the mix, which was added for workability, being driven out as water vapour during the hydration or hardening of the cement and condensing at the form face and saturating the outer skin of the concrete. When formwork is removed after 24 hr or so in the hot summer months this condition does not occur and the concrete colour is much lighter.

Solution

Do nothing! Allow the excess water to evaporate from the concrete naturally. It may take several months, but as it does so it will carbonate by absorbing CO_2 which will form calcium carbonate as it combines with the free lime and will return to a light colour. It is

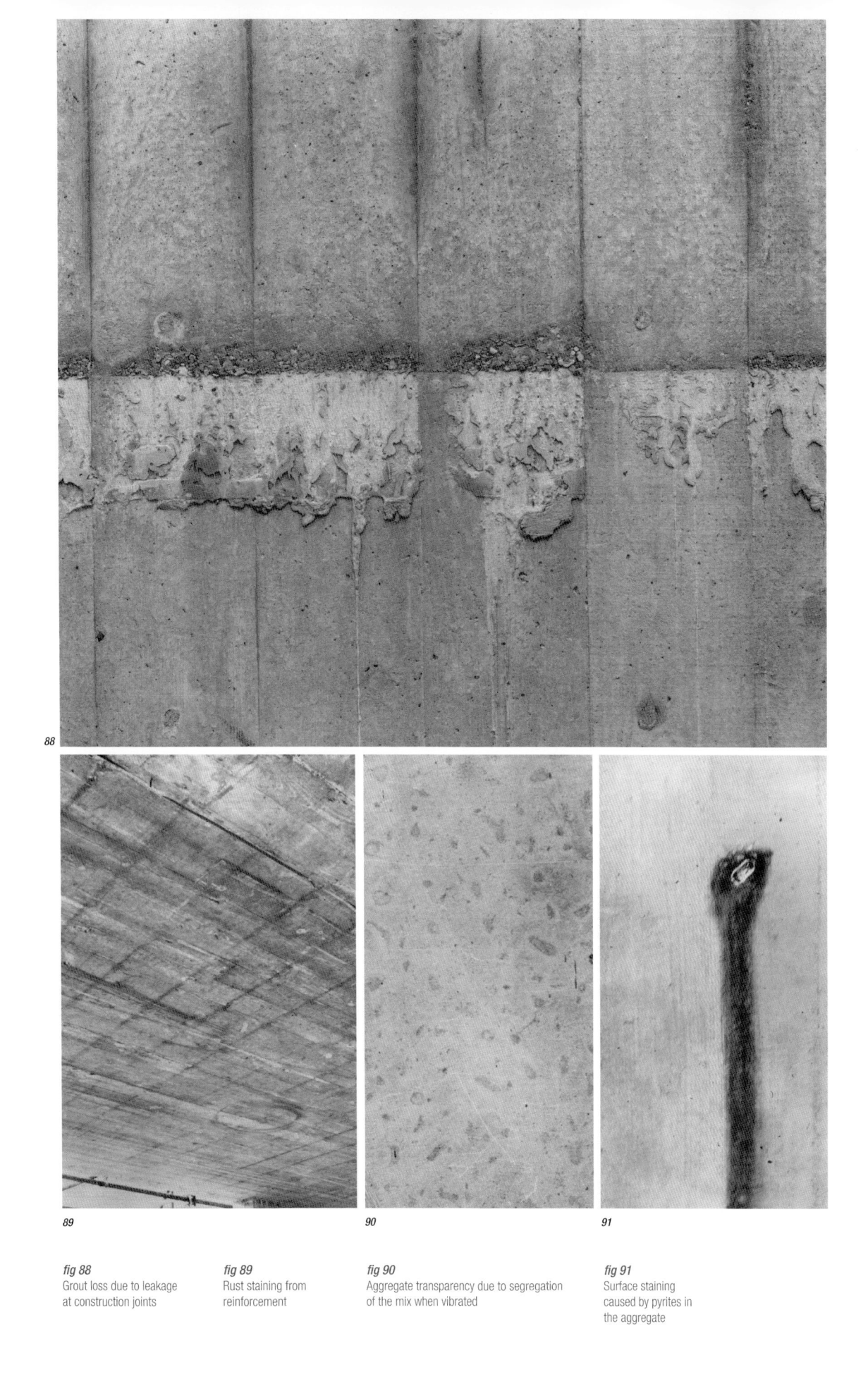

fig 88
Grout loss due to leakage at construction joints

fig 89
Rust staining from reinforcement

fig 90
Aggregate transparency due to segregation of the mix when vibrated

fig 91
Surface staining caused by pyrites in the aggregate

69

advisable to ensure that the surface does not become re-saturated with rainwater and so the surface should be coated with a water-repellent but vapour-permeable coating such as a dilute silane or siloxane.

PATCHY SURFACE DISCOLORATION

This can happen when the concrete is poured directly onto the fair face formwork near the top and is allowed to travel down the face to the bottom of the pour. Whitish patches and flow lines appear on an otherwise smooth, dense, blowhole-free surface finish.

Solution

It is important never to pour against the fair face formwork as this may leave an excess of cement particles on the surface at these points and may even scour the form face. Use a tremie or direct the flow to the non-exposed side.

REMEDIES

SURFACE DISCOLORATION

To remove surface discoloration caused by rust stains, lime streaks and patchy superficial marks, the surface should be washed with dilute hydrochloric acid. A test patch should be carried out first.

Alternatively, the face can be lightly sanded down with fine sandpaper, leaving no scratch or circular score marks. A test patch should always be carried out first. Sponge blasting may also give the same result.

If this does not give a good result a light grit-blast can be tried, removing 1 mm of the surface laitance without exposing the aggregate. It will leave a slightly textured surface. Initially, the surface colour may be a tone darker, but once it begins to carbonate it will lighten up.

CONGEALED SURFACE GROUT RUNS

Thick grout runs and fingers of grout can be carefully tapped and removed by a bolster as they are not very hard and will chip away from the background concrete.

HONEYCOMBING AND NIBBLED CORNERS

Where possible it is best to do nothing if, overall, the honeycombed area or nibbled edge does not look unsightly. The more repairs are made to the surface, the patchier and more unsightly it is likely to become, and then the best remedy may be to paint it. It is possible to repair a honeycombed area well, but the procedure requires skilled workmanship and plenty of time.

If wall or column corners have suffered grout loss and the edges look like nibbled biscuits and have lost definition, then a concrete repair can de done.

A concrete repair of uniform colour-matched concrete is virtually impossible to produce. The best that can be achieved may fall short of the finish of the surrounding unblemished concrete. Acceptance of this fact and a willingness to compromise on the part of both the architect and client are therefore desirable. If the repair is still visually unsatisfactory but the surface profile and edge definition has been restored well, it may be advisable to colour-wash the whole surface to mask these visual differences using appropriate materials.

REPAIRS

Repairs should be carried out as late as possible, when the original concrete has fully hardened and started to carbonate and lighten in colour.

A skilled person with a steady hand and a patient disposition should be selected to carry out the work.

MORTAR MIX

For filling of blowholes, honeycombing and repairing chipped, nibbled arisses, a mix of 50/50 Portland cement and white cement with silver sand in the proportion one part total cement to two parts silver sand by weight should be used. A styrene butadiene rubber dispersion agent (SBR) should be mixed with an equal amount of water and this gauged liquid should be added to the dry mixture of materials until a mortar of just-moist consistency is achieved.

MAKING GOOD

Blowholes

The surface must be wire brushed to remove any loose material and to ensure a good key for the repair mortar. The substrate should be thoroughly damped down to control and minimise suction and, while the surface is still just damp, large blowholes should be punched in and filled with mortar mix. When the mortar has just set, using a wooden float with a dampened sponge rubber face the surface should be rubbed over to remove all excess mortar and leave the surface flush.

Honeycombing

The surface should be bolstered, chiselled and wire brushed to remove loose material. The surface of the honeycombing should be damped down with a dilution of one part SBR with three parts water and, while still damp, the mortar mix punched in with a suitable tool. It is vital to ensure that all corners are filled, there are no air gaps and a good tight finish is obtained around the edges. For restoring a board-mark finish, the face should be rubbed with a sponge-faced wooden float and the board placed on the surface and tamped with a hammer to obtain the board-mark effect. The board should be left in place for 3 days. The face of the boards must be coated with release agent beforehand.

A steel trowel may be used for a smooth phenolic face type finish and a wooden float for a matt surface finish. The patch should be covered with polythene and kept damp for 5 days.

Arisses

The surface must be wire brushed and damped down with one part SBR and three parts water. A board former should be applied to one face to hold the mortar in place and then the mortar mix punched in with a small trowel or putty knife. The surface may be rubbed with a wet sponge float when the mortar has just set or a smooth trowel as appropriate for either a board-marked or smooth cast finish and covered with polythene to cure for 5 days. The shutter face should be removed after 1 day and the surface rubbed as appropriate with a wet sponge float or steel trowel and covered with polythene to cure for 5 days.

Concrete surface sealer

One coat of brush-applied silane or siloxane water repellent should be applied in accordance with the manufacturer's instructions to provide a non-staining, water-repellent, vapour-permeable seal to the repaired surface when it has cured. A trial area should be tested before work commences.

fig 92
Problem wall with resin and tannin staining

fig 93
Close-up of the wall after dilute acid wash

figs 94, 95
Views of the cleaned walls

PART 2
CASE STUDIES
IN CONCRETE

Thames Barrier Park   075
Patel Taylor (2000)

Persistence Works   084
Feilden Clegg Bradley (2001)

The Art House   097
Fraser Brown Mackenna (2002)

The Anderson House   107
Jamie Fobert (2002)

Aberdeen Lane   117
Azman Owens Architects (2003)

One Centaur Street   127
de Rijke Marsh Morgan (2003)

85 Southwark Street   137
Allies and Morrison (2003)

The Bannerman Centre   147
Rivington Street Studio (2004)

The Brick House   157
Caruso St John (2005)

The Collection   167
Panter Hudspith (2005)

Playgolf, Northwick Park   179
Charles Mador Architects (2005)

E-Innovation Centre   189
BDP Manchester (2005)

The Jones House   199
Alan Jones Architects (2005)

Spedant Works   209
Greenway and Lee Architects (2005)

Central Venture Park   219
Eger Architects (2006)

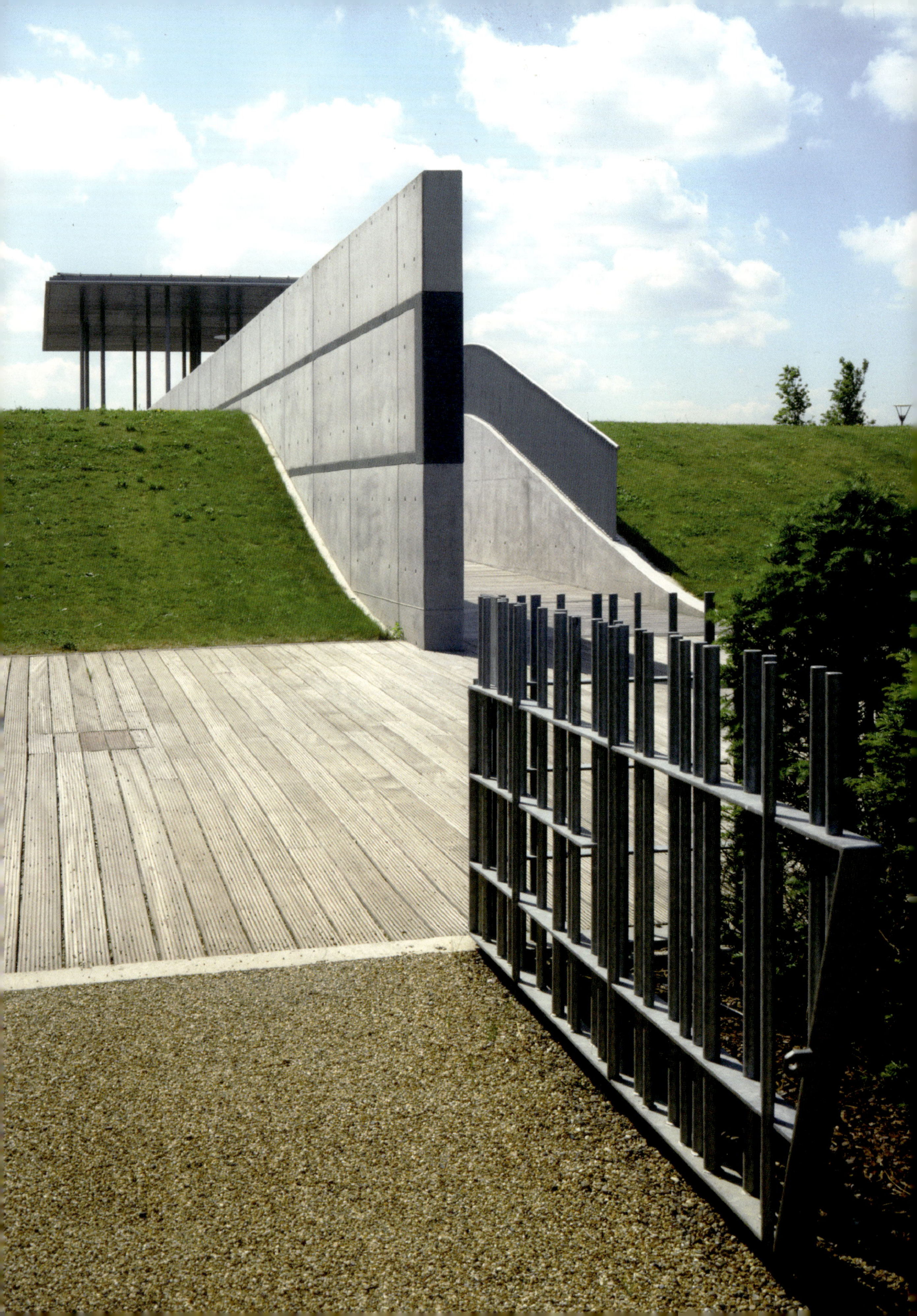

THAMES BARRIER PARK
NORTH WOOLWICH ROAD, LONDON

Architect: Patel Taylor

Location

The park is on the north side of the River Thames, adjacent to the Thames Barrier, and is approached by North Woolwich Road in Silvertown. By Docklands Light Railway (DLR) it is at the terminus of Pontoon Dock or a pleasant walk from the DLR station at the City Airport. Car and coach parking is free.

Introduction

Wondering where to go to this weekend with the kids or just looking for a place to relax, watch the river and enjoy a walk among 22 acres of imaginative planting? Head for Thames Barrier Park, it's got it all – play areas, picnic areas, river walk, seating, lawns, trees, contoured yew and maygreen hedges, five-a-side football pitch, a fountain plaza with 32 water jets to provide a cooling and natural entertainment for everyone.

One of the most imaginative and attractive features of the park is the Green Dock. Renowned international horticulturalists Alain Cousseran and Alain Provost selected colourful flowers and shrubs which reflect the river's ever-changing spectrum of tints, shades and shapes, creating a micro-climate where varieties of plants and butterflies abound.

A circuitous stroll along the 1 km boundary pathways will bring you to the Visitor Pavilion Coffee Shop where you can indulge yourself with gourmet coffee, tea, cold drinks, hot snacks, sandwiches, salads, pastries, confectionary and ices.

ARCHITECTURAL INSITU CONCRETE - CASE STUDIES

Architecture discussion
Andrew Taylor

We won the competition with our collaborators Group Signes and Arup. The site was 22 acres, located in a prominent position next to the Thames Barrier itself. An aerial view of the site shows the vastness of the man-made topography of Docklands and reveals a very level, dry area where the planes of water read like plateaux of blue under open skies. In a way, the urban park wanted to reflect that topography and we wanted it to look like a machined piece of green space.

When we first started looking at the area it was in complete transition. There was the Tate and Lyle building bringing in huge ships to their dockside refinery, a working scrapyard next door, piled high with rusting metal, that was incredibly noisy, an oil refinery to the west of the site and the remnants of Ronan Point to the north, near Custom House Square, which was being demolished. There was no visible linkage through such an infrastructure but the London Docklands Development Corporation (LDDC) had the vision to transform it with road networks, build the DLR and impose bold planning ideas. The site itself fronts onto the main road which was once called the Silvertown tramway because there were so many railway sidings that were the feeders for this once busy dockyard. We have a new DLR station built here called Pontoon Dock, which is part of the new spur from Canning Town to the City Airport.

The site was covered in 150 mm of crushed concrete as a defence against illegal vehicle entry to the Thames Barrier. Between the river edge and the road there was a 5 m drop, with the road being the lowest point and at river level height. When the docks were dug out they created bunds of high level plateaux. The LDDC had

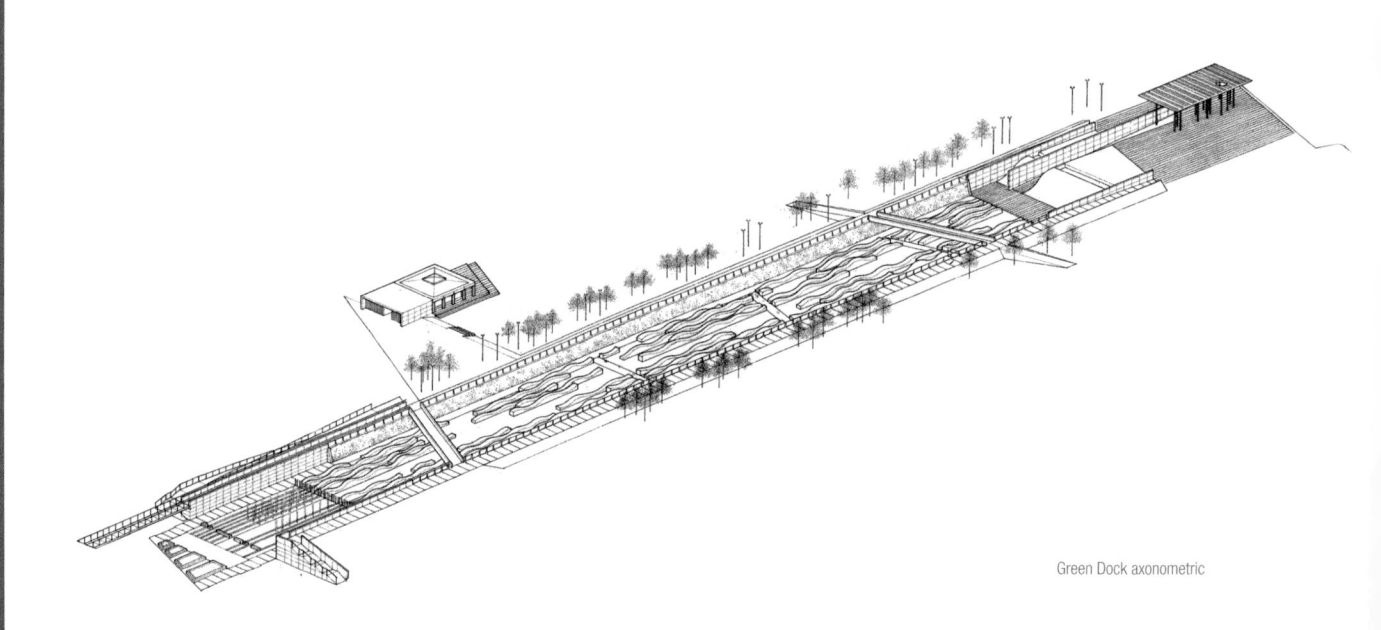

Green Dock axonometric

THAMES BARRIER PARK

the initiative to make the urban connection north–south across this vast plain leading from the Park in the south, then hopping across the waterways by bridges, slicing through Millennium Mill, to end up at the Excel Centre. The park idea was an initiative to bring value into the area and to complement the residential building that has mushroomed around the park.

As well as making urban connections with other parts of Docklands and filling the space with greenery, we had to invent a context for the park as there was no existing community. There were no details given in the competition brief, there was nothing around the park to make reference to. This was a scheme where the end result was truly a team effort, where everyone had an input. The engineering strategy with Arup was to make the land formation and in doing so we broke one of the few rules of the competition which was not to disturb the land profile that existed on the site. We cut a slice into the ground to create the Green Dock, allowing us to use the fill to raise the north of the site to level. We also had to excavate pits for the trees and we used the fill for further landscaping. The Green Dock is the machined landscape and within it are fountains, formal planting, concrete retaining walls, the Pavilion and Visitor Centre and overhead bridge links.

The architectural concept was to enclose the park and frame it like a composition. Halfway through the contract, the scrapyard nearby was operating as noisily as ever and we often experienced strong, bracing winds blowing across the site, but when we went 7 or 8 m down into the cutting of the Green Dock there was silence and stillness.

The axis of the park and framing of it mean that as you walk out of it going north you can take the new footbridge over North Woolwich Road which is at park plateau level and continue towards the Excel Centre. The plateau of the park is framed by concrete elements such as seating making perimeter thresholds; the cutting is defined by the concrete blade walls, the front area by the ha-ha with gates and river frontage – a manicured piece where we did some repairs to the river wall with sheet piling and concrete capping beams.

The plan appears quite simple but it is very detailed topographically. The main datum of the park is 6.50 m AOD; the walkway at the back and along the edges is 7.50 m AOD. The river side is 6 m AOD. There are pathways between the formal rows of planting on the plateau and a diagonal path that runs from the north-east corner, crossing the Green Dock by bridge to reach the

top: Green Dock seen from the Pavilion
bottom: Green Dock and Rainbow Garden

77

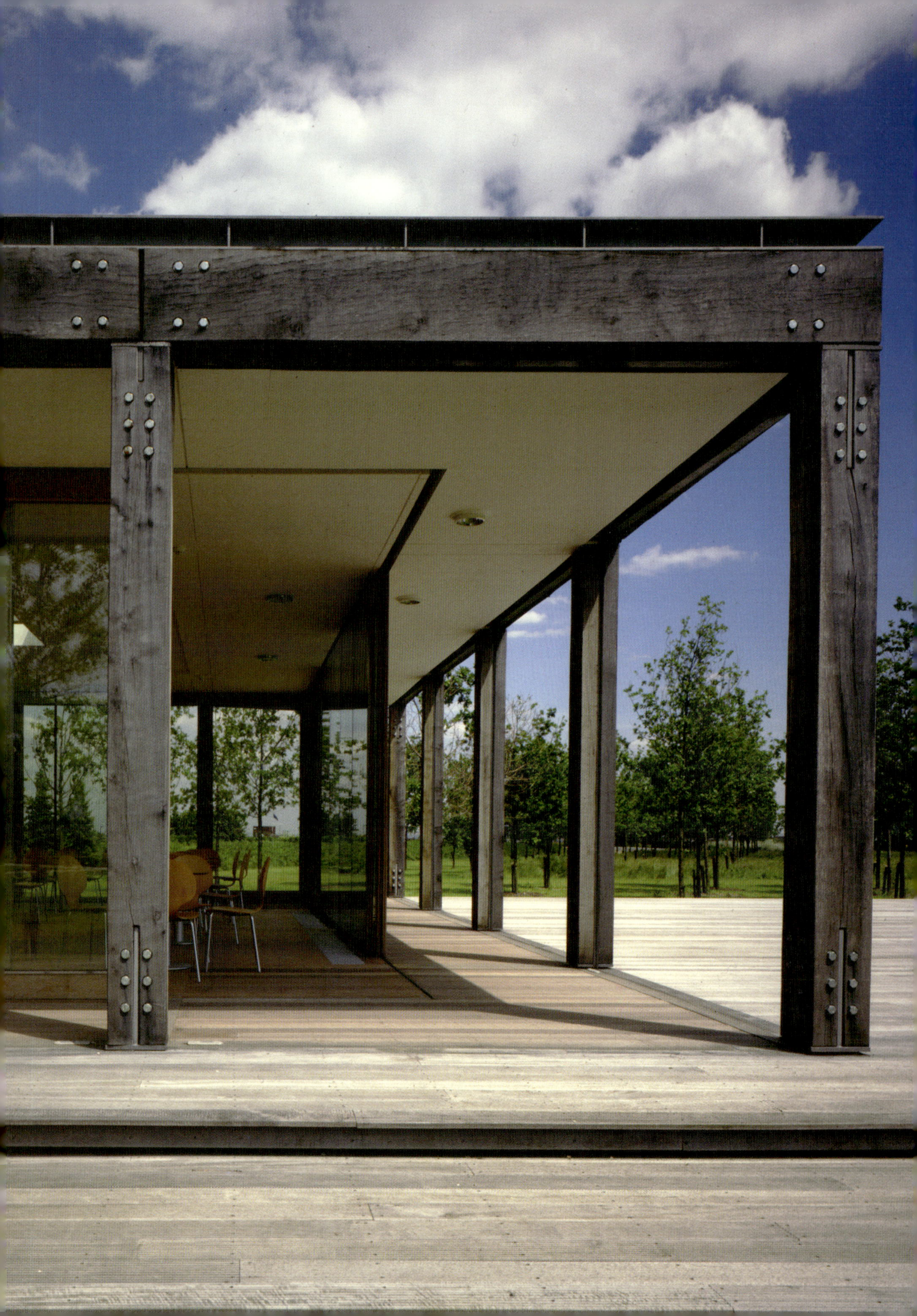

south-west tip of the park facing the river and, with the 300 m long Green Dock, these create the axial links that tie space together. The plateau is 250 m square, which is a kilometre in total along the perimeter. The plateau has a distinctive character; it is planted with meadow grass and tree lines of silver birches, oaks and specimen trees. Also, with so many landscape and scale changes, with a cutting 7 m deep and a canopy 14 m high, there is a lot of visual interest and depth.

All the level changes are made with small or large retaining walls and handrailing. The idea of Green Dock was to have concrete walls framing the 30 m-wide formal planting, the granite paving and the fountains. When we were considering the detailing of the paths and walls, we used photographs of the pontoon dock. What we liked about it was the robustness of its construction, the sheet piling and the concrete.

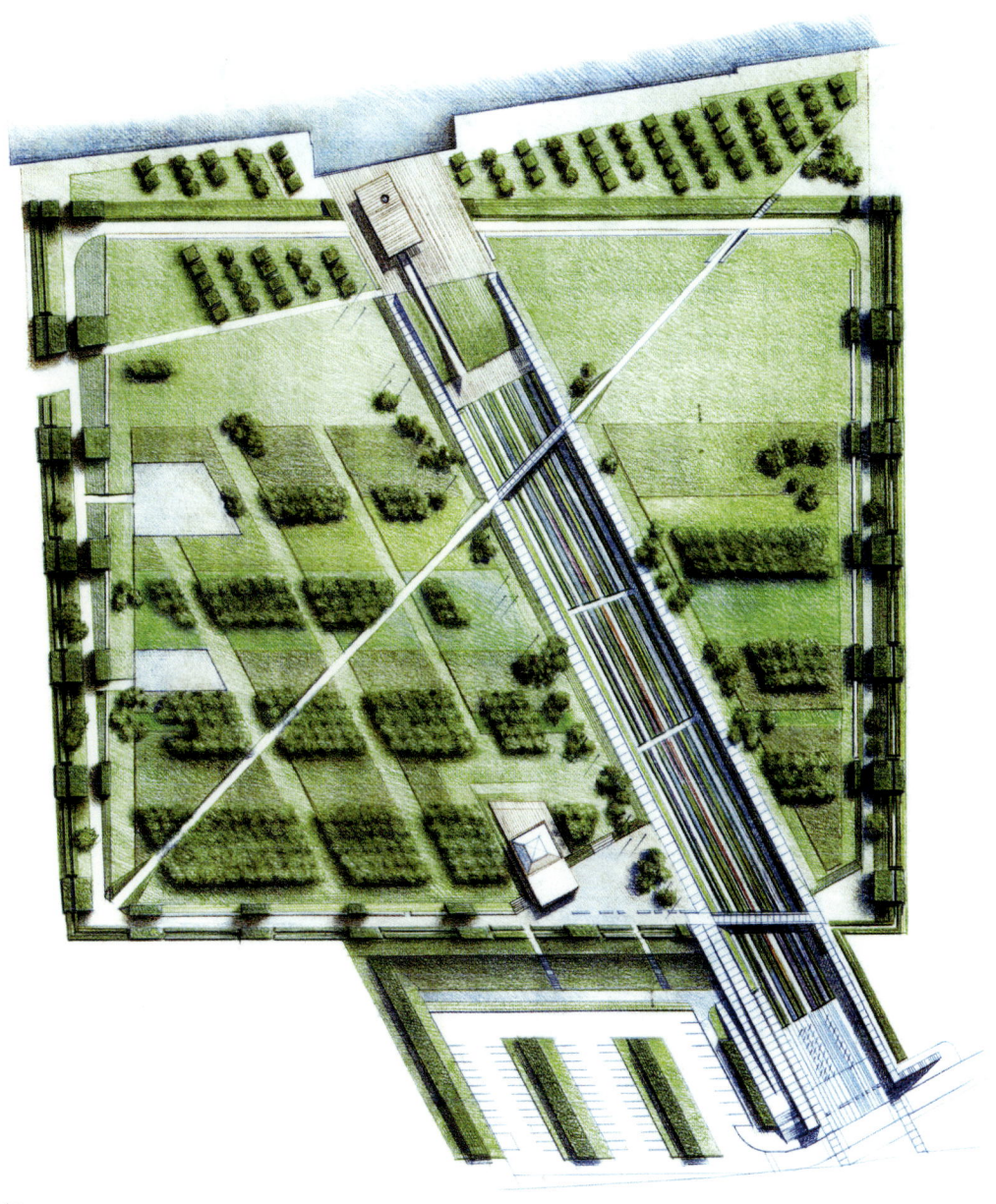

above: Site plan
left: Visitor Centre

ARCHITECTURAL INSITU CONCRETE - CASE STUDIES

We kept the granite strips that covered some sections of the concrete the same unit sizes as the individual sheets of ply formwork that made up the concrete wall. The concrete was rebated back 25 mm to create inset panels into which we glued the 20 mm-thick granite slab panels to maintain a continuous front line. The exposed concrete was cast with a positive ariss running on the horizontal line of the butt joint between the 1,220 mm-wide ply sheets. These lines were continued but rebated on the darker granite panels. The concrete contractor used a film face ply, which was fitted to a rigid support system that was moved as one piece to make several castings 6 m long and up to 7 m high. When we were considering the detailing of the wall, the client was concerned about the large areas of concrete, so we added the positive ariss detail to break up the flatness. Carlo Scarpa is a recurrent theme in our work, from whom we drew inspiration.

We quite like rough cast with smooth surfaces and the contrast of materials. We used green oak with concrete with no banding on the Visitor Centre, which works well. We felt that the concrete walls had to be simple and efficient and built as smooth-faced slabs of concrete, and we would contrast that by hanging banded panels of grey granite in places. On the whole, the concrete quality was very good.

The entrance to the park has a ramp designed as a box culvert where people can circulate and which is used to house the pumping gear for the fountains. In essence, it is a free-standing structure backfilled in parts. In the cutting there is a walkway with mesh spanning above the planting, which you walk over before you enter the gardens. The next element is the Visitor Centre which you can access via the bridge, which picks up the north walkway of the site.

We had to solve the problem of a building that included a café, toilets, a kitchen and store rooms. Some areas don't want windows; others, on the café side, are all windows and the building sits in the middle of a park! The concept was to have a concrete box for the kitchen, toilets and store rooms and a green oak framed café with the glazing set back 1 m, around the perimeter. Metal shutters slide down inside the green oak frame to secure the building. We drew inspiration from the recessed joints on Louis Kahn's Dacca assembly building. The contrast between two materials is echoed by the concrete and greenery in the park.

The v-joints were formed by routing them out of two sheets of ply, which worked very well. The bolt holes were set out to a grid that

Pavillion and canopy

we had drawn; some of them were dummy holes. The panels were laid out on the ground after placing the strong backs and metal walings and face fixed with nails, which were set out. You do see the nail holes and sometimes, where they damage the surface, patch repairs with plastic body filler have been made and you see the difference, but these were in small, isolated areas and were acceptable.

Along in the pontoon dock, we designed an insitu edge slab rather than precast panels to which the handrail was bolted. Insitu created a slightly weathered, coarser finish which was more like stone. The slab had a saw cut running parallel with the handrail about 450 mm in from the front edge. The pathway along the top edge of the Green Dock was sand blasted and the handrail section left smooth. On exiting the Green Dock we introduce banding into the walls to reflect strata as you walk towards the Pavilion and waterfront.

The Pavilion canopy is 7 m high and supported on slender steel hollow sections in a random pattern and was quite complicated to analyse structurally. The plaza is timber-lined and the canopy soffit is the same timber and so, as you look out over the Thames, you are sandwiched between timbers. On the river front there is a gravel pathway that is bonded with resin. The balustrades are cast iron with stainless steel handrails on the river edge; elsewhere in the park they have hardwood handrails..

Concrete material

One of the first things we tell the contractor is not to touch any of the concrete if it has blemishes or faults. To wait until we have come to have a look to agree what's to be done. Too many times they bag it up or patch up the surface in the hope that it will be OK and it rarely is.

We did samples, tested out formwork and spacers and preferred a smooth finish. We wanted a white concrete but that was expensive. The next best was GGBS blended with grey cement, but in the end the grey OP worked best. Sometimes, when they first opened up the shutters the concrete looked patchy and disappointing, but we knew from experience that in 6 months it would be fine. It's hard to imagine when it's your first time, hard to be certain that doing nothing will improve it! Arup's R&D helped us to draft the concrete specification and prepare the construction detail. And I like the analogy that David Bennett uses, that a good piece of concrete is about good joinery.

On this project we were on site virtually all the time and checked every bit of formwork before the contractor was allowed to pour. The contractor cast the first wall on the Pavilion, and that was condemned and they pulled it down. There were lots horizontal pour lines, surface blemishes. The formwork was well crafted but we think they did not get good service from the ready mixed company and there were delays between the pours.

They then started on the walls of the Green Dock and that went very well. We had a good working relationship with the concrete contractor.

CONSTRUCTION NOTES

Anthony O'Connor, Konform Ltd

Our involvement was to provide formwork, install reinforcement and cast concrete to all the retaining walls, seating areas, Pavilion building, the capping beams along the Thames and the four short span pedestrian bridges within the development. The value was £750k and we finished in December 2000, having been on site for 2 years.

Design information came through to us quite slowly – these were the rebar schedules, detail for formwork panel layouts and we were made to start the work out of sequence so that we ended up using more ply than we had priced for. The original programme had been to start with the capping beams next to the Thames and work our way out towards North Woolwich Road, leaving the area behind us free for other trades to follow, such as the landscape gardeners and so on. The actual information issued was not in line with that, and we had to work on the smaller walls first and build the larger, taller ones afterwards, which meant we wasted a lot more formwork and materials and our labour was not as productive.

The supply of concrete at times did cause us a few problems. The main contractor had selected Tarmac Topmix as their batching plant was just up the road, less than 5 minutes from the site entrance. Because the batching plant was so close to the project we found that some truck mixers had not mixed the ingredients thoroughly before they arrived on site. On one occasion we were pouring some oval-shaped columns to the bridge. It was a Friday, the mix was dark and it appeared very wet to begin with but had passed the slump test. When we poured the other columns from the same truck mixer we noticed it was very stony and harsh. When we struck the four columns the next day, two looked fine and two were awful. We were instructed to carry out remedial repairs to the honeycombed column surfaces.

With regard to the walls, we used Ainsworth Pourform 107 ply which has a medium dense paper overlay. For the backing system we used aluminium waling beams at 350 mm centres horizontally, supplied by RMD, and used our own strong backs at 1 m centres. We assembled the system for wall pours that were up to 7 m long and up to 7 m high. The top line of the wall had to be level while the bottom reduces as it follows the rising slope. It was logical to start with the biggest panel and work our way towards smallest, but the main contractor had us working the other way. In addition, the base of the wall had been cast with stepped levels to account for the slope, which made it necessary for us to completely dismantle the wall pour and reassemble it to cast the adjacent wall. It would have been much better to have kept the base at one level and imported fill to create the slope, as it would have been faster and cheaper.

We used nails to fix the ply to the backing system to a regular pattern and sealed the nail heads with resin to give a clean strike with no rust marks. The sheet size shown on the drawings was 2,400 × 1,200 mm, which meant we had to cut every sheet to trim one edge by 20 mm and other by 40 mm since they are supplied as 1,220 mm × 2,440 mm panels. It was too late to remind the architect that sheets are 8 ft × 4 ft.

We trimmed the sheets on site by hand-held electric saws and for the nail patterns we made up a template. The assembly was laid out on the ground and brought together. As we were using aluminium beams to link across the sheets, the whole unit would be quite flexible when picked up by crane and 'sprig' causing the sheets to twist and warp. We inserted cross-braces to provide additional rigidity. The shallow rebates on the face ply for the granite inserts were made of timber 20 mm deep with a width of 90 mm, which was splayed for easy release. Detailing like this reminds you that the walls were originally designed as precast units and were later changed to insitu but they left the granite insets in place. When we struck the shutter, the rebate remained in place as they had been fixed with only panel pins and mastic. We waited a few days so we could pull them out easily. If they are fixed rigidly to the shutter, the ariss could get damaged. Every time we reused them they had shrunk a bit, even though they had been varnished and oiled. So the rebate became narrower. Later, when the granite strips came to site, we had to diamond grind and chamfer the rebate to get them to fit.

Other problems we had came from red staining due to iron pyrites in the coarse aggregates. They were using marine dredged aggregates and we were told it was unavoidable and we had to make good. Also, we were getting differences in the sand colour and that affected the shade of the grey colour. Part-way through the project the main contractor changed concrete supplier. The new mix was easier to place and trowel than the previous mix, which was harsh at times and difficult to trowel due to the microsilica, which made it sticky, and the GGBS.

The dwarf walls to the balustrade around the perimeter had a splay with 1:8 falls and when we used the mix with GGBS included, we had problems. It was stiff to place but once it was vibrated it became very sloppy with a lot of bleed-water rising to the surface and very difficult then to trowel. After placing the concrete that afternoon, many hours later we were still trying to trowel the surface, which had not set. The next day the surface was powdery, delaminating and crazed. The technical adviser from Tarmac took one look at the surface and said it was the wrong concrete mix for trowelling to falls. They changed the mix to a C40 fair face with OP and 20–10 mm single sized rounded aggregates, which gave an excellent finish and we had no further problems with bleed-water or trowelling.

When we were placing GGBS/OP blended concrete for the main walls we were told, some 2 hours after compacting and filling the wall, that we had to go back to revibrate the top 500 mm to reduce the excess bleed-water that had risen and which could cause a discoloration. This was extra work which we had not allowed for and we also had to strengthen the formwork supports. Despite all this, the finish was never the best, leaving a dusty surface every time. When they changed to a C40 OP mix we did not have that problem.

Rebar was so congested in the main 7 m high retaining walls that we could only get a 75 mm tremie down past the lower half of the formwork. The top half of the tremie tube was 150 mm wide and so long as the concrete had good workability it would flow down the tube. We only used the one 7 m tremie tube located centrally in the wall pour with the flexible hose of the pump connected to it. We positioned internal poker vibrators each side of the tremie to cause the concrete to flow and distribute along the formwork. It proved impossible to move the tremie pipe and thread it past the rebar cage to move it along the formwork.

We achieved at least seven uses of the ply using Adolease chemical release agent. The nail heads were covered with 'form fill' which did not leave any rust marks on the concrete surface as plastic filler sometimes does. We covered the formwork overnight with plastic sheeting to reduce the likelihood of any waterlogging and swelling of the timber due to rain.

PROJECT TEAM

Client: London Development Agency
Architect: Patel Taylor
Landscape designer: Group Signes
Structural engineer: Arup
Main contractor: May Gurney
Concrete subcontractor: Konform Ltd
Completion date: December 2000
Project value: £12.5 million

PERSISTENCE WORKS
SHEFFIELD

Architect: Feilden Clegg Bradley

Location

The monolithic structure of the six-storey Persistence Works buttresses the top end of Brown Street, the art hub of the city. It's a good walk from the railway station and not far from the shiny steel façade of the defunct National Centre for Popular Music, now the student union building of Sheffield Hallam University.

Introduction

From street level, wherever your eyes take you the building is clean, unmarked and refreshingly crisp. Persistence Works is a large insitu concrete structure conveying a freshness and vigour in appearance that seems everlastingly new. The pale-grey concrete that cloaks the building and is exposed internally does not give any hint of an overbearing, dirt-encrusted, brutal presence. It is quite the opposite. This is no-frills architecture with an Ando-like concrete finish that is honest, bold and on an impressive scale. The manner and determination with which the architect, engineer and contractor went about saving cost without diminishing design integrity, has been both imaginative and bombastic in execution. Having to halve the frame cost to keep the overall building cost within budget and still have a structure, takes some doing.

The new arts building had to offer 68 high quality and affordable studio spaces for crafts people and artists – jewellery designers, metal bashers, sculptors, potters, painters, weavers and illustrators. Some studios had to be large enough to contain 6 m-high sculptures, while others were the size of a spare bedroom for working with jewellery items. The proportion of high ceiling spaces, conventional design studio space and small work spaces was predetermined by the mix of artists that had occupied the cramped old cutlery warehouse, rented by the Yorkshire ArtSpace Society.

The ground floor has the high ceilings that contain the large studios, while the upper and lower levels of the taller six-storey building facing away from Brown Street contain the smaller studio spaces.

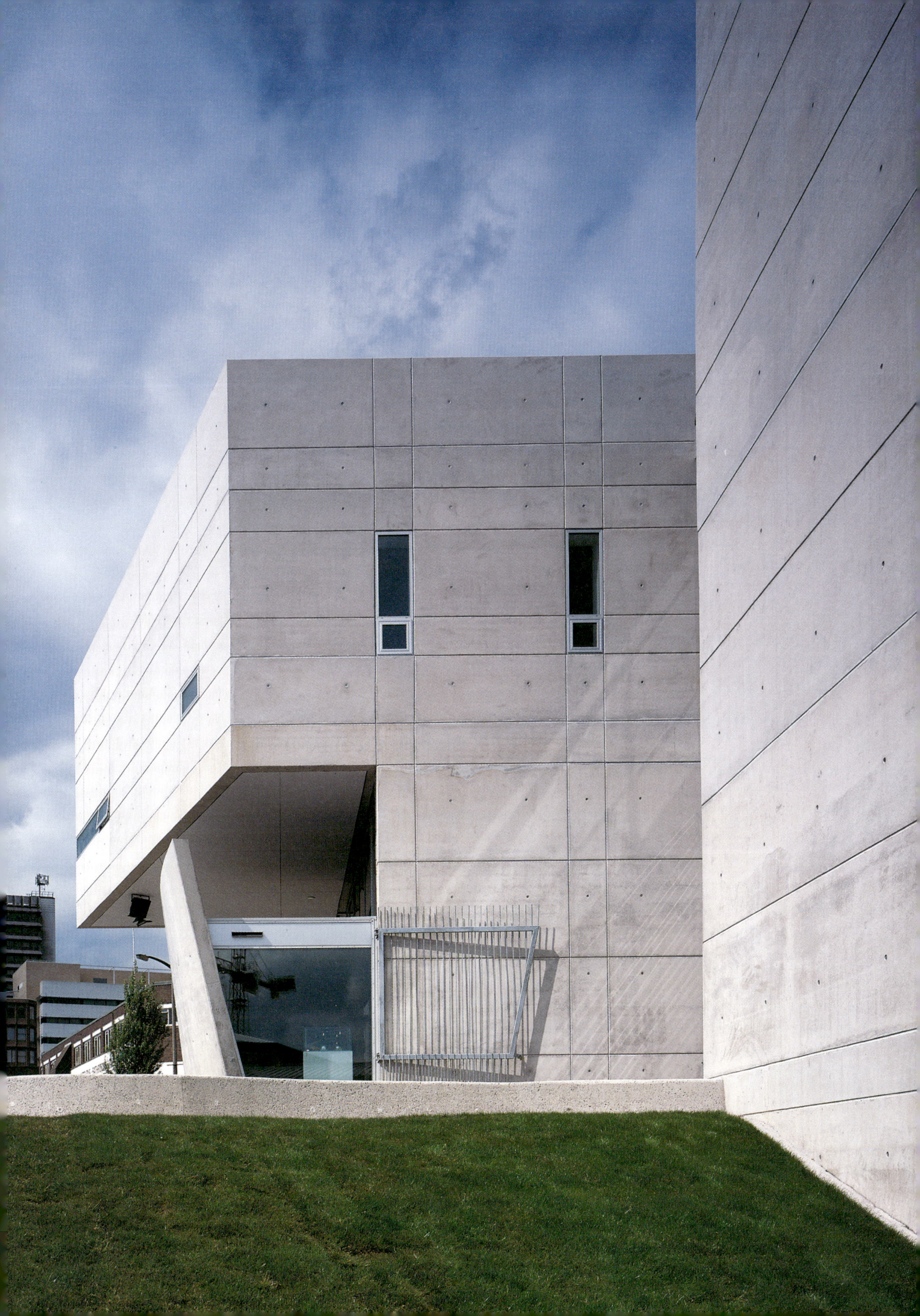

ARCHITECTURAL INSITU CONCRETE - CASE STUDIES

Architecture discussion
Julia Kashdan-Brown, Partner

The ArtSpace project as a new building type was a real challenge for us; we wanted to create something of its time, but echoing the industrial past of the Cultural Industries quarter in Sheffield, where the building was located. Insitu concrete seemed a natural choice for the building. It was a material that was attractive as a raw material, that could be moulded into forms and it seemed appropriate symbolically for an arts community. It was also robust, durable and economic – reflecting the ruggedness of the brick-built cutlery warehouse the artists had outgrown.

The challenge we set ourselves was to ensure the concrete would not darken or become stained with prolonged exposure, and convince the client that concrete was visually pleasing! We showed Kate Dore, the director of Yorkshire ArtSpace, examples of what we considered to be beautiful concrete buildings, like Kahn's Salk Institute and some of Tadao Ando's work. We visited seminal buildings like the National Theatre and the former David Mellor cutlery store. We also researched the internal white concrete at the Museum of Scotland in Edinburgh and talked to the concrete subcontractors, O'Rourke. We had long discussions about modern materials and the more we talked about concrete, the more appropriate it seemed and the more enthusiastic the client became.

The success of the project and use of insitu concrete throughout came about through strong support from the client, and her belief in our ability to deliver, the commitment of main contractors MJ Gleeson and, of course, a lot of careful research and trial sample panels to set the required standards and finish treatment. It has been extremely gratifying that the concrete looks so fresh and that it seems to improve in appearance as it ages. It feels like a building that has been there for a long time.

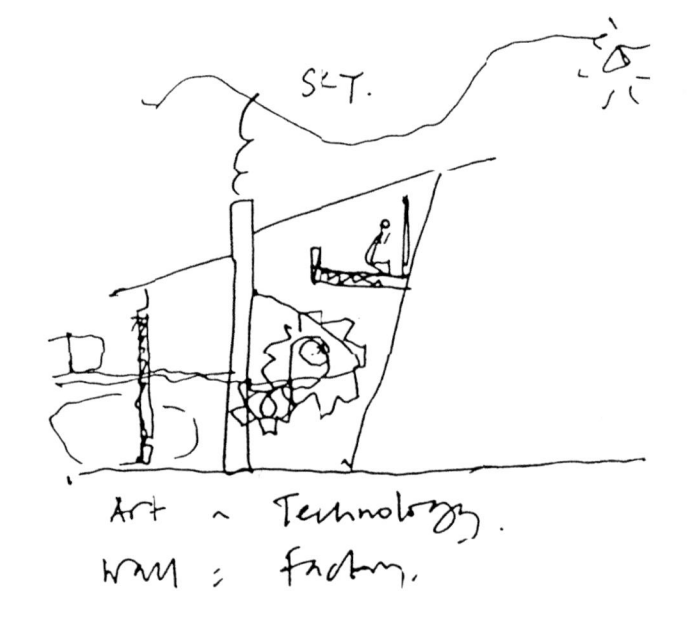

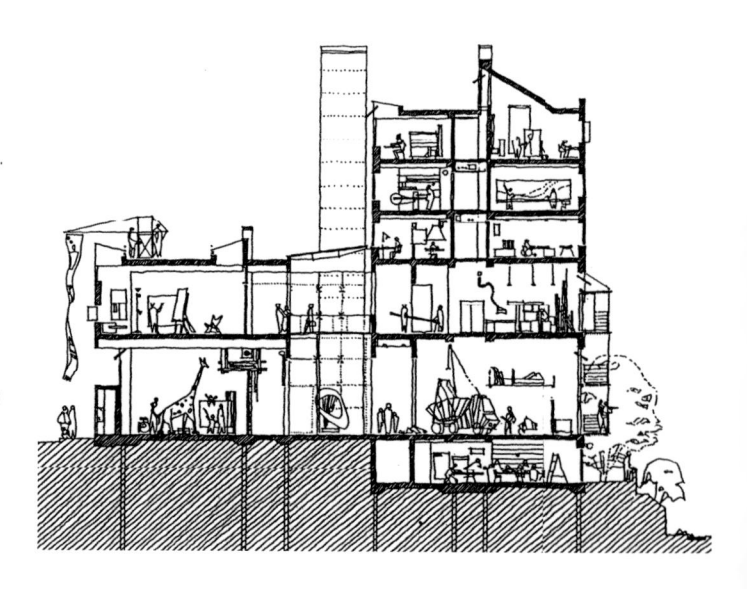

Concept sketches

PERSISTENCE WORKS

Architecture considerations
Toby Lewis, Project Architect

It was a delight to have such a committed contractor to build our project. MJ Gleeson, the main contractor, declared their intention to bid for the concrete work at the outset. In contrast, many of the other concrete subcontractors who bid had either not read or not understood what was specified and were judged non-compliant.

A main concern in the design was cold bridging where the floor beam meets the external wall. Calculations of the heat loss at these points showed that it was fairly insignificant and there was a low risk of condensation. Concrete has an advantage over steel because its thermal mass, lower conductivity and vapour permeability makes this form of construction possible.

Modelling the surface colour

For the main surfaces we considered a concrete mix using white cement and took our reference from the Museum of Scotland in Edinburgh. Having talked it over with the contractor, O'Rourke, this proved to be too expensive. Instead we looked at pale-grey concrete colour mixes made locally in Sheffield using 40–80 per cent GGBS (ground granular blastfurnace slag) cement replacement. A number of samples at specification stage were made and we noticed a significant colour change, to a much lighter and slightly warmer grey, between 40 and 60 per cent replacement by GGBS. Beyond 60 per cent replacement there was not a significant further change in the colour. We chose a 60 per cent GGBS, 40 per cent OP cement blend. We discussed the performance of this concrete with Gleeson to ensure that there was enough heat of hydration for the concrete to set in winter without freezing. In addition, we agreed that the formwork should remain in place for 48 hours before striking to compensate for the possible slower rate of strength gain. The contractor assessed that this would not affect their construction programme. We had one incident in the middle of winter when a slab had not gone off after 4 days. It eventually cured and a test core drilled out of the concrete proved it was just on the limit of the required strength.

We initially tried to create a blue surface colour for the stair cores and a red for the curved wall to the entrance. We wanted to use integral pigments, but this proved very costly.

top: Rear elevation
bottom: Corner detail on front elevation

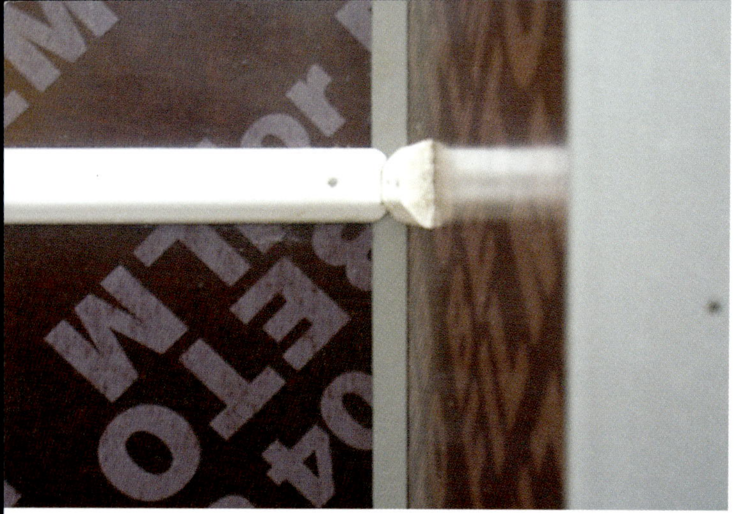
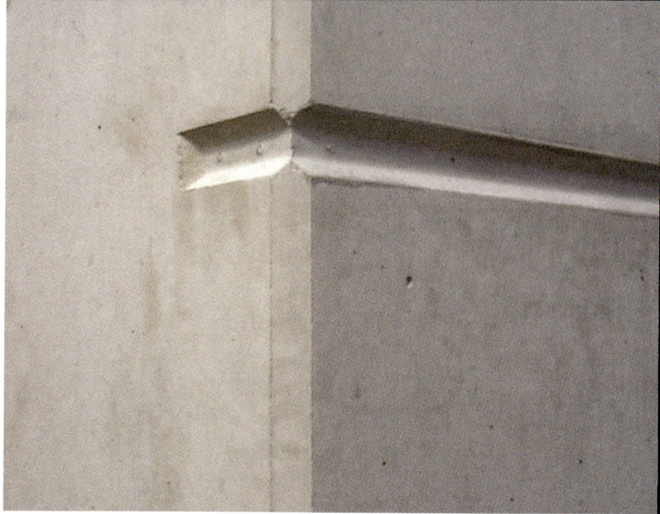
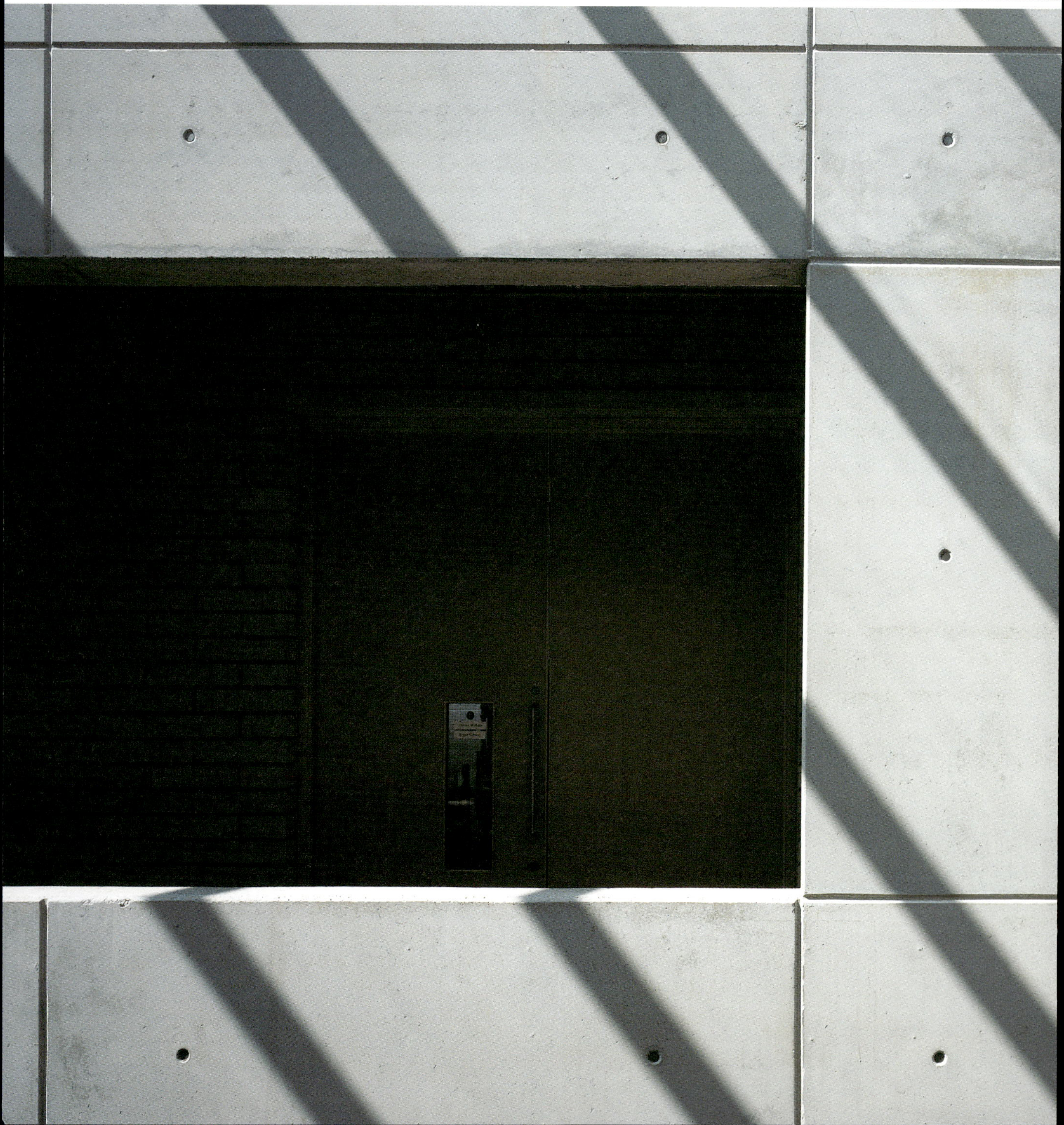

Alternatives considered included applied finishes and renders, blue tiles and cladding panels. In the end, the use of colour was omitted because of the cost.

There was an environmental benefit of using the GGBS, which allows a reduction of the cement content and saves money. It is a waste product which would otherwise go to landfill. By having 60 per cent GGBS cement replacement we halved the embodied CO2 of the concrete and achieved a lighter, warmer colour. Although it can make the hardening and finishing of the concrete more erratic in the cold winter months, by understanding the principles it is possible to work well with it.

Pre-contract samples

One of the aspects that made the project so enjoyable was the attitude and enterprise of the contractor. They recognised that their concrete expertise was based on work done some time ago. They put considerable research time into updating their knowledge and understanding how to achieve a fine finish. They carried out samples off site to look at formwork materials, release agents and to assess the choice of profiles to form feature rebates. They tested proprietary polyurethane faced, phenolic film faced and melamine faced plywood panels. In the end they used a phenolic film faced panel which gave a consistently good finish. The main difficulty was

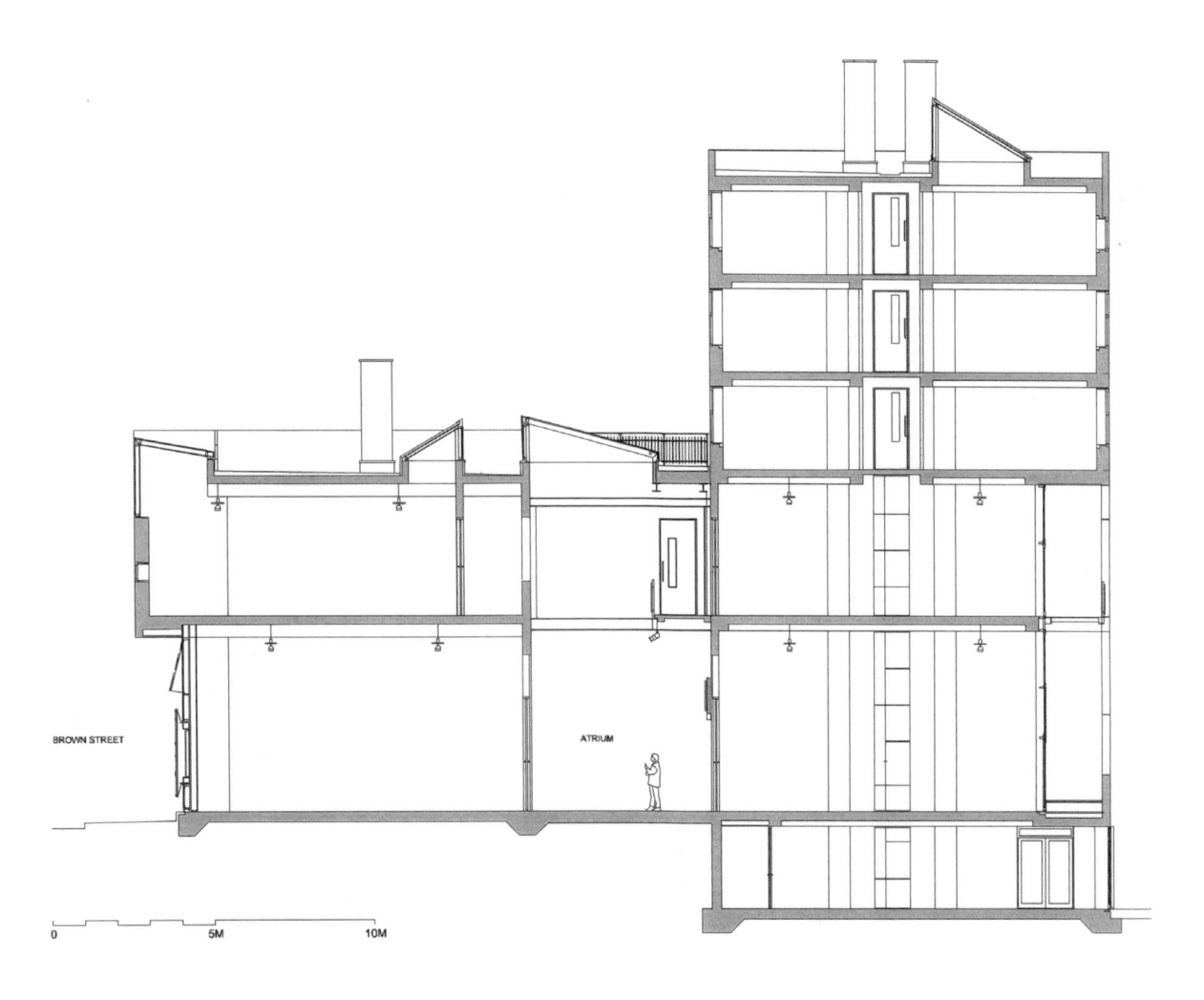

Section

opposite:
top left: Plastic rebate fillet
top right: Cast finish of rebate feature
bottom: Internal wall

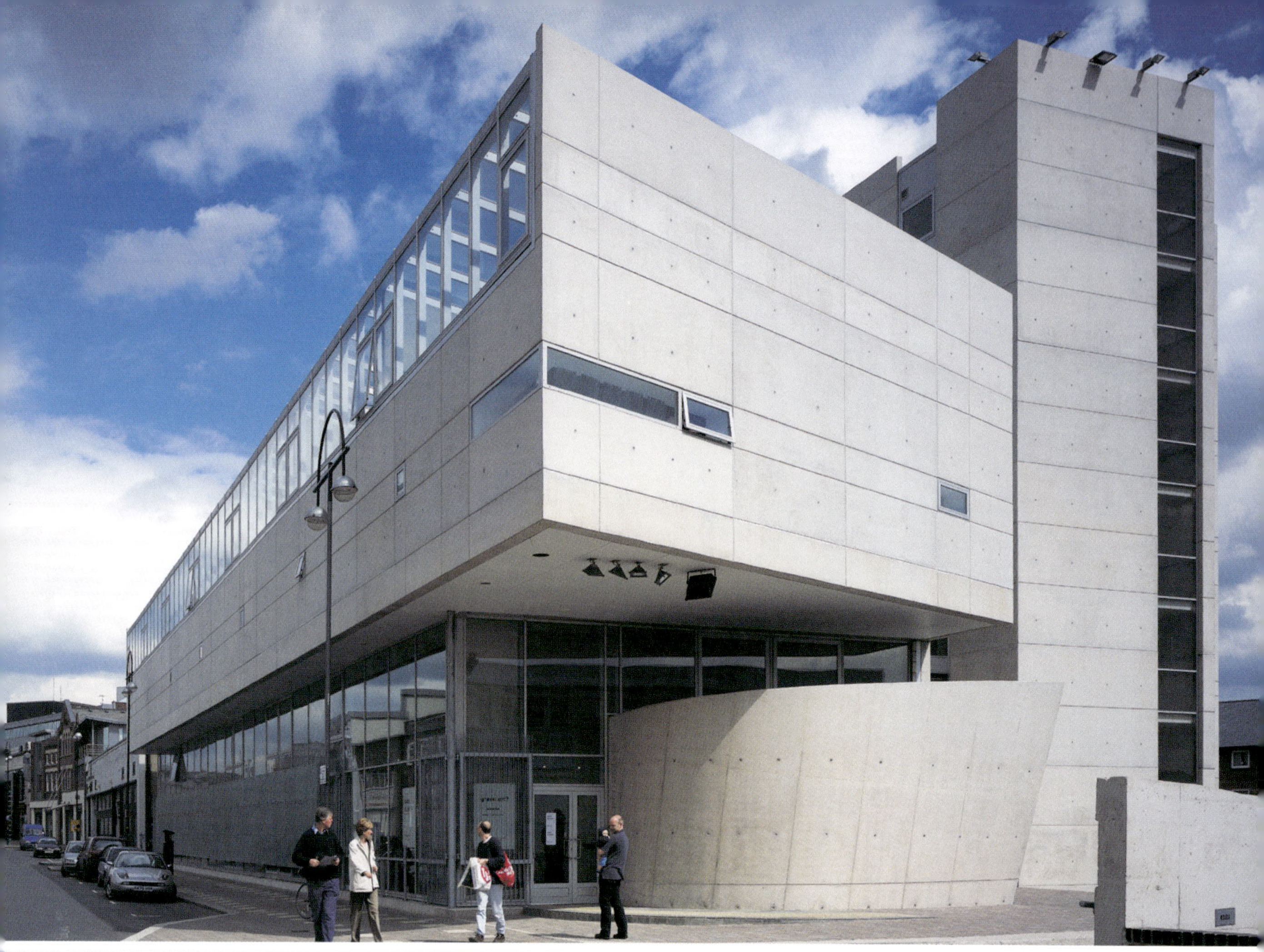

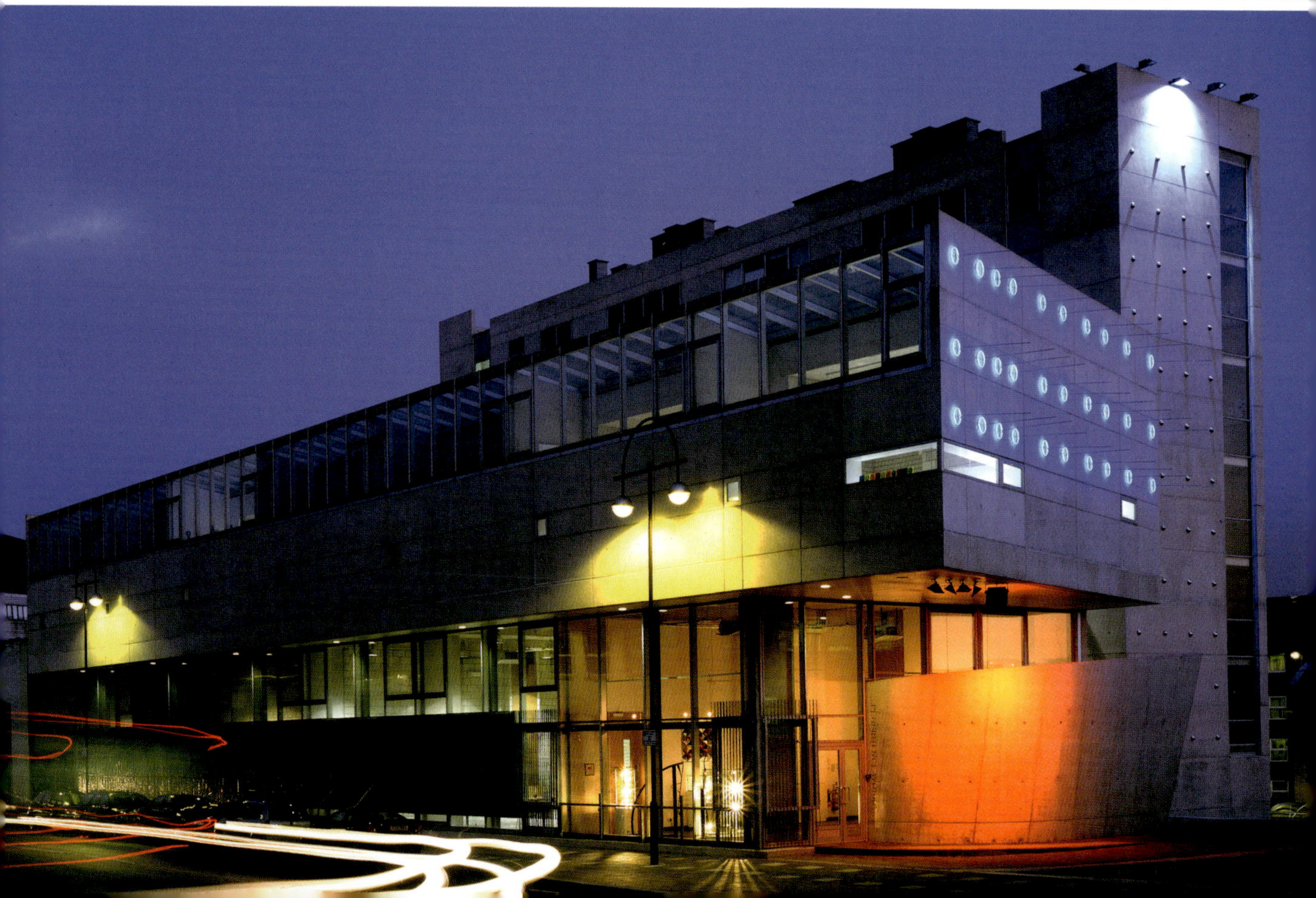

in forming the horizontal rebates on the building elevations, both in shape and material. In trials we adjusted the profile of the rebate to quite a shallow slope to minimise air bubbles being trapped on the lower face. For the material the contractor tried timber, steel, aluminium and plastic. The timber did not give a good finish as it spalled when it was pulled out due to the timber swelling. With the steel rebate the grout would squeeze behind the angle causing the front ariss to break off on release. The aluminium reacted with the cement to create a fizzy mush and that was also discarded. In the end they used a plastic tiling quadrant, cut flat on the face in their joinery workshop, and reversed to form a shallow V depression. The plastic quadrant was bedded in an acrylic caulk so that the fillet made a seal with the formwork to eliminate any grout loss. Later, our clerk of works did find a plastic extrusion purpose-made for rebates but it was too late to incorporate as the job had already started. It was a very expressive and effective feature.

We considered several measures to reduce surface staining, which has blighted many modern concrete buildings. We initially designed projecting cornices at particular levels on the façade to reduce dirt staining by throwing rainwater off the main elevations. These were later omitted on cost grounds. We also looked at staining caused by window ledges from which settled dirt can steak down the façade. We resolved this by placing the windows and their frames flush with the façade, eliminating ledges altogether. In considering surface treatments we initially tried a pigmented product made by Keim. They prepared colour samples and we also visited Ponds Forge Sports Centre in Sheffield where they had treated a concrete surface with a heavily pigmented Concretal Lasur coating. We considered this unsuitable as the pigmented coating took all the colour variation, patina and life out of the concrete. Keim then suggested their clear Lotexan which was a siloxane. This inhibits surface wetting and reduces dirt adhesion. Unlike clear epoxy sealants, which look like varnish, the eye cannot see where the Lotexan has been applied. The contractor challenged me to tell the natural surface from the coated surface. I couldn't, until water was splashed on the surface and the untreated concrete soaked it up. It was a magic product.

In the concrete finishes specification we initially asked for an ambitious enhanced grade C finish (ref. BS 8110) which called for an 'as struck' surface with very low defects and did not allow any repairs. Gleeson stated that they had put in an allowance for 25 per cent of all panels being condemned to achieve that standard. As we needed to reduce the construction costs and make it easier for the contractor to achieve a certain standard of finish we were prepared to accept a lower class 'as struck' finish, provided more skilled repair techniques where used where necessary and this approach was adopted.

Construction

We designed the panel layouts as part of the architecture. MJ Gleeson discussed how they wanted to cast the walls, the pour heights and where they wanted to put day-joints. We detailed and developed the panel layouts and the feature joints from that requirement and issued detailed drawings to the contractor, showing the feature rebates. We detailed it so that the floor slab thickness could be expressed on the façade and ensured that the same mix was used for the slabs as the walls to give a consistent colour match.

We inserted stainless steel discs over the grouted bolt holes on the façade. For the cavity weep holes on the façade, we used a plastic pipe which was colour matched to the concrete and located in unobtrusive places. For movement joints, we wanted to avoid the 'smeared on' look of mastic sealants and found a compressible filler strip that matched the colour of the concrete and created a neater appearance. The building was constructed using kickers for the internal columns and kickerless construction for the external walls.

The contractor started off using an Ischebeck Titan formwork system to build the walls, but it proved to be a less adaptable system, so Gleeson changed to a standard SGB system. This comes with a ply board fixed to the units, which was overlaid on site with new phenolic film faced plywood to give a good finish. Early on they found that any nails driven in through the film face spilt the surface and the panels swelled and began to delaminate. They changed to drilled only face fixings and filled the screw heads with plastic body filler and coated all the cut edges of the ply panels with a fast-drying marine paint. Despite that, they experienced some rippling of the surface after the first casting when the boards were new. On the second use this effect disappeared as the moisture content of the ply behind the film face began to even out. They started leaving the new boards out in the rain to prime them so that they did not ripple on first use. Panels were replaced after four uses.

top: Front elevation by day
bottom: Front elevation by night

ARCHITECTURAL INSITU CONCRETE - CASE STUDIES

One mysterious phenomenon that occurred was a blue stain that appeared on the concrete surface after casting. Our initial thought was that it may have been a stain from a pigment in the release agent. The blue stain was worse when the concrete section was thick or the shutters had remained in place for longer. In the end it was found to be a temporary condition that can occur with GGBS cements and faded away completely after about four weeks.

When the concrete was cast on cold, damp days in the winter the surface looked mottled and blotchy in colour, like dead flesh, although it was smooth. This did not happen in summer. We were initially alarmed but after some months the blotchy surface colour also began to disappear, possibly as surface carbonation set in.

The difficulty with the waterproof clear coating was the conditions under which it could be applied on site. It required a continuous period of dry weather, and a minimum of 3 days with no rain. We kept postponing the occasion when we could apply it and waited an inordinate length time. The fact that we waited those extra months helped the concrete colour to improve and stabilise.

A lot of the interior face of the building was blockwork and we were fortunate to find a standard Tarmac block from their Maltby works that was almost the same colour as the insitu concrete wall. However, when the first batch of blocks arrived on site they were much more pink. The supplier generously took the blocks back, though they were only a standard concrete block, and agreed to batch our blocks from one mix and store them so that we had a consistent colour.

The only bit of concrete that was not of the same colour was the stair flights, which were precast. The manufacturer would not, for some reason, use a GGBS mix to match the colour of the walls. The lintels over the blockwork door openings were cast on site with the same mix to keep the colour consistent.

The concrete floors were power floated in the open which was a remarkable achievement. There was some hand-finishing around starter bars. The circulation areas were stained black and saw cuts were made in the concrete surface to stop the stain bleeding to other parts. Generally, the power floated floors were a success, except for two areas in the upper studios where it started raining and the surface could not be levelled. Here a proprietary levelling compound was applied after the surface had hardened. The power

Internal walkway

floated floors were sealed with a SIKA product called 'Purigo S', a water-based product designed to reduce surface dusting. The atrium walls internally were also treated with Lotexan.

The curved wall of the entrance

We had looked at prefabricated moulds to obtain the double curvature using glass reinforced plastic (GRP) but that was expensive. By remodelling the form to a canted single curvature we were able to form the shape by bending plywood panels. We made CAD layout drawings of a number of parallel templates, which were then fabricated on site into a large form nicknamed 'the pigeon holes'. This was the last bit of concrete that was cast. It was a poetic way to sign off the completion of the building.

Structural considerations
Steve Fisher, Buro Happold

It is common at the time of tender to produce reinforcement estimates only because we have not completed the detailed design of the building. On a building with an unusual structure such as Persistence Works it can contain an element of guesswork. We use a rule of thumb which gives a density of rebar tonnage/cubic metre of concrete. It is generally a conservative value as we tend to err on the side of caution. We don't want to be in a position where the tender price exceeds the cost estimate. At tender stage the actual contractor was given general arrangement drawings, structural section sizes and the weights of bar sizes in the bills of quantity. For slabs this is about 150 kg/m^3, but we put in a slightly higher figure at 175 kg/m^3. No reinforcement drawings were issued with the tender.

In this building the walls which are exposed are load-bearing elements – thin, slender beams or long, narrow columns that carry the floor loads and create diaphragm action. The cantilever on the north elevation over the entrance lobby is composed of these blade walls which were designed as deep beams. We were concerned that cracking could occur on the exposed face so we tended to use close centred reinforcement near the surface to limit the crack widths. The outcome of all this is that by the time we had detailed all the reinforcement drawings and issued them to the contractor there was a reduction in the tonnage used, which was about 20 per cent less than the tender figure and this helped to reduce the overall cost of construction.

All the external walls contribute to the structural stability and load carrying. There is no redundancy. The reason for this was the need for column-free spaces in the interiors and this was made possible by using the perimeter walls as load-bearing elements and stiffening diaphragms. This created 10–12 m of column-free space within the building on each floor. There are transverse floor beams within the floor slab that are supported by the perimeter walls and usually a central column to reduce the overall spans.

The building is divided structurally into two halves separated by a central atrium, the southern half of which is six stories high and contains a basement and acts as a moment frame with no shear walls. We were concerned with the overall length of the building and also concerned that we should not get any shrinkage cracking on the walls, especially around the window openings on the rear elevation. Dowel connections on the spandrel panels and the adjacent columns allow for expansion and contraction at these locations. The dowel allows the façade to give frame action. The crack lines, if they occurred, were hidden in the shadow of the featured rebates on the façade. These were a structural device to induce the cracks along these channels and so hide them also in the construction joint lines.

On the northern block there are shear walls that continue to the ground floor so that frame action is derived from the columns and floor slabs with resistance to sway enhanced by the perimeter wall line. The foundations of the buildings are supported on continuous flight augur (CFA) piles. There was a demolition and excavation contract prior to commencement of the main structure. The main complications were the brick vaults and the tanked storage vessels below ground as this was the site of a former tannery. All the large obstructions and storage tanks were excavated. The materials were crushed and mixed with the excavated spoil down to a depth of 2.5 m, and any hot spots in the subsoil due to contamination were removed and disposed of. It was uneconomic to dispose of the entire waste fill off site. Intermixing the crushed bricks with the subsoil contained and controlled the quality of subsoil. There was an original methane problem so we had to install a capping membrane with ventilation ducts around the perimeter.

The most interesting aspect of the design on the northern block involved the first floor walls and floor cantilever over the ground floor entrance. It made predictions of the stress in the walls quite difficult. We did a lot of work on finite element analysis to check the stress in the walls, which we corroborated with manual calculations. That allowed us to examine where the peak stresses occurred and to contour the reinforcement density appropriately. We predicted that the crack widths would be 0.3 mm and would appear within the first 4 months if they did occur. The floors are flat slabs spanning between primary beams; these span between the external walls. Because we used the external wall as the structure, there had to be a discontinuity between the external wall and the floor slab to prevent cold bridging. That is the reason why we used beam strips to minimise the contact area with the external wall.

CONSTRUCTION NOTES
Sean Quinn, MJ Gleeson

By their own admission Gleeson have never before taken on a challenge on the scale of Persistence Works. They have plenty of expertise in formwork assembly, concrete handling, concrete frames and fair faced civil engineering construction which would give them the essential skills to do the job. But it was the fine tuning, the careful selection of formwork, release agent and concrete mix that was going make the telling difference.

We spent 3 months reading through technical books and press articles on fair faced concrete, then made a series of sample panels working with RMC in Sheffield, using three different concrete mixes, a number of different formwork faces and even tested several different materials to form the recessed V joints around the panels, to come up with the optimum solution.

Finding a cost saving of £600,000 on the £1.6m price for the frame alone was just as time consuming. Out went the blue concrete staircase and the cornices to the exterior walls, the texturing and architectural modulations. Engineer, architect and ourselves sat down together to run a value engineering exercise over the entire structure. The specification was rewritten to allow the contractor the option of repairing minor surface defects. Section depth of walls, beams and slabs and size of columns was standardised and all rebar accurately scheduled and quantified. Curved or architecturally shaped elements were eliminated as far as possible. It's surprising just how much you can shave off a building cost, when you know the life of the project depends on it.

Following preliminary trials, phenolic resin faced plywood Wisa – Betofilm panels were preferred for the façade walls. The wall panels were laid out in 1.2 m × 3 m long sheets to an agreed plan and then screw fixed to an SGB Logik metal-backed panel support system, with feature recessed 'V' joints formed between adjacent panels. Screws were countersunk on the fair face panel, filled with car body filler and then rubbed with wet and dry papers. Every cut edge and tie bolt hole was coated with fast-drying acrylic paint, all joints between panels were silicone sealed and wiped clean to reduce the risk of formwork panels warping and any grout leakage. In the architect's search for a very light-coloured concrete a 60/40 GGBS/OP cement mix was settled on and after preliminary trials with local 20 mm limestone aggregate, a 10 mm single sized rounded gravel was preferred with a target slump of 125 mm. To reduce the risk of trapping air voids during placing, the skip had a nozzle attachment which tremied the concrete through a length of flexible hose pipe to the bottom of the lift.

Wax oil made by RFA was used for the release agent, which smelt just like the 'dubbin' you use to waterproof leather walking boots. By keeping the strike times in summer to 48 hr and winter to 72 hr, a non-dusting, homogenous surface finish was achieved which had a slight blue hue to start with, that faded with time. The 750 mm × 250 mm columns were cast using Ischebeck Titan metal forms with a ply faced lining. They were easy to handle and quick to erect. A pair of joiners could assemble six 7 m high columns for the large studios in a day.

One statistic sticks in the mind when Gleeson were asked about remedial work and repairs. Of the 303 separate wall pours they had to cast, only seven came out poorly and had to be taken down.

The Epilogue
The last word must go to the dynamic director of the Yorkshire ArtSpace Society, Kate Dore who has been the inspirational driver behind the funding, organisation and management of this facility. 'Feilden Clegg Bradley was our choice for architect and was coincidently the only team to bring a woman architect, Julia Kashdan-Brown, to the selection interview. Art people want a substantial, hard wearing material not some flimsy lightweight cladding for the building. You will be surprised what art people can get up to in their studios! Stone was too expensive and bricks hark back to the cutlery building which we wanted to move away from. Concrete was ideal; it has industrial might and an intrinsic beauty. I like working within the spartan restraint and limited palette of pale grey concrete and buff blockwork walls and take perverse delight in looking up at the 'orange peel' marks staining the bare concrete ceiling in my office. It reminds me of the time I spent on site climbing up ladders looking at the work in progress while 7 months pregnant. The ceiling speaks honestly to me about the way the structure was built, warts and all, and the rusting tails of tie wire create an abstract visual pattern, which I wanted left in place.'

PROJECT TEAM

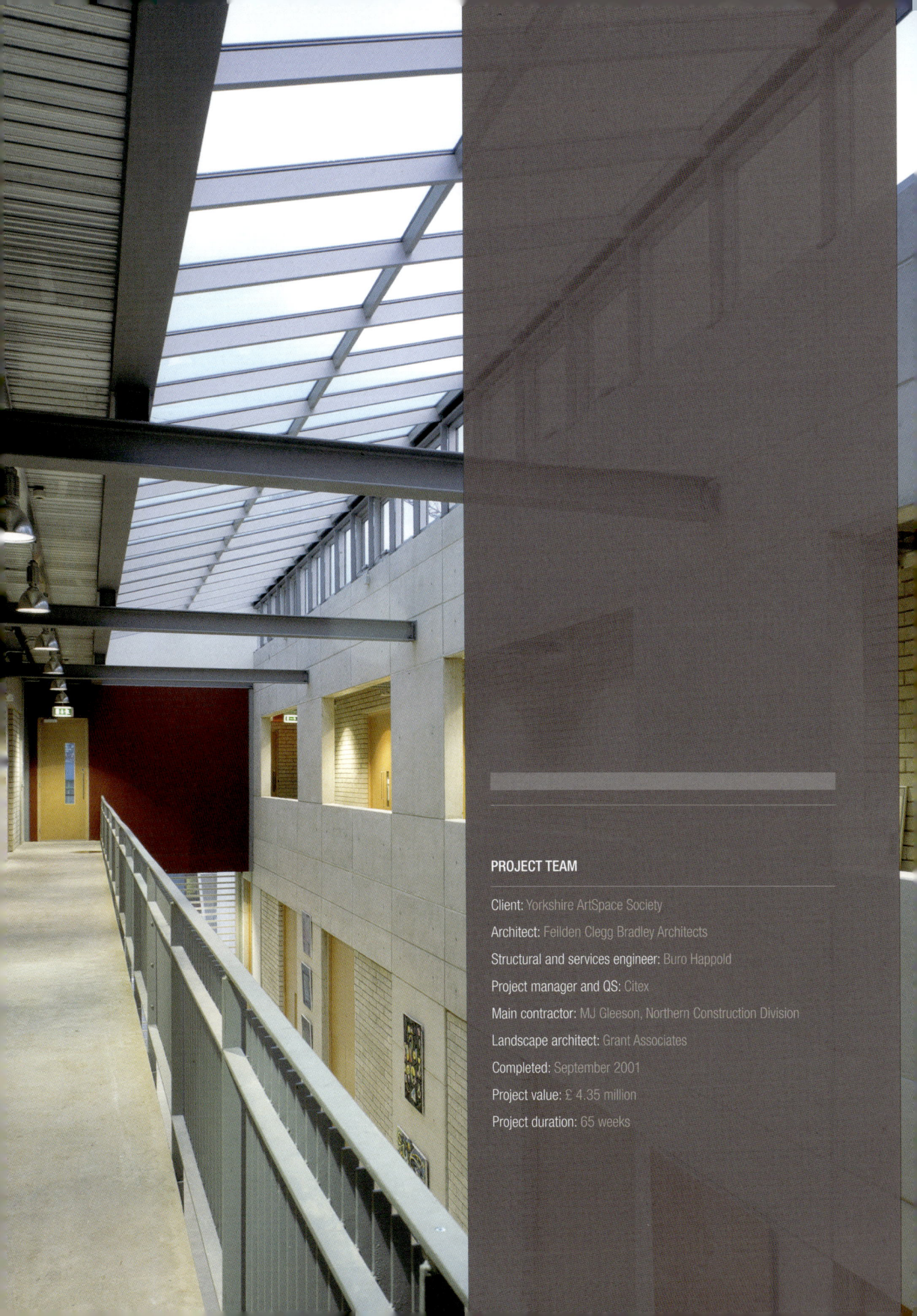

Client: Yorkshire ArtSpace Society

Architect: Feilden Clegg Bradley Architects

Structural and services engineer: Buro Happold

Project manager and QS: Citex

Main contractor: MJ Gleeson, Northern Construction Division

Landscape architect: Grant Associates

Completed: September 2001

Project value: £ 4.35 million

Project duration: 65 weeks

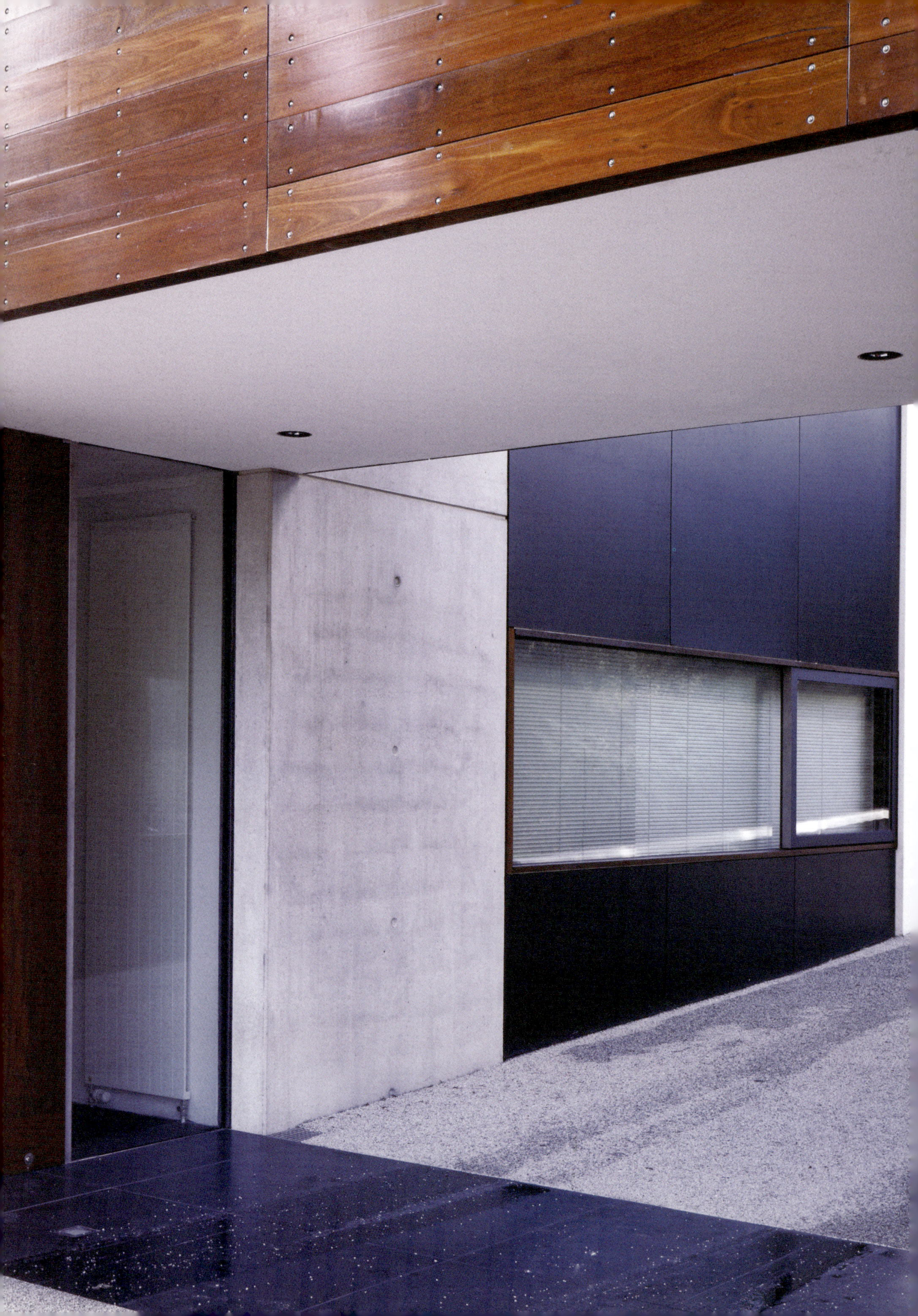

THE ART HOUSE
HERTFORDSHIRE

Architect: Fraser Brown Mackenna

Location

The house is situated in the plush West Hertfordshire countryside where there are good train links into Central London. The location of the house must remain confidential.

Introduction

The concrete superstructure of the house has been constructed on a reinforced concrete raft that sits on piled foundations. Many of the concrete walls and columns are exposed in the interior and have a high-quality fair face finish.

The long, open-plan, two-storey rectangular house has external elevations that boast a variety of finishes – concrete, render, glass, hardwood weatherboard and resin lamina boarding. The weatherboarding is made from Jarrah, a Western Australian hardwood, just one of the many materials to be specially imported to meet the project's requirements.

Internally, the bespoke design has made the detailed setting out and construction plan meticulous and exacting. There is no skirting where the floors meet the walls and the full height doors have no frames and no door handles and shut snugly under the concrete slab soffit. Apart from certain items of sanitary ware, very little in this building could be bought 'off the shelf'; they are 'one off' specials to ensure the production of a very original building.

The landscaping has incorporated the use of Jarrah decking and granite paving within a large water feature that is accessed from one of the sliding windows on the ground floor. A smooth, blue 5.5 tonne ovoid sculpture is supported above the wild lawn on a single pin.

ARCHITECTURAL INSITU CONCRETE - CASE STUDIES

Architecture discussion
Angus Brown

The client was a sponsor of the Slade School and when we designed a new exhibition structure for the Sculpture School, he contacted us and said 'I want a house just like that'. The Sculpture School was about rationalising existing space and inserting a mezzanine floor into a listed building to improve the circulation space and create more usable volumes.

In the beginning we were looking for suitable sites to build a new house but after fruitless searches, the client decided to have his own pleasant, circa 1900 vernacular house knocked down and rebuilt as he did not want to move out of the area. We approached the local authority to seek their views about a modern concrete house being built in a conservation area. They thought the proposal would be totally unacceptable. The client was quietly determined that he was going to have his modern concrete house and asked us to undertake a full design including artist impressions, rendered images and working models to show the planners what it would look like.

On seeing the scheme the planners did a complete u-turn and agreed wholeheartedly that the project would make an interesting house. Their comments were 'This house will make an outstanding contribution to the built environment of the area' and 'The town needs more quality architecture of this calibre'. The planners and conservation officer totally supported the scheme right through the planning process, for without their support the scheme would not have been built.

The concept of the house is that it has large rectangular volume with punched holes and wide glass elevations held in place by blade walls, through which the landscape comes into the house. The concrete blade walls and recessed window frames and sliding glass doors on the front elevation allow openings to be maximised and the inner structure to be expressed.

The client was keen on racing and cars. The driveway through the house is one of the voids, the other side of which is an indoor garden with a flowering tree.

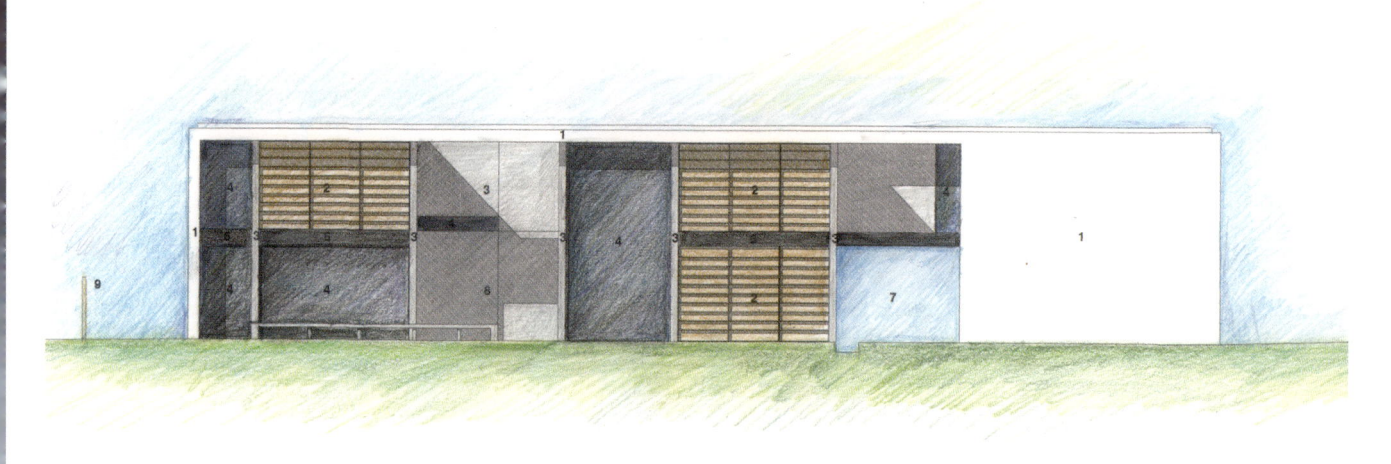

Sketch of front elevation

THE ART HOUSE

The house plan is divided around three zones – the habitation zone to the front, the circulation zone and gallery space along the spine wall at the rear and the services area with toilets, walk-in wardrobes and electrics contained in the middle zone. On the ground floor the driveway separates the habitable area into two separate spaces that are reconnected on the first floor by a glazed bridge. John Aiken was commissioned to install the two ovoid sculptures to the front lawns and the two mirror pools in the back garden, all visible from the living room, and George Papadopoulos was commissioned to provide a beautiful double height cracked glass window. Bruce MacLean and John Aiken designed and supplied the dining room table and Janet Stoyel provided a set of sliding screens made out of a woven mesh of stainless steel rods on which are patterns of leaves and ivy to provide privacy to the master bedroom.

In terms of the materials, we tried to keep the palette simple; in essence it was just concrete and glass. There was some difficulty in achieving the u-values with so much glass, particularly the re-entrant areas. The large openings are glazed with low emittance, argon-filled double glazing units sourced from East Germany. The glass is set into the double storey high blade walls of lightweight concrete, which are thermally broken but with no risk of condensation as a long thermal pathway was created by the recessed frames. The glass elevation is shaded by wooden external louvers suspended on stainless steel cables. The ground floor is covered in granite slabs imported from Portugal.

We went through several options for the concrete mix after Liapor decided not to supply the lightweight precast wall panels that we had specified. We took the decision to use a lightweight aggregate concrete with a low u-value to cast the perimeter walls and blade walls – Lytag, a sintered PFA aggregate that reduces the concrete density and can overcome issues of thermal bridging. The air entrained Lytag aggregate allowed us to create a C40 concrete with a 0.9 u-value. This material enabled the provision of a wide thermal break behind timber lining at the glass junction to avoid condensation forming on the inside face of the walls and columns. This was achieved without adding any additional insulation. Charter Construction, the main contractor, cast the double height gallery wall and the blade walls with the lightweight concrete mix using steel shutters faced with a phenolic film faced ply. There were no tie bolts permitted to avoid thermal bridging. Internally, the concrete blade walls are always naturally expressed and are combined with

top: Blade column and roller blinds
bottom: Internal corridor and staircase

load-bearing blockwork walls finished in white painted plaster. The first floor and the roof is an insitu flat slab with standard concrete mix. We used a white render on the external face of the 200 mm thick concrete gallery wall and the blade walls as a means of waterproofing the single skin thickness.

We felt, in terms of flexibility of the house, that we did not want an exposed soffit slab and the constraints of having to determine all light and service penetrations. We used skim coated plasterboard for the ceiling to hide the services within it. The large window elevations facing the garden have electrical roller blinds and the blind boxes are framed in plasterboard and painted white to mask their location. There is a Sarnafil roof covering to the concrete which is insulated and the rainwater is collected for the planters. Black ashphalt was used to line the mirror pools to the rear of the house.

The door details run to the underside of the ceiling and there are no door handles; they are opened by applying pressure to the door face. I did not believe a British contractor could build a

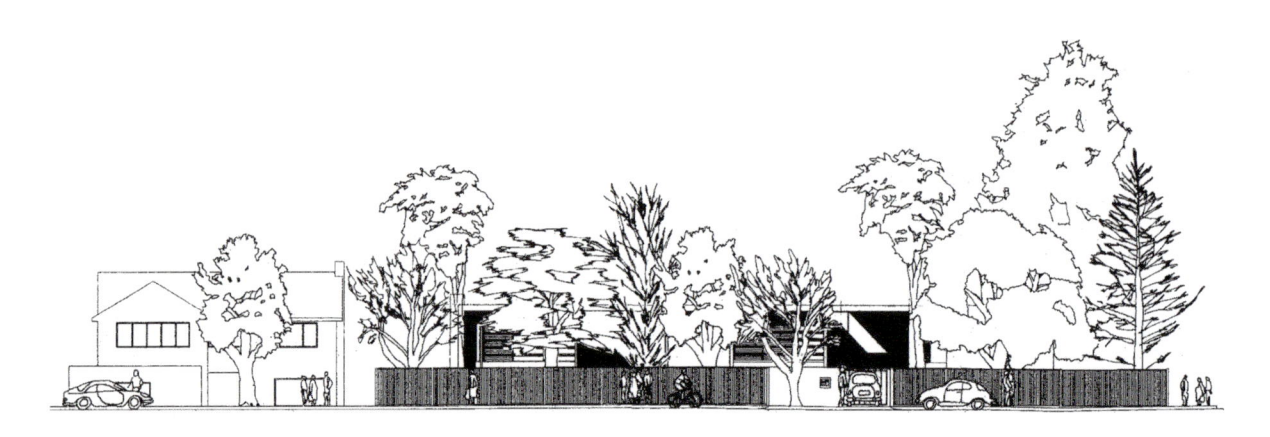

top: Construction progress
bottom: Sketch of street elevation

opposite:
top: Living room
bottom: First floor gallery

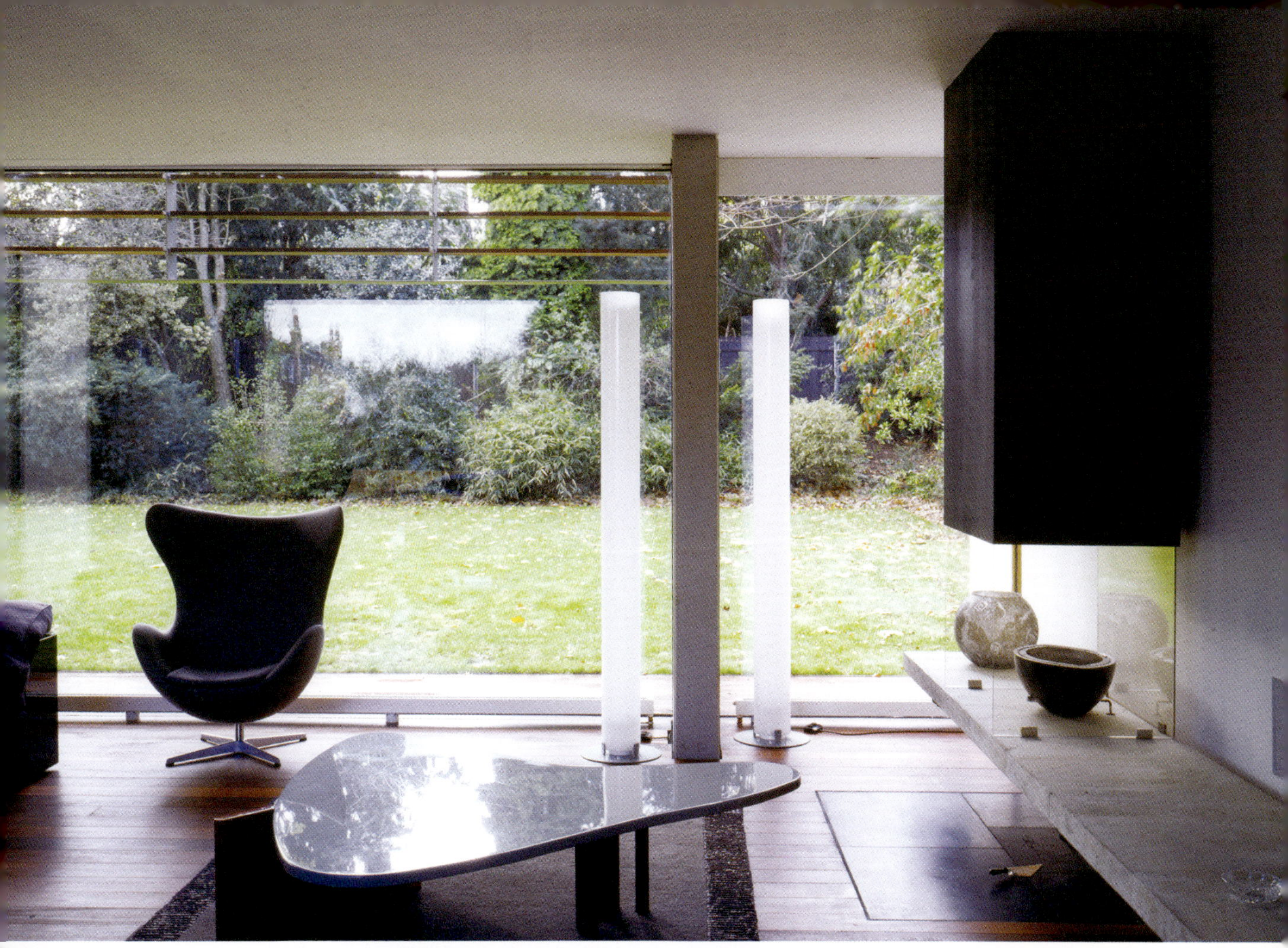
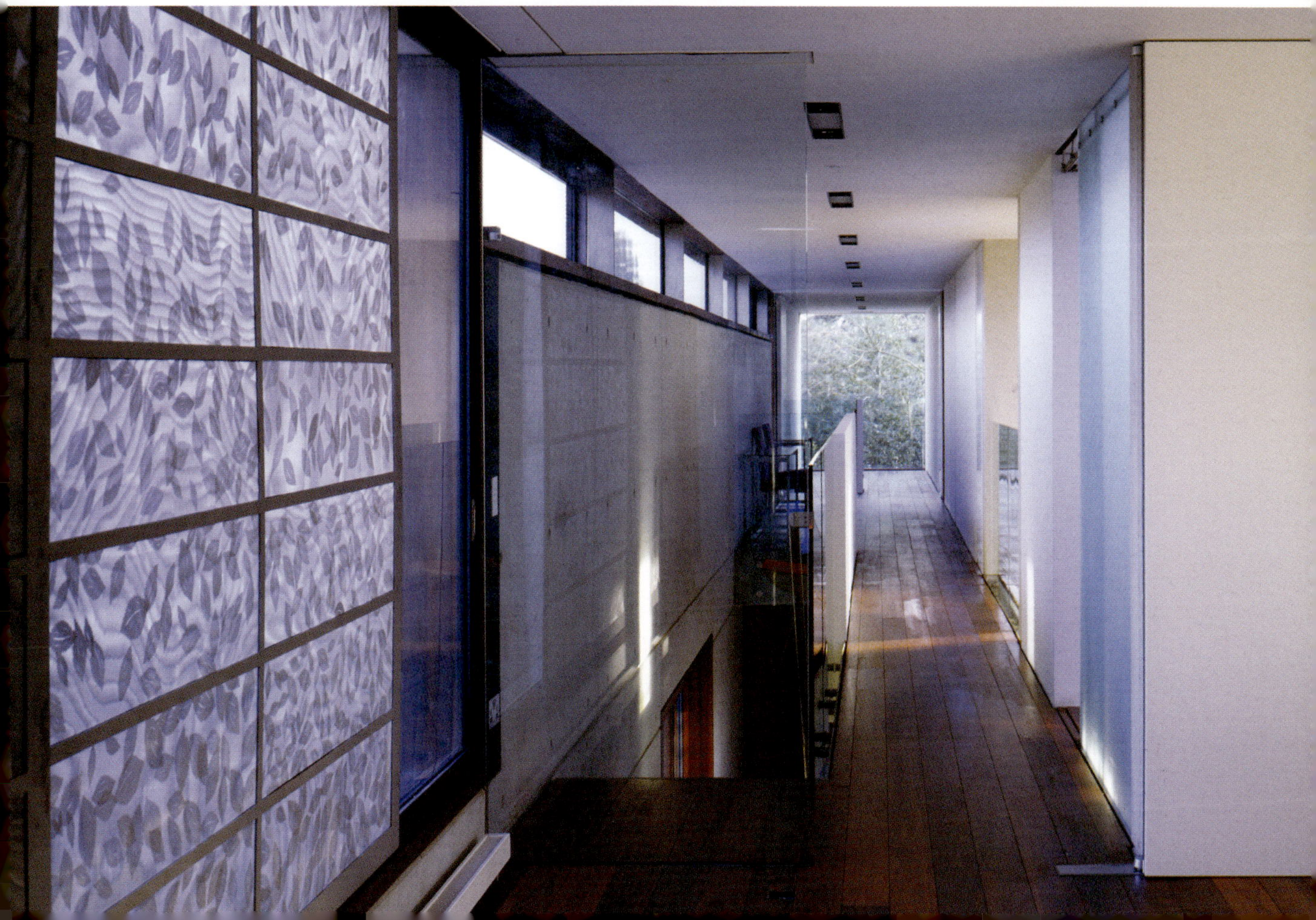

concrete structure with the tolerance of +/- 3 mm to meet the door tolerances between the ceiling and the floor; but they did. All the doors were shot into place.

There is a large log fireplace with a concrete bench in the lounge, the dining table was jointly designed by Bruce MacLean and John Aiken, both colleagues at The Slade School of Fine Art. The gallery was intended as an art show with a collection of contemporary paintings and sculptures by former students and staff from the Slade School but the client in the end preferred the simple surfaces of exposed concrete and therefore the artwork has yet to materialise.

No wonder *The Sunday Times* called this house the 'Art House'.

CONSTRUCTION NOTES
Paul Jenkins, Project Architect

The design of the house maximises its connection to the garden with large glazed openings, courtyards and sliding doors and clearly required a non-traditional structural solution. Concrete fin walls with recessed window frames allowed the window openings to be maximised and the structure to be expressed. The desire to keep the visual structure as thin as possible posed structural and insulation problems.

It was proposed that the primary enclosure should be constructed from precast lightweight concrete panels. This would give consistency of finish and predictable thermal and structural performance. Discussion with various precast suppliers in the UK proved fruitless and the design was eventually developed with Liapor in Germany. They make bespoke lightweight concrete panels for houses in Germany and Poland. The decision to expose a significant proportion of the concrete structure both internally and externally meant that their 200 mm thick mid-density panels, with a u-value of around 2, were selected.

Regrettably, Liapor pulled out at the last minute citing UK contract and regulation problems as their reasons. This left the client and design team with a serious dilemma. The planning permission called for fair faced concrete for a number of the external structural elements and time was pressing. An insitu option was considered and, following investigations into mixes with Tarmac Topmix and the shuttering system with the concrete contractor, it was accepted as the appropriate alternative.

The concrete had to provide a thermal conductivity no higher than 0.9 W/m · K and that required it to have a density of around 1,850 kg/m^3, a strength equivalent to C40 and a finish and colour as good as precast. The final design mix consisted of OPC cement with white microsilica to give a light colour matrix. The fine aggregate was a light-coloured washed sand and the coarse aggregate was a graded 12–4 mm Lytag. Cast against phenolic film faced ply and steel moulds, this gave a silky smooth surface and a consistent pale-grey colour.

Tarmac Topmix was the preferred supplier but the concrete mix could not be transported far because of the quick setting properties imparted by the microsilica and the tendency for the mix to segregate. The only batching plant they had within a reasonable distance was at a gravel extraction site with a planning restriction preventing the use of imported aggregate. A temporary waiver had to be obtained to allow Lytag to be used.

The mix was easy to pour into the 200 mm-wide shutters and around the very dense reinforcement. Although designed as a pump mix it was placed using skips and consolidated with minimal vibration to avoid the segregation and flotation of the lightweight aggregate experienced with normal poker vibration.

The 28-day cube strengths ranged from 65 to 90 MPa, far exceeding the design specification. This resulted in a higher than specified density, nearer 2,000 kg/m^3 with a correspondingly higher u-value. Detailed assessment showed that the exposed concrete would still fulfil statutory regulations regarding cold bridging and the higher heat loss could be offset by alterations to the insulation specified elsewhere.

The shuttering needed to give a smooth, fine finish, be economical, quick to move and reassemble and adaptable to a number of different configurations. The final system involved using a limited number of different sized panels and returns with neoprene seal strips which could be combined to create all of the panels and walls required. Detailed drawings showing configurations of panel sizes were prepared by the architects. The system worked well and the quality of the resulting fair faced work was as good as the Swiss projects that were used as the benchmark.

top right: Main elevation
bottom right: Cracked glass window by George Papadopoulos

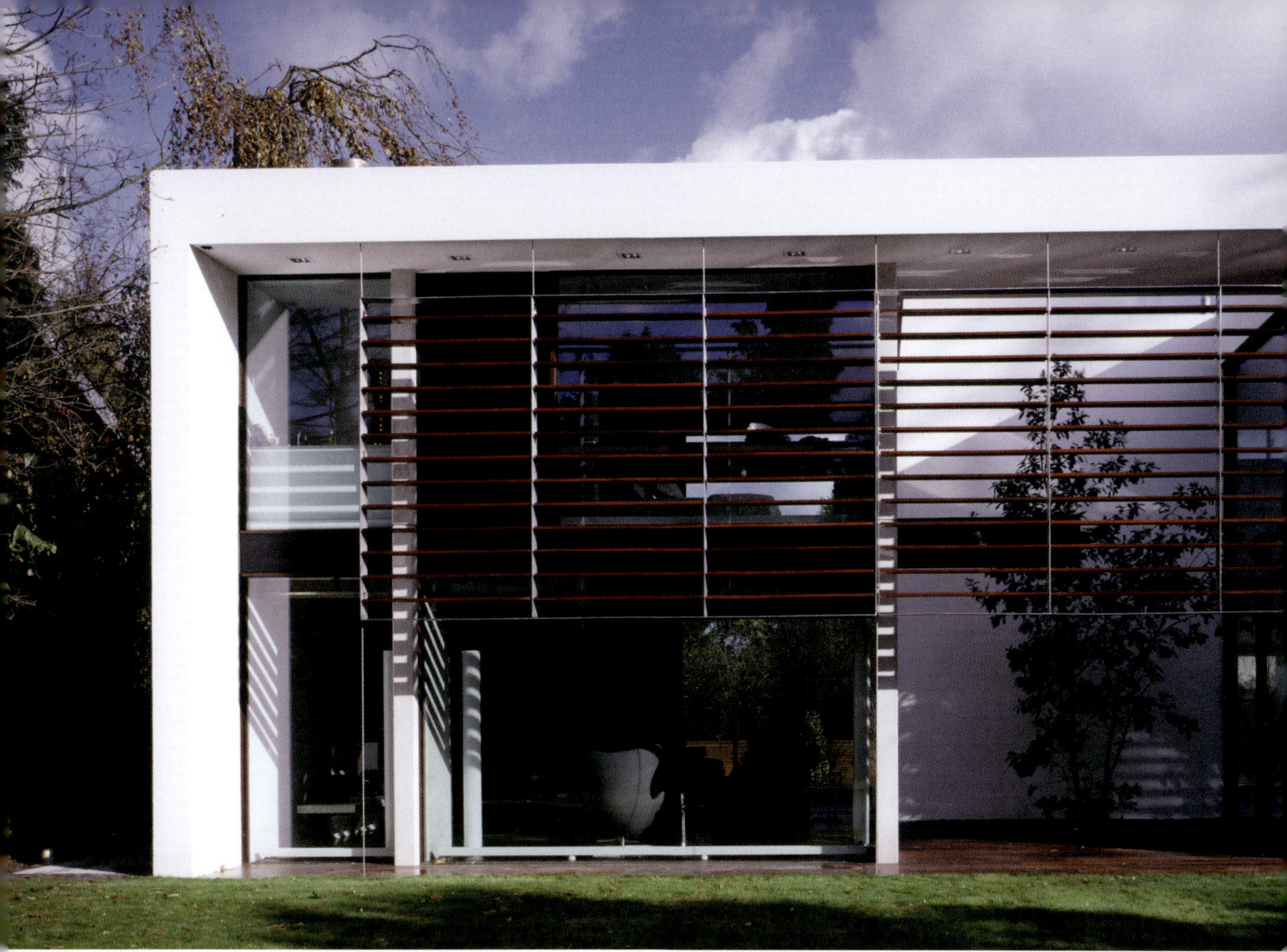

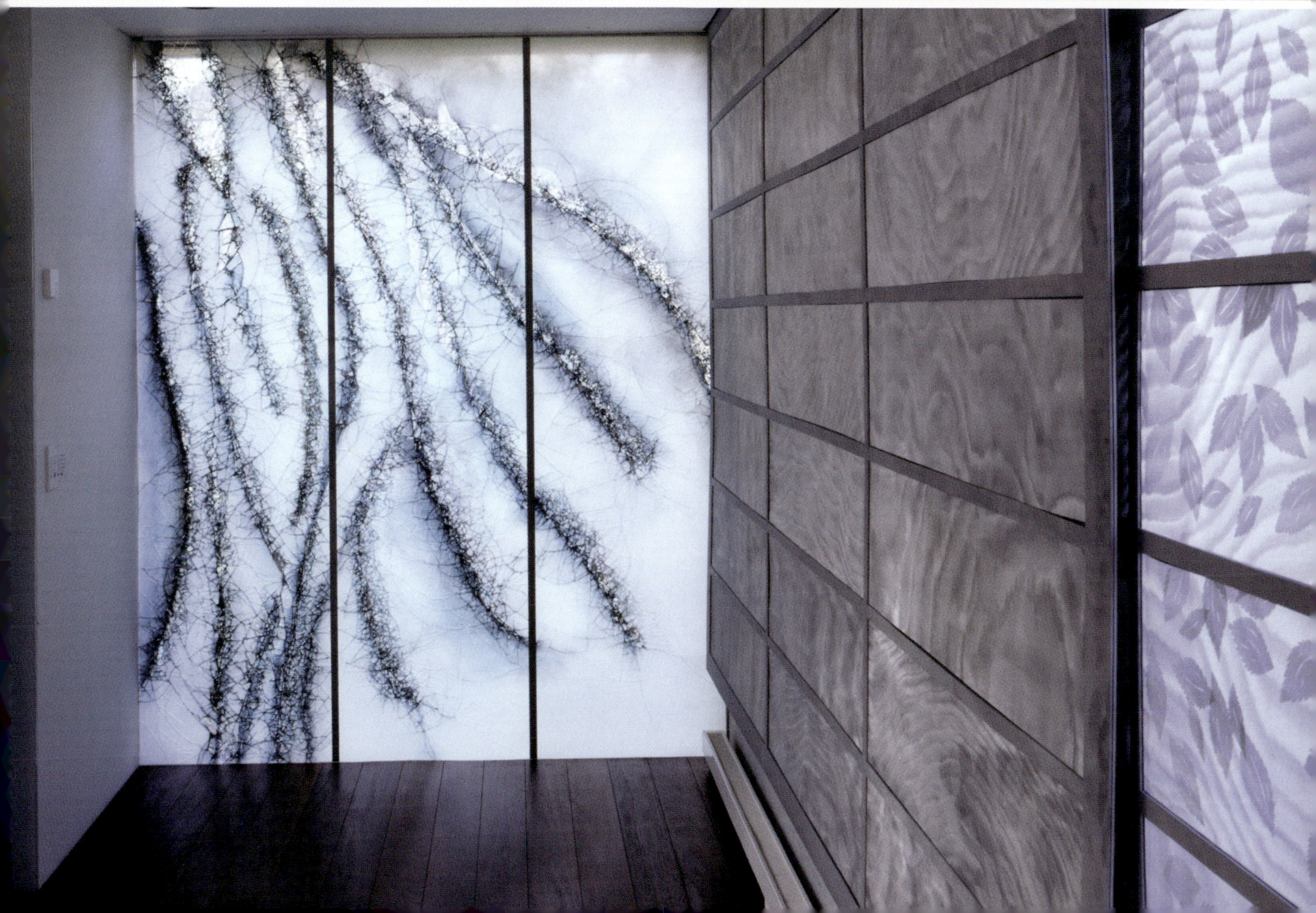

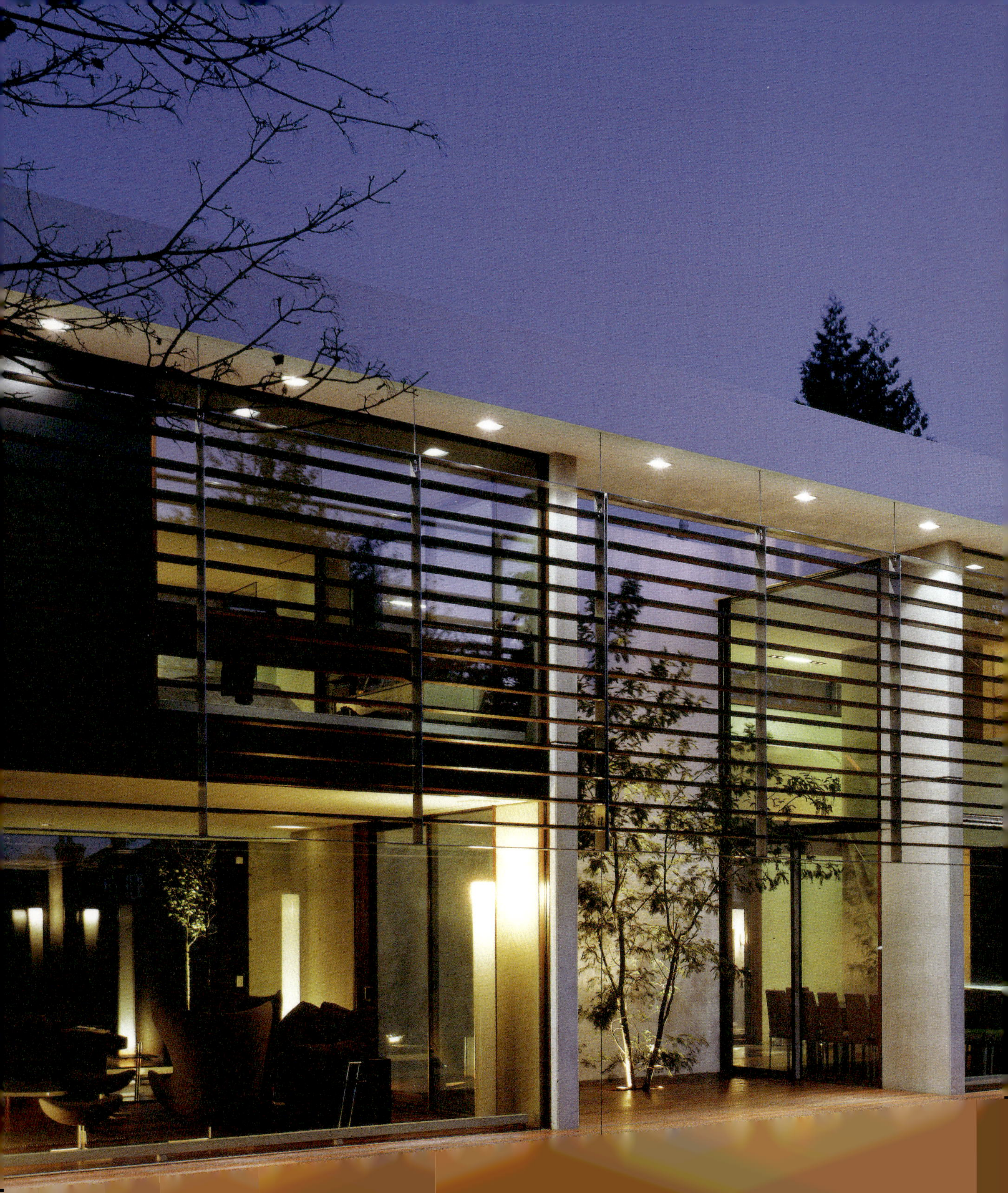

PROJECT TEAM

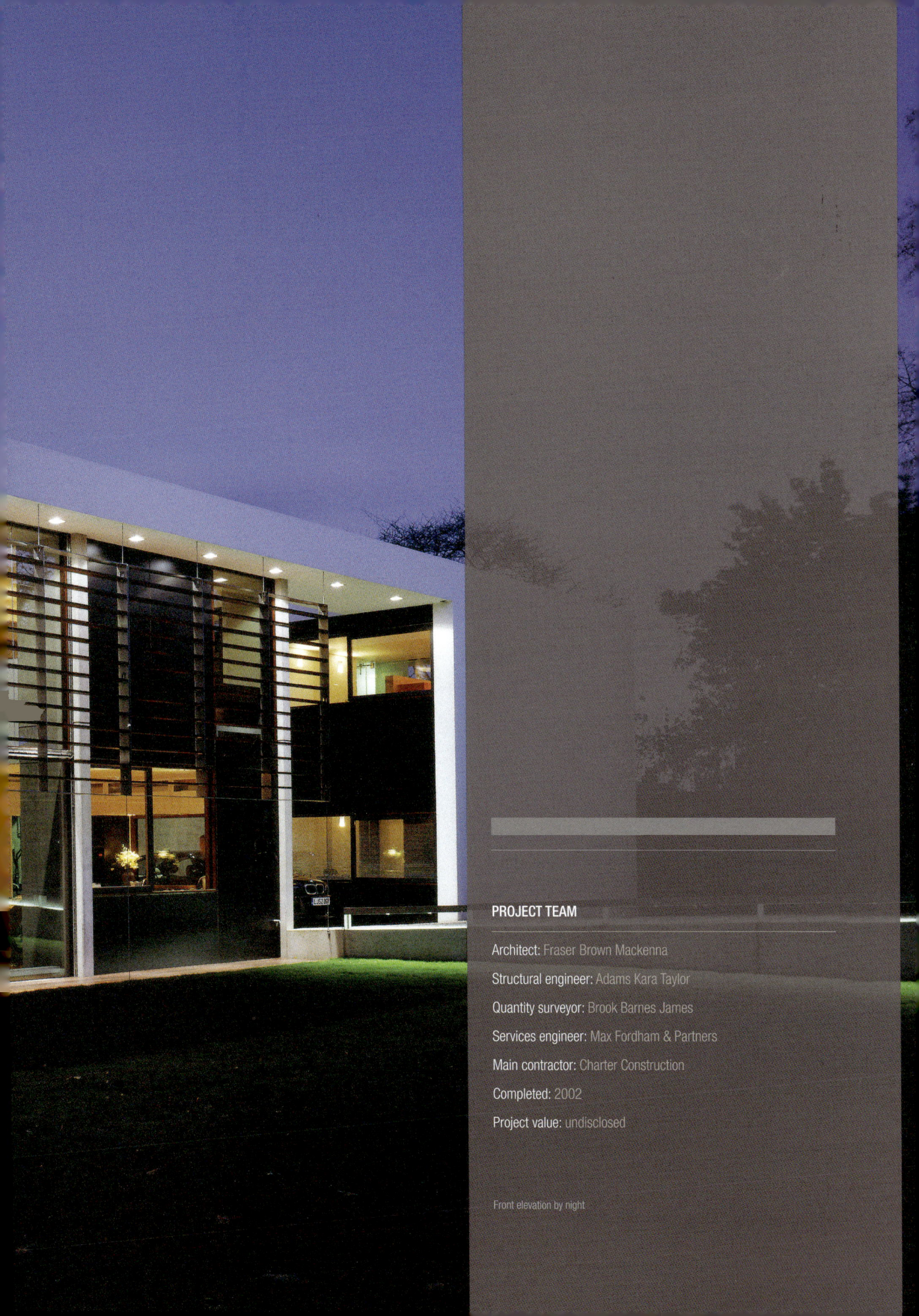

Architect: Fraser Brown Mackenna
Structural engineer: Adams Kara Taylor
Quantity surveyor: Brook Barnes James
Services engineer: Max Fordham & Partners
Main contractor: Charter Construction
Completed: 2002
Project value: undisclosed

Front elevation by night

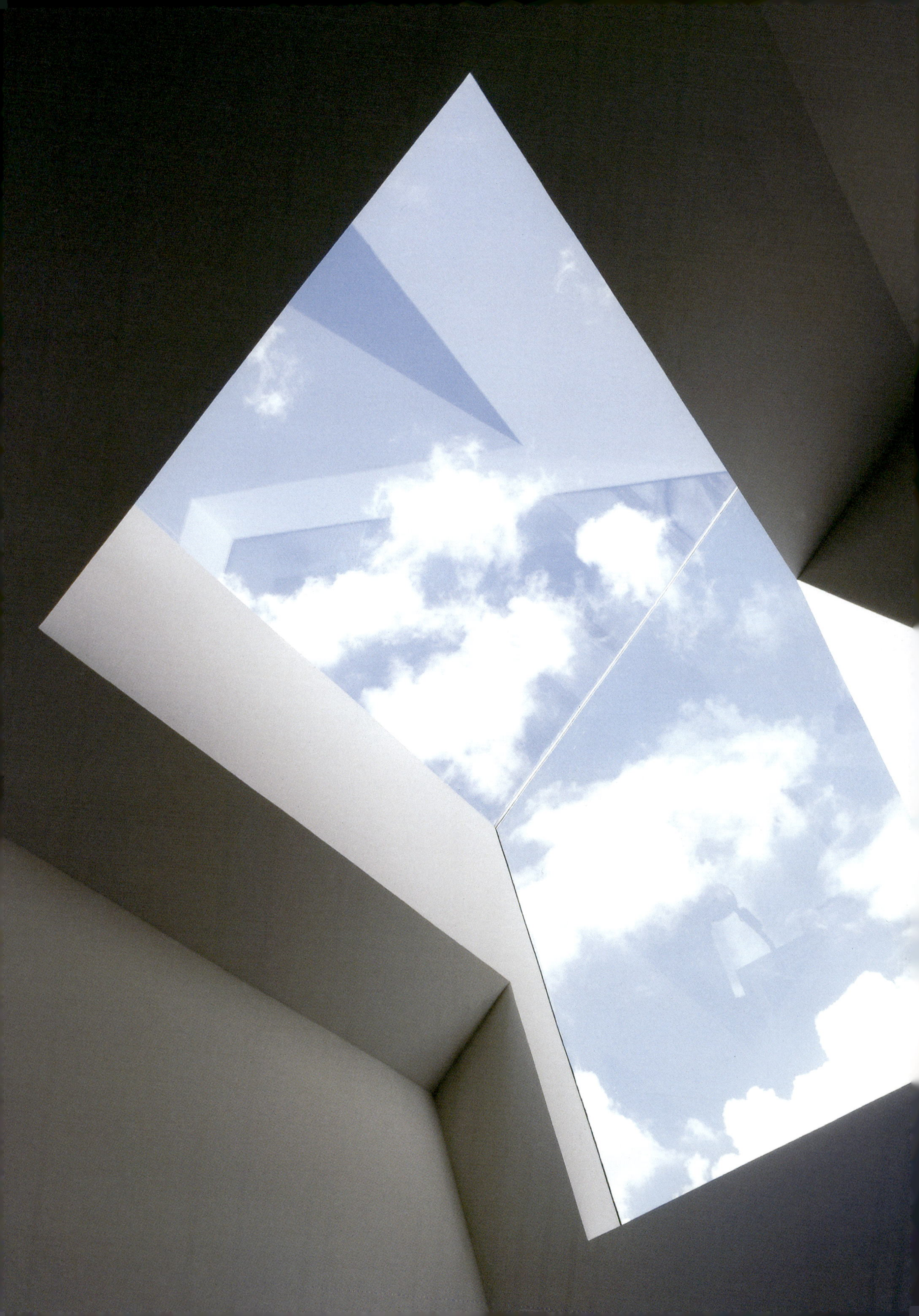

THE ANDERSON HOUSE
LONDON

Architect: Jamie Fobert

Location

The front door to the property is on Gosfield Street, not far from Oxford Street, in the heart of the West End of London. Entry is made through a 1 m wide passage between adjacent buildings and the house occupies a site enclosed on all sides that had once been a bakery.

Introduction

Architectural critics have admired the concrete 'ashlar' finish to the walls, the imaginative use of space for such an unprepossessing site and the way the many party wall agreements were handled and resolved for this cocooned, courtyard house in concrete.

The main feature wall in the house, affectionately known as the 'T-wall' because of its shape, forms a double height living room space. It is 3.6 m high and 350 mm thick at the base reducing to 275 mm at the top of the wall. It has two return walls that form short corridors through to the kitchen and entry. The beam section at the top of the 'T' spans the openings to the kitchen and landing and runs across the 5 m width of the room. Concrete had to be placed through the narrow 250 mm opening at the top of the wall to fill the 350 mm thickness of the wall below it. There was reinforcement, tie bolt sleeves and polythene-lined formwork panels to negotiate before the concrete could find its way to the bottom of the pour. When the formwork was struck there was hardly a blemish or blowhole to be seen on the exposed concrete face.

This is what the judges said of the winner of The Manser Medal for Best One-off New House in 2003:

'This is an extraordinary house, but one that is the worthy winner of this year's Manser Medal. It is a house that can only be understood and appreciated by visiting it, for its architect, Jamie Fobert, has solved what many other architects and most house builders would regard as an intractable problem – that of building a house that is almost completely underground, with almost no external walls and virtually no windows, only skylights.

It does so, not only in an innovative way, but with vigour and architectural rigour, eschewing traditional or pastiche interiors in favour of solutions that are dramatic, contemporary and well constructed. In particular, the quality of the concrete finishes in the house are remarkable.'

Architecture discussion

Jamie Fobert

I had worked with concrete before, when I worked for 9 years as a senior associate at David Chipperfield's offices. During that time I learned a lot about concrete while travelling in Japan, where I saw some of the work of Tadao Ando. Since the early years of my own practice, we have been experimenting with concrete. Before we began work on the Anderson House we had first used concrete in a series of shops for the American cosmetic company, Aveda, and in the Cargo restaurant/bar/nightclub.

The Aveda shops sell organic products. Our approach was to echo the simplicity of the products by designing the structure, fittings and furniture with solid materials. Insitu concrete was utilised to create freestanding furniture elements throughout the shop. All the surfaces were left plain and we used ply shuttering throughout. In terms of mix, our approach was 'whatever is in the truck will do'. There was to be nothing precious or special about the mix; colour variation and patination was part of the process. Unlike wood or bricks, which come in modular sizes, concrete is both monolithic and plastic. We explored its potential for sculptural form, cantilevers and angles. The shop included an organic café, which is where David Anderson, who used to lunch there, first became aware of our work. He was attracted by the simple, tectonic and unfussy concrete of the Aveda shop and he asked us if we could design a domestic building, using the same approach.

About the same time we were approached by a restaurant owner who wanted to convert railway arches into a new venue – Cargo. We approached the three large arches as a kind of landscape, within which elements of the brief were considered as small buildings. We wanted to keep an industrial feel to the venue. Against the vast surface of red brick we decided to build these smaller elements in concrete. We experimented with a number of surface treatments, casting concrete against various surfaces readily available from builders' merchants. From these

top: Sketch of T-wall and main room
bottom: Section
opposite: T-wall and double height living room space

experiments, we chose two concrete finishes. Some elements were cast using chipboard ply and for others we experimented with a polythene liner to create a special, polished finish. The idea of using polythene had originated by accident when, on a previous job, the builder had used a thick blue builder's plastic sheeting to line the foundations and filled it too high so part of it was visible above ground. When the plastic was removed and the edges of the foundation were exposed, it had an incredibly smooth, rippled surface, almost like polished marble. The polished concrete was expressed and very much liked by the client. Four years later we returned to this smooth finish for Cargo. In Cargo the walls were cast in long lengths. Big sheets of polythene were draped from the top to the bottom, just like a curtain, held tight but allowed to ripple. The sheets were tough enough not to tear but flexible enough to be able to fall into small folds.

Another accident had resulted in the finished surface of the chipboard walls. When we had made the sample for the chipboard finish, the contractors first waxed the board thoroughly, and the chipboard pattern was mirrored on the concrete, leaving no particles of the board behind. But when they came to cast the large pieces, they did not apply the wax so consistently and some chips of wood remained ingrained in the concrete. The contractor thought we would reject it, but we thought it was acceptable in the rough context of the venue. Overall, the concrete resembled the pattern one finds in galvanised metal, but enlarged.

While we were designing the Anderson House, the client came to see our work at Cargo. He liked the ripple effect on the walls, and felt it would work well in a residential building, particularly as it was smooth, shiny and reflected so much light.

Design development

In designing the Anderson House, we were preoccupied with the volume of space and the materiality of the finish. The initial decision to build with concrete was made not for aesthetic but for functional reasons. The engineer had advised that we could not rely on the perimeter walls to carry the loads. We had considered using steelwork, but that would have meant craning steel sections up and over the residential apartment blocks that surrounded the building – a high-risk proposal. Building with concrete was a logical, though not a cheap solution to the problem, as you can easily pump it in some 20 m from the street.

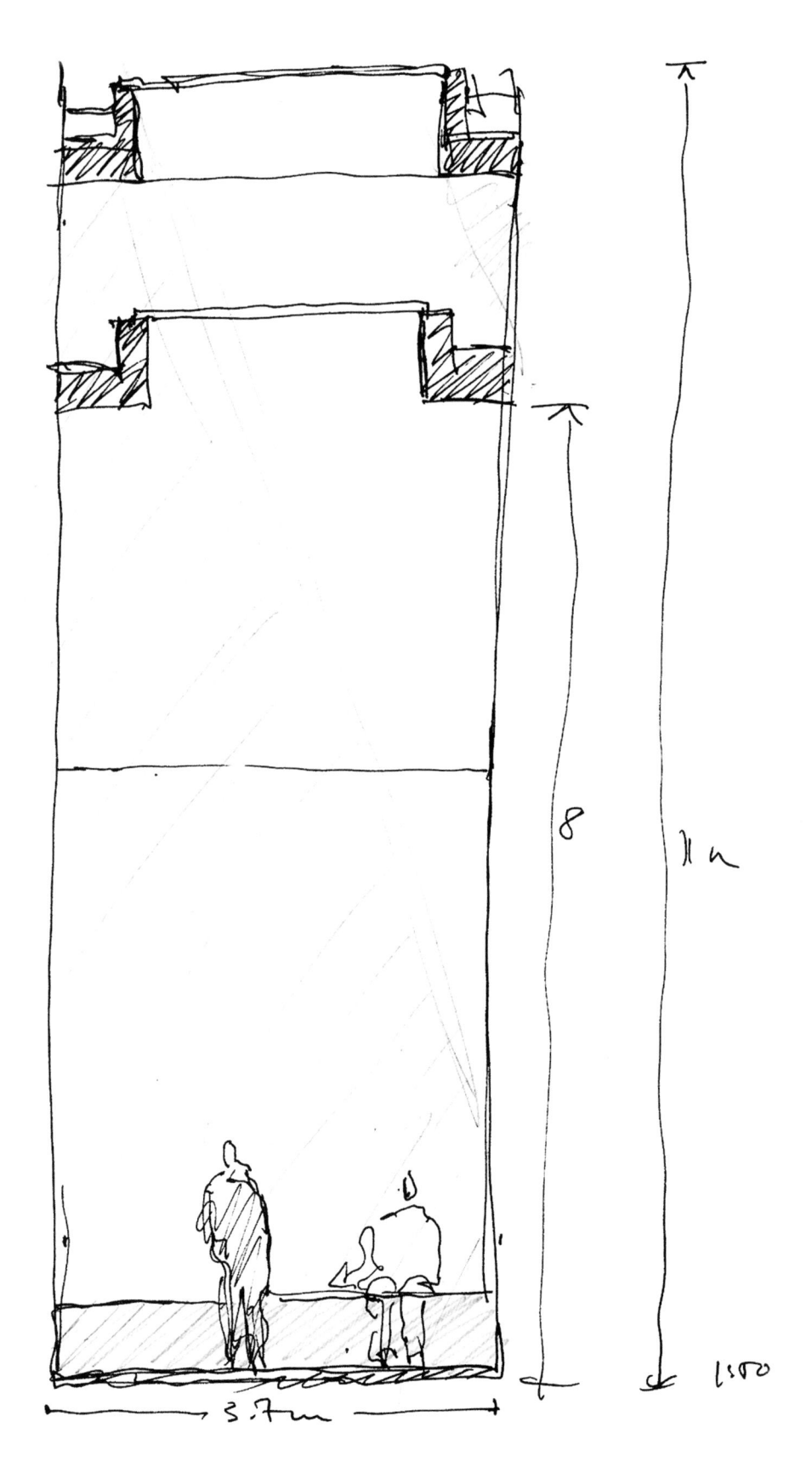

Sketch of internal spaces

opposite:
left: Living room
top right: Roof frame
bottom right: Bathroom

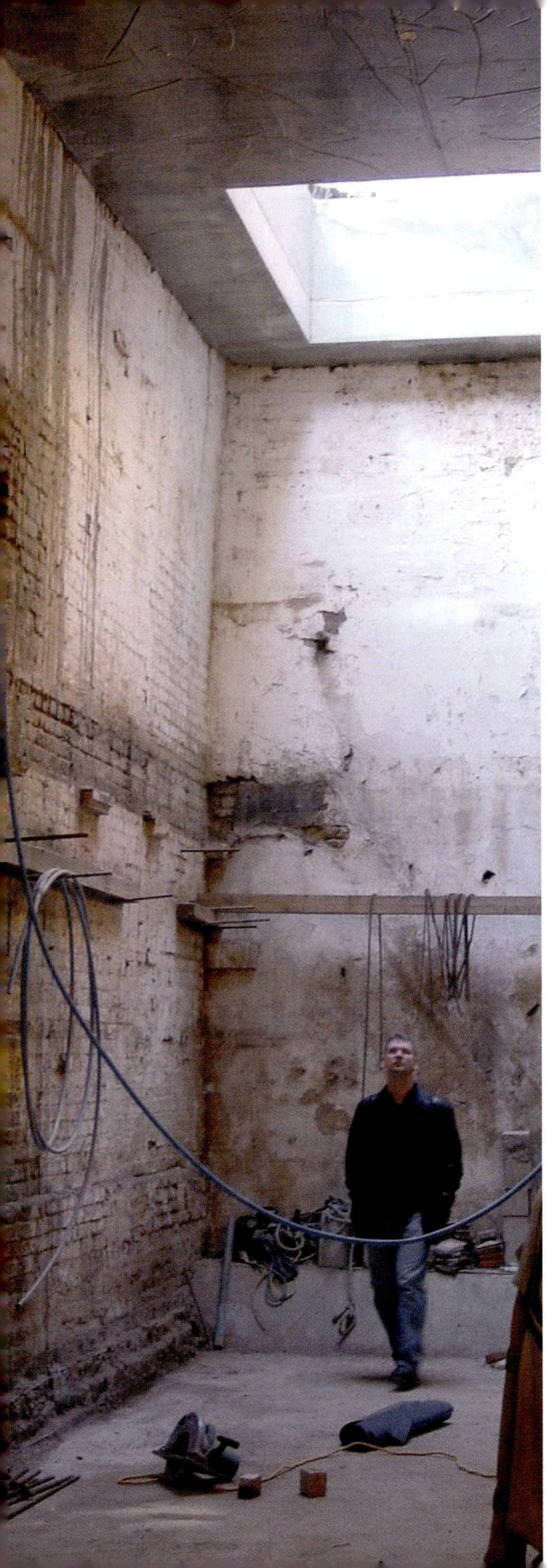

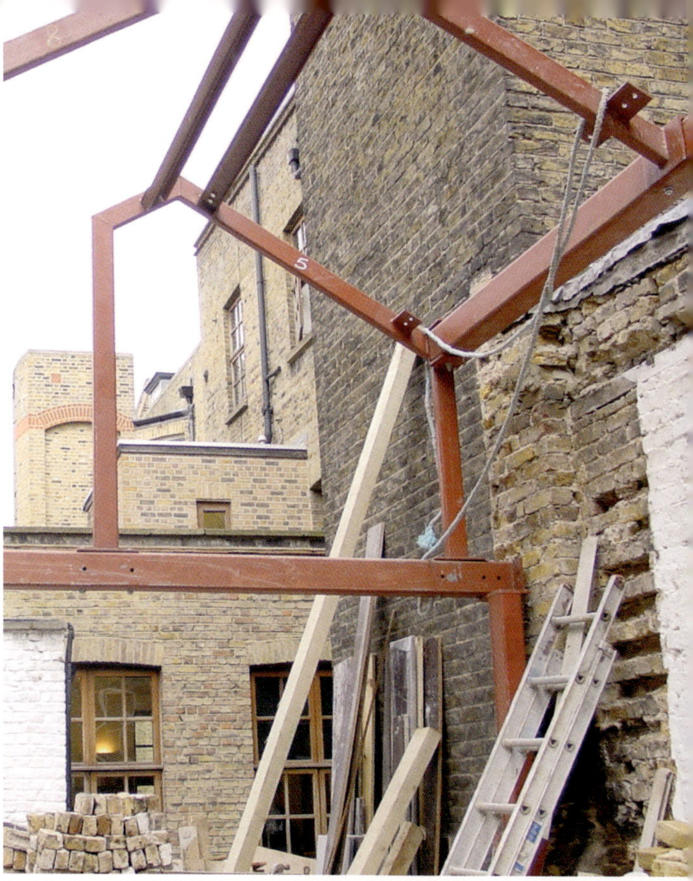

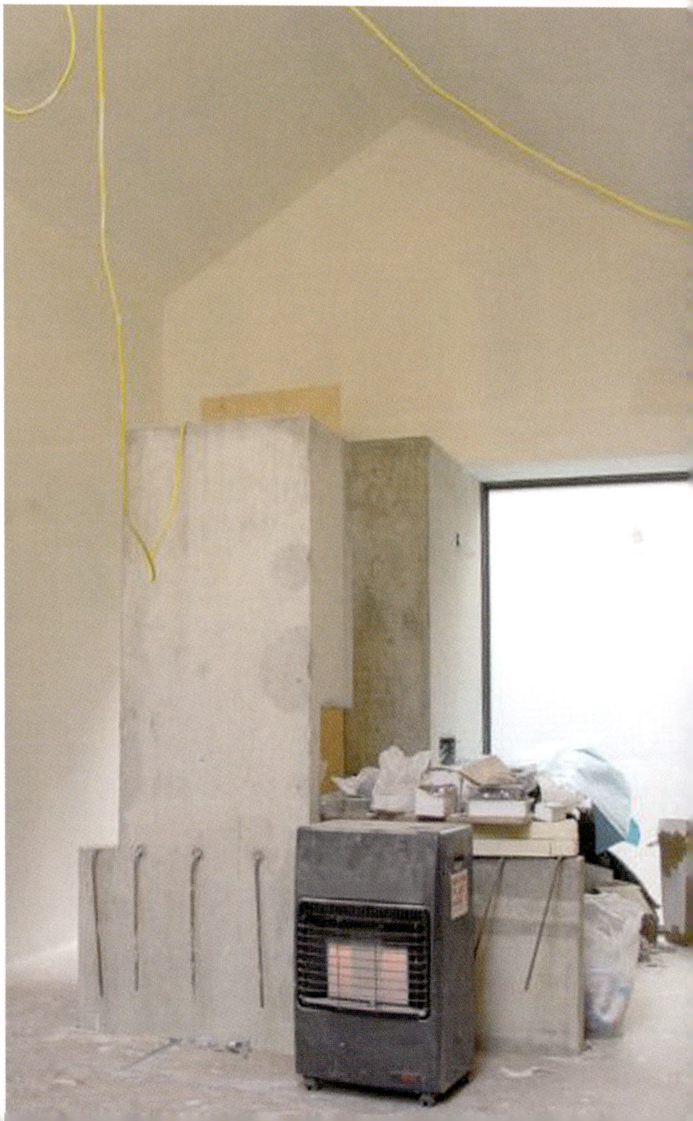

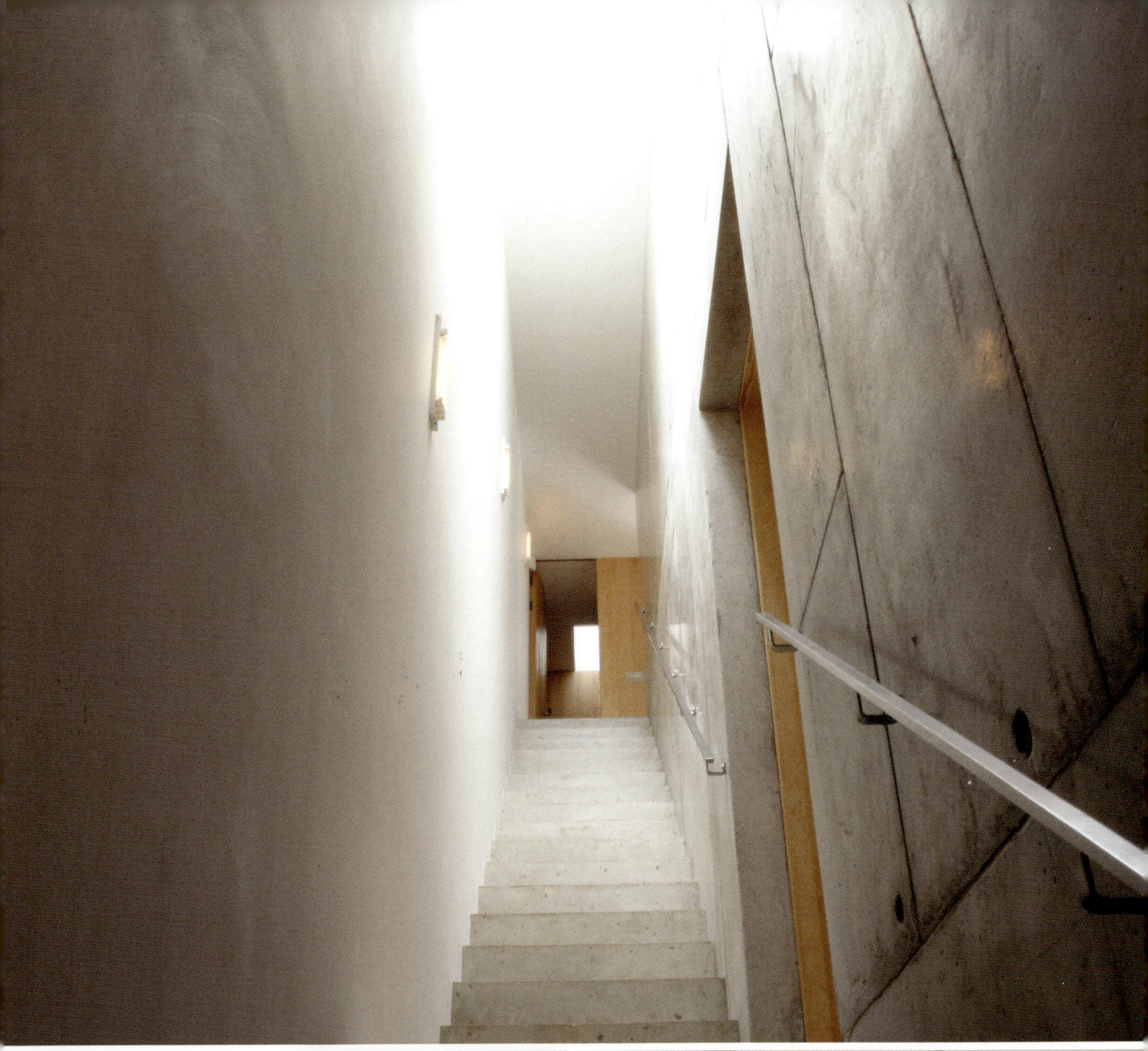
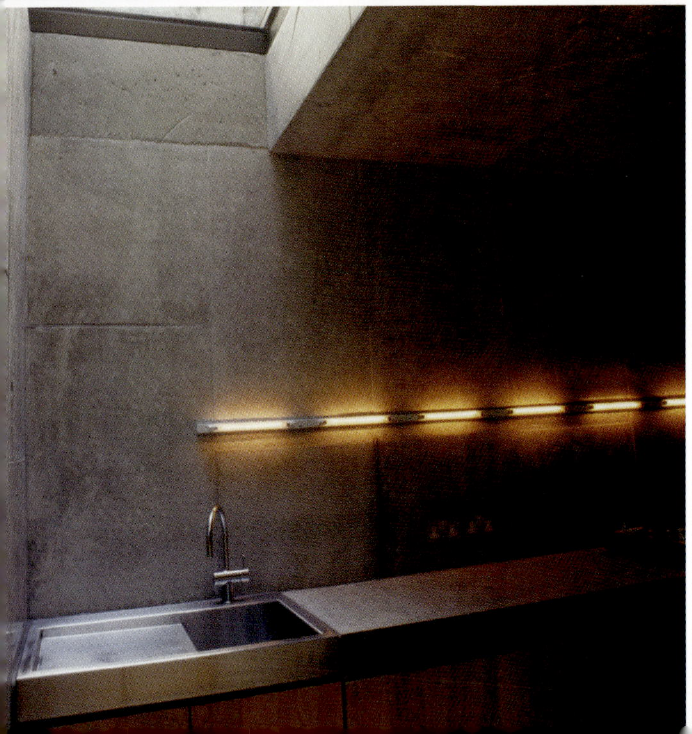
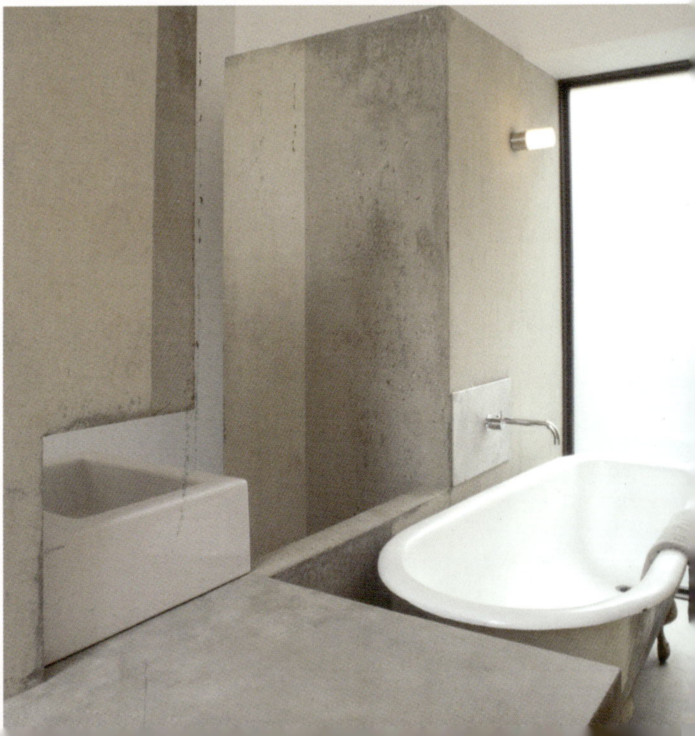

Volume and light

The planners had stipulated that the existing envelope could not be exceeded. They allowed the site to be infilled but required a sloping roof towards an existing residential light well, while the parapet to the same light well could not be raised. This restriction forced the first critical decision: to descend to basement level from the street, passing below the fixed parapet.

Within the constraints of the site a series of spatial volumes were inserted, each responding to an aspect of the brief and developing distinct proportions. The long descending passage from the street gives the house a distance and sense of removal from the city. The main body of the house was primarily organised in section. The main living space took the maximum volume the site could afford and became generous beyond any expectation when entering the house. As you enter the living room, light falls from above through the large window to a small court. At the far end light pours in through the roof light. The generous light sources negate any feelings of confinement in the restricted, underground space. Rather, light has been treated as a material around which the house has been formed.

The stair to the upper levels follows the perimeter of the site leading to a guest room and the master bedroom. In two locations the vertical volume of the stair drops to follow the constraints of the site. Here, folded roof lights were placed to both light the stair and dissolve the corner. At the end of this progression in the uppermost reach of the house the rebuilt pitched roof of the original workshop forms the master bedroom.

Construction specification

As the new house could not rely on the party walls, the majority of the load is carried on a central pile cap, with 18 piles each 10 m deep. From this, a central concrete structure rises forming the principal walls of the interior, stairs and ceilings of the first floor. All three processes of structure, wall and finish are resolved in a single material. The steel frames of the upper floors rest on these concrete platforms. The concrete was cast in vertical shuttering panels, each wrapped in polythene. This surface treatment gives the concrete a highly reflective and polished finish. The concrete also reduces in scale to become furniture, virtually emerging out of the larger surfaces. This was explored most fully in the bathroom element, which provides a framework for the loose porcelain fittings of bath and basin. This element was a re-creation of the concrete tables, plinth and separating walls that we had created in the Aveda shop.

For the Anderson House we had started to think about a domestic scale for the sheets of ply and to create an ashlar effect for the sheets of polythene. We decided to cut the 8 × 4 ft sheets to make panels that were 2 × 8 ft with staggered joints, rather like stonework. We then drew all the walls with 2 × 8 ft panel layouts and detailed how the joints would work. We thought we would wrap each panel with thick gauge polythene before they were fixed. Where the polythene-covered sheets butted, joints were created as positive creases, protruding from the wall. Next we drafted a specification for the contractor to make the panels, wrapping the polythene around each one and fixing them to a backing formwork. We made sure that the polythene was kept tight to avoid any ripples, as here the client wanted a marble-smooth finish. He was happy to accept the ripples on the ceilings by way of contrast. To hide the untidy polythene at the top of the wall we cast the floor slab onto the wall with the polythene tucked behind a rebate or shadow gap.

The concrete colour

At first we were happy to accept the concrete as it came from the truck mixer with no particular requirement, but once we noticed the variations in colour in the basement lower levels we asked for the colour to be consistent. Tarmac Topmix developed a concrete mix that gave us a colour consistency.

Acoustics

Sound insulation was an important issue to resolve in the lining of the existing volume. Originally you could hear the televisions from the apartments through the party walls on all sides. So we lined the brickwork perimeter walls with acoustic insulation fixed to the stud framing. What we did not account for was quite how uncomfortable the echo and the acoustics would be in a confined concrete space with 6 m high walls 4 m apart. When the house was finished, we re-lined one wall in a grey fabric to deaden the echo. The softness of the grey fabric was well balanced against the richly toned concrete.

The T-wall

The T-wall was designed as a single monolithic construction that we wanted built without any construction joints. The wall was to be the central support for the whole structure and the main feature of the double height living space. The upper floor loads of the house are transferred down the T-wall and it was a significant structural element. There are 24 different party walls on the perimeter of

top: Staircase corridor
bottom left: Kitchen
bottom right: Bathroom

the building, so 80 per cent of the load of the structure has to be fed along the central wall from the floor slabs and various transfer beams, with the remaining 20 per cent carried by the perimeter walls. The loads on the centre of the building are carried down the central core and the T-wall, which sits on a piled raft with a pile depth of 22 m. At the site meeting, it was clear that the contractor was planning to separate the T-wall into a number of pours which would produce a series of joint lines that would affect the reading of the wall as single element.

The contractor agreed to cast the wall without any tie bars or construction joints. However, after a site inspection it was apparent that tie bolts had been used, exacerbated by the asymmetrical and haphazard placement of the bolts. After this we had to specify tie bolt positions and carefully consider pouring sequences and joint positions before any formwork was erected. There was a lot more rigour and cooperation after that as we were able to lay out the tie bolt positions and the construction joints to suit the architecture. The client was quite happy to accept the T-wall with the tie-bolt positions once he had seen the finished wall, as the casting of the wall in one pour had produced subtle marble-like rippling patterns on the finished face of the concrete.

The T-wall also forms part of the kitchen, which has an exposed concrete slab to the soffit. When the slab was laid the lighting and heating conduits were inserted into the pour. One of the most difficult design coordinations was to have all the service layouts fixed before the walls and floors could be concreted. The different trades had to be disciplined to lay out the lighting and heating ducts into the walls and slabs with complete precision. Any errors in the setting out would have resulted in redrilling into the walls post-cast, which would have ruined the finished face of the concrete walls. The lights hidden in the walls of the kitchen were florescent tungsten tubes – simple, low-tech items that you can buy at any DIY store – placed in clusters of five or six.

CONSTRUCTION NOTES
John Saville, Horgan Brothers

We had carefully chosen an industrial type of 'shrink wrap' polythene as thick as 2,000 gauge, which would not tear when it was stretched across the ply or rupture when stapled to the back of the ply for the polythene faced 'ashlar'. The ashlar panels were then fixed to the backing panels of formwork by back screwing to avoid puncturing the polythene. The ashlar panels were butted up to each other to create the jointed layout that the architect had detailed. As the minimum of tie bolts were permitted to be seen, the bottom set, just above the 75 mm kicker line, was hidden below finished floor level while the top ties were kept above the wall height and sleeved through the RMD Kwikform strong backs. The tie bolts were pushed through the polythene and ply backing. The plastic cones trapped the edges of the polythene to make a grout-tight joint as the tie bolt was tightened and clamped to the external shutters. Only the middle line of tie bolts was visible on the wall surface.

It took nearly 4 hours to pour the T-wall and the other linking walls, using a static pump line and a series of flexible rubber pipeline extensions. This brought the concrete to the top of the formwork and along the length of the walls. The rubber line was broken back and uncoupled as the pour furthest from the static line was completed. Internal poker vibration and movable external clamp-on vibrators were used to compact the concrete. A combination of 7 m^3 and 9 m^3 truck mixers despatched from Tarmac Topmix's Kings Cross yard supplied the 40 N concrete to the site. It contained 40 per cent GGBS cement replacement and was a pump mix with a water/cement ratio of 0.55. No release agent was used on the polythene in case it should etch or degrade it. Only one use of the polythene-covered panels was possible as every wall was different. The perimeter walls in the other rooms, the long staircase corridor and the floor slabs were polythene finished in much the same way.

However, the internal walls of the shower room and the worktops in the bathroom and kitchen were conventional fair face concrete. A phenolic film faced Russian ply was used for shuttering these walls; there was no polythene. A smooth hand-trowelled finish was given to the worktop surfaces. The chippies who built the formwork, the gang that placed the concrete, the joiners, plumbers, electricians and finishers were all in-house staff employed by Horgan Brothers. The spirit of teamwork and good workmanship is there for all to see. We completed to budget and programme, although handicapped by having to bring materials through an access corridor from the main road that was a doorway just 1 m wide and 2.8 m high. Through this tunnel we carried away the rubble of an existing three-storey building as it was demolished piece by piece. The piling rig was too big, so it had to be dismantled and pushed through the narrow entrance slot. The shutters, the props, the scaffold support and all the reinforcing bars were manhandled through this small aperture. The architect had the ideas, and we like to think we made it work.

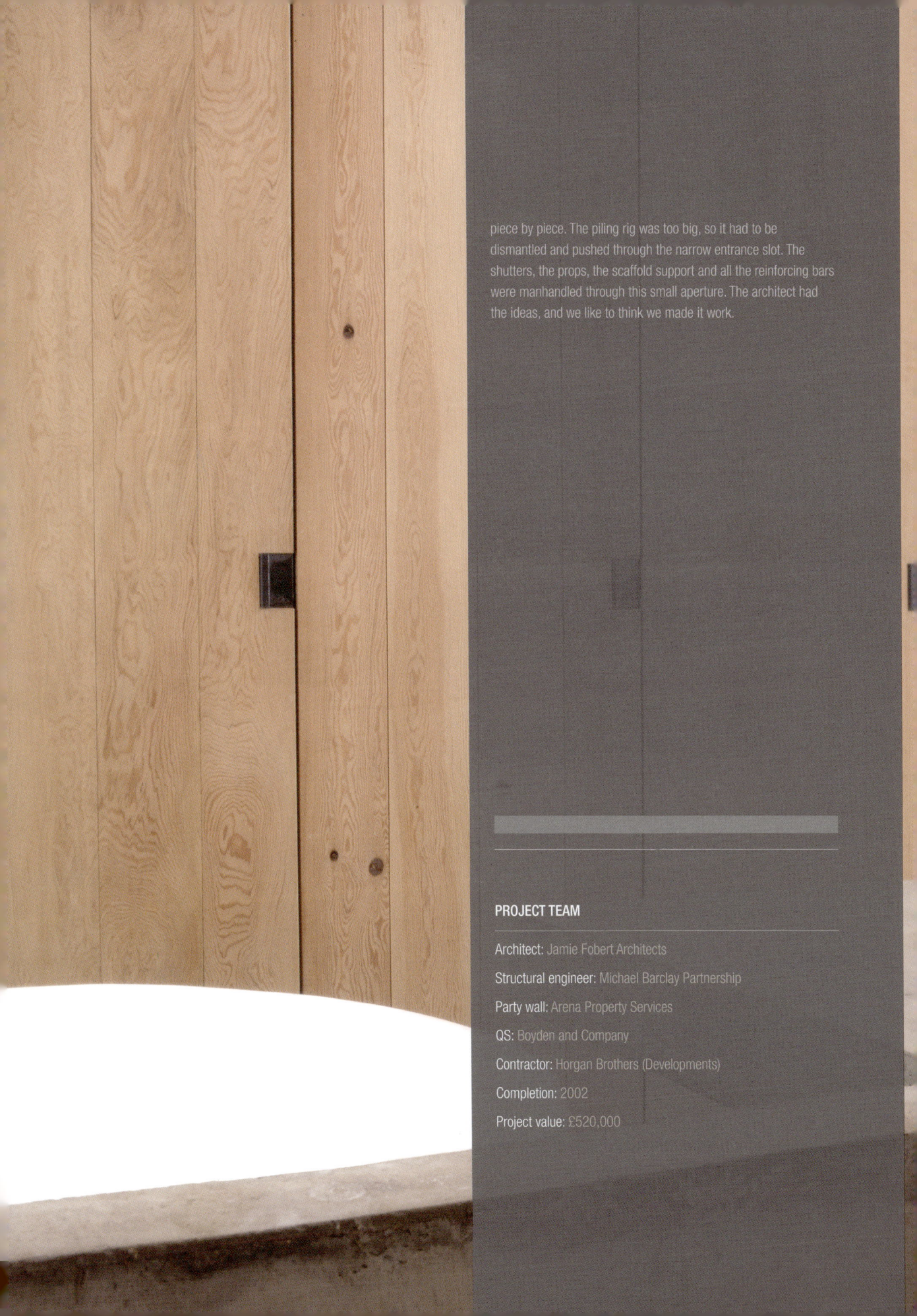

PROJECT TEAM

Architect: Jamie Fobert Architects

Structural engineer: Michael Barclay Partnership

Party wall: Arena Property Services

QS: Boyden and Company

Contractor: Horgan Brothers (Developments)

Completion: 2002

Project value: £520,000

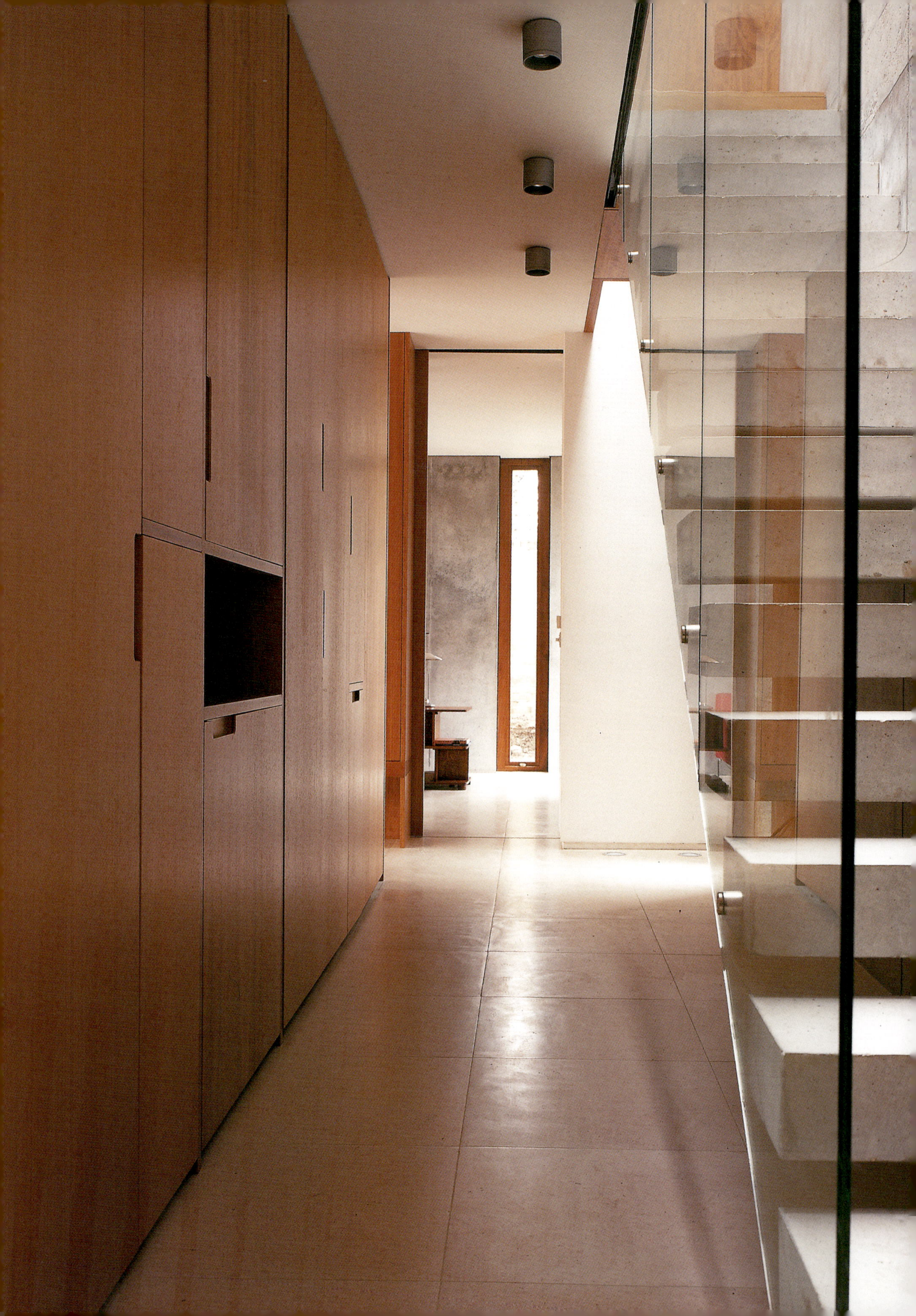

ABERDEEN LANE
ISLINGTON

Architect: Azman Owens Architects

Location

The house is located in Highbury, London, at the end of Aberdeen Lane which is an unmade road off Highbury Grove. The tube station of Highbury and Islington on the Victoria Line is a 15-minute walk away.

Introduction

This new build house for a family of six is located on an unadopted lane dominated by workshops, studios and light industrial sheds. The site backs on to an established residential neighbourhood. This trendsetting modern house in Islington is a statement in wall-to-wall insitu concrete. What is so surprising about the concrete is its marble-smooth finish, its flawlessly clean surface and seamless appearance.

'This is a concrete, timber, glass and limestone house with each material expressed to shape its character and architecture. It is modernist, cubist and the planes of structure and texture are skilfully overlapped and interlocked. A demonstration of a labour of aspiration and love by its owners, architect [and contractor]'.

Michael Manser *citation for the Manser Medal*

Architecture discussion
Ferhan Azman

The focus of the brief was to create a house with sufficient space and comfort for a family of six. The site is bordered by a detached house on the west side, terraced mews houses on the east side and a garden wall at the rear. The starting point of the concept was the decision on the orientation of the house. The decision to face the house inwards to create a courtyard rather than facing the lane was followed by a series of decisions that led to the choice of materials and method of construction. It was decided that the north and south walls of the house (the walls facing the lane and the house at the rear) would be treated as solid without many openings, in order to maintain a definitive edge to the lane and avoid being overlooked. The other instrumental parameter was the depth of the mews house at the east side of the site. With these two critical influences, the house was designed as two interlocking cubes of internally and externally exposed reinforced concrete walls. The choice of concrete was driven by the desired solidness of the north and south walls. We wanted these walls to be seamless planes and proposed concrete to the client. The house was designed like a doll's house with a west-facing 'courtyard' façade left transparent with large panes of glass in timber framing. This was a response to the clients' brief for a house accommodating the varying needs of a large family. Timber louvers were installed at first floor level for a degree of privacy.

We chose to use the same materials throughout the house, internally and externally, which are limited to concrete, limestone, timber and glass apart from the quaint guest bathroom which has stainless steel walls and a red rubber floor.

Aberdeen Lane in Islington is a bit off the beaten track, along a dirt road, past a collection of single-storey warehouses, back gardens and drab garage lots. Why build something so ultra modern and spanking new stuck down a long, uninviting industrial lane, in the shadow of a 1920s brick-built mansion? For the client this

First floor landing and stairwell

site is perfect – it's a 5-minute walk from the tube station, 10 minutes by car into the city, it's in a very sought-after location but, more to the point, it was a plot of land within their budget. The client thought the site interesting and private. The location gave us the opportunity to create something dramatic and monolithic in appearance – there were no buildings close by to relate to. Instinctively, our thoughts went to insitu concrete and when we suggested it to the client they were surprised but not shocked. What was causing the greatest anxiety to both architect and client was finding a contractor who would price the job within budget and also execute the concrete beautifully; a notion that many designers think are mutually incompatible objectives. Varbud Construction in Perivale had worked for the architects before on refurbishment projects – they made furniture, fitted out retail units, built handmade kitchens, did all the plastering, plumbing and brickwork, even cast concrete floors and walls and were always very competitive in their pricing. They were keen to take on Aberdeen Lane and we trusted their integrity, knowing how carefully they work, but fair face insitu concrete was going to be a new challenge for them. This bold project was going to establish their reputation.

The double skin concrete walls of the box structure act as bookends, retaining the open glass elevations that look westwards over the courtyard garden, the large ash tree, the iroko panelled timber garage and garden room. The concrete inner walls support the concrete first floor slab, the flat roof and staircase. The living spaces are all on the ground floor – the kitchen, lounge, TV room and dining area – with the bedrooms on the first floor. They are accessed via an open plan precast staircase that runs along the double height east wall, whose roof light floods the staircase with daylight. Single skin blockwork walls divide the internal spaces of the house into their functional uses – bedrooms, bathrooms, walk-in wardrobes and so on. All the 200 mm-thick inner load-bearing concrete walls and the floors were cast, then the 150 mm outer skin of the concrete cavity walls were cast as a second lift. The finished concrete surface is marble-smooth to touch, light-grey in colour and full of subtle abstract flecks and variations depending on the angle of the light.

We wanted a smooth-faced, light-grey concrete that would not get dirt-encrusted or stained. What has been achieved has come up to all our expectations. The surface is sensual, cool and very tactile and makes a strong contrast with the limestone ground floor tiles, the wooden framed windows and the elm-covered first floor panels.

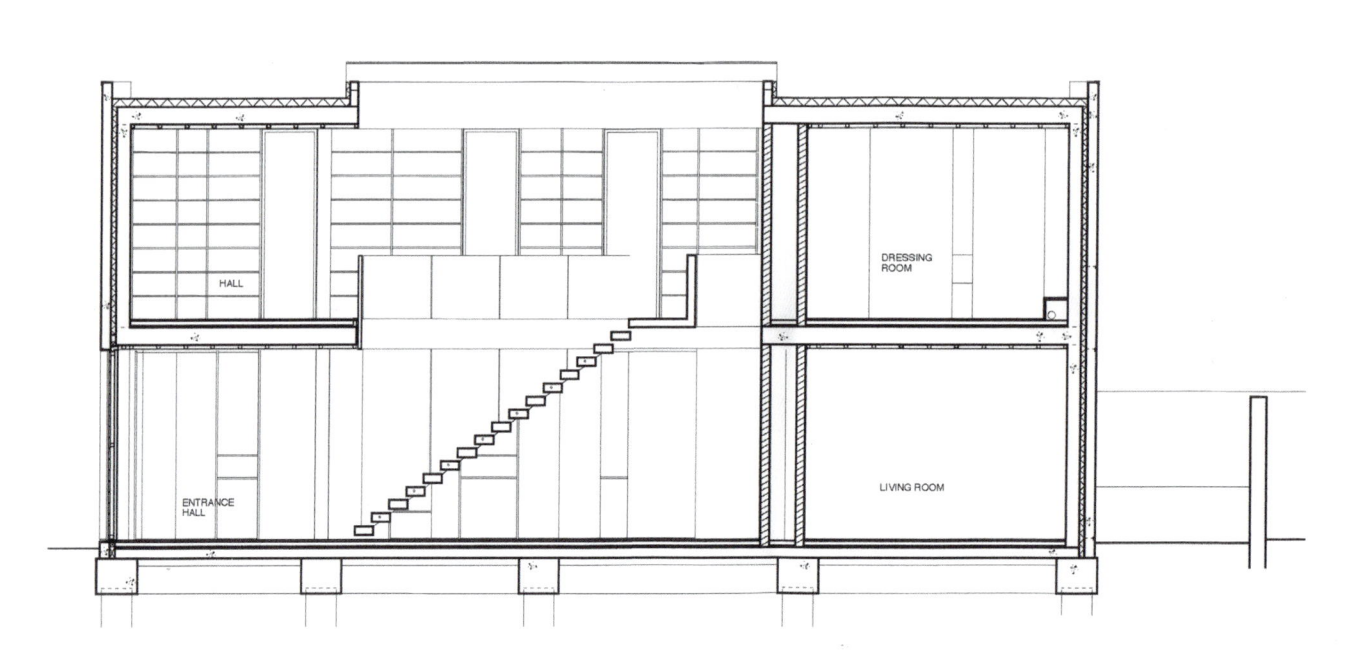

top: Front elevation
bottom: Longitudinal section

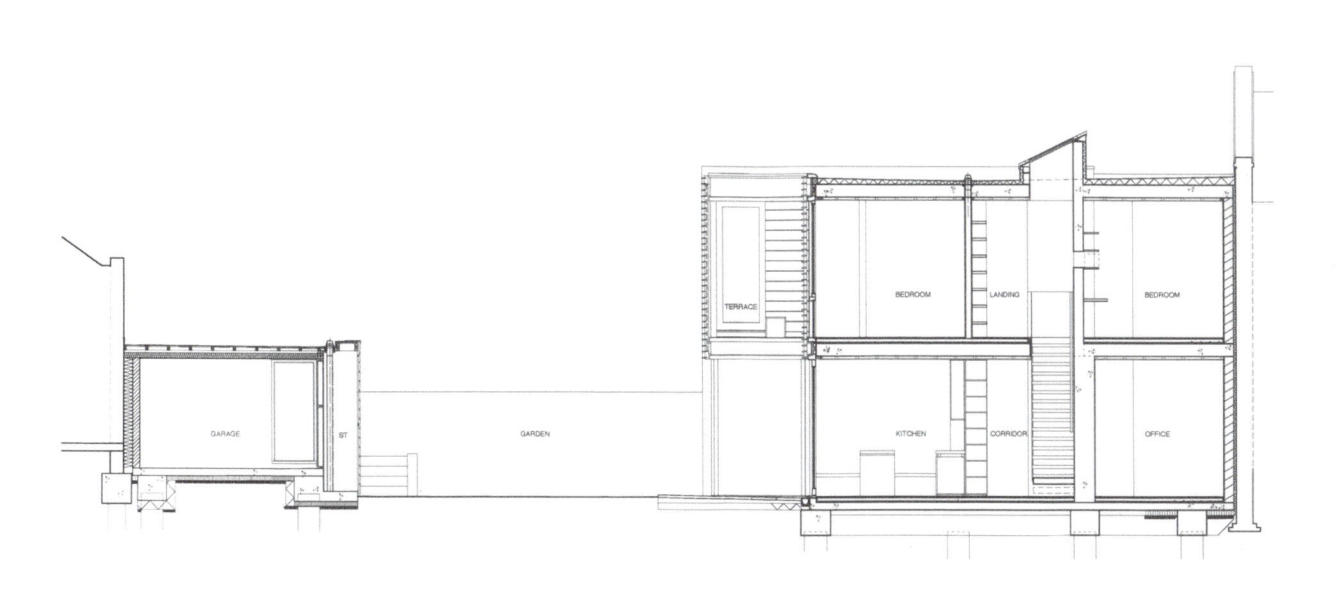

top: Front elevation by night
bottom: Transverse section and garage

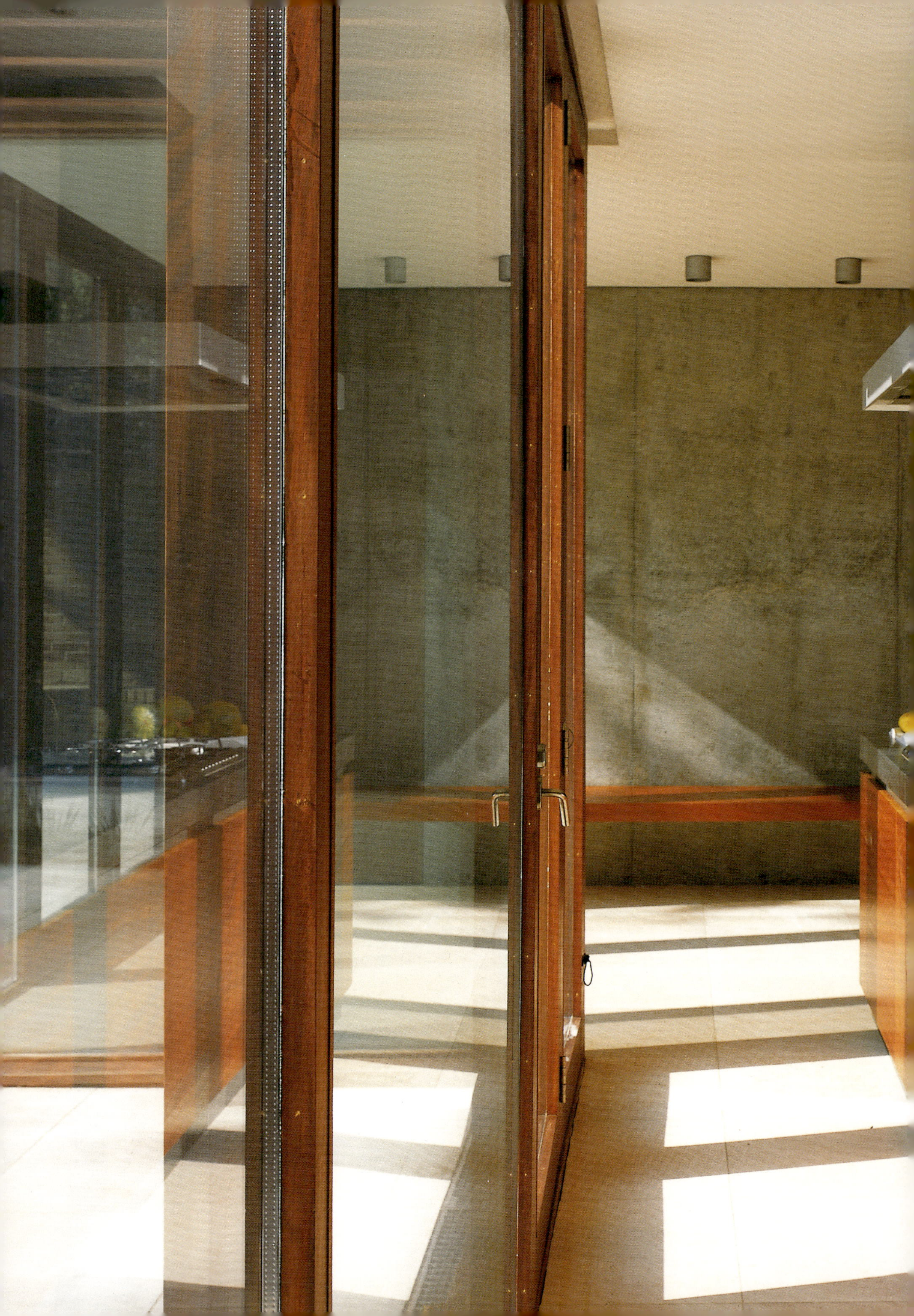

CONSTRUCTION NOTES
David Bennett

After the contractor had completed the sample panels to check the efficiency of different release agents, they machine cut the large birch faced ply sheets in their workshops and prepared the surface by sanding it down, then coating it with two coats of lacquer.

Once on site, the birch faced panels were lightly oiled with a high-performance chemical release agent manufactured by Nufins, before they were screw fixed to the backing ply. They were not allowed to screw fix on the fair face side of the ply. Everything had to be screwed from the back of the panels. In addition, the strong backs and waling to support the forms were designed with no tie bolts over the body of the formwork. A-Plant, who supplied all the props, had never designed temporary works with quite this degree of sophistication and control, but it worked very well. The push–pull props in the mid-span, the close centred walings over the lower half and the double row of strong backs, kept the shutters rigid and true under the 3 m head of liquid concrete. The concrete walls are as straight as a pole, perfectly plumb, with no lipping or bowing over their height. Wacker UK Ltd arranged a training workshop to show the concrete ganger how to use their constant amplitude pokers effectively. We were supplied with the best concrete we have ever seen for consistency of mix and workability by Hanson Premix. Their service to the site has been first class.

Varbud Construction have proved to the architect that a caring, listening contractor willing to work with you – but with no previous fair face concrete experience – can make achieving a fine cast concrete construction a painless process.

Kitchen

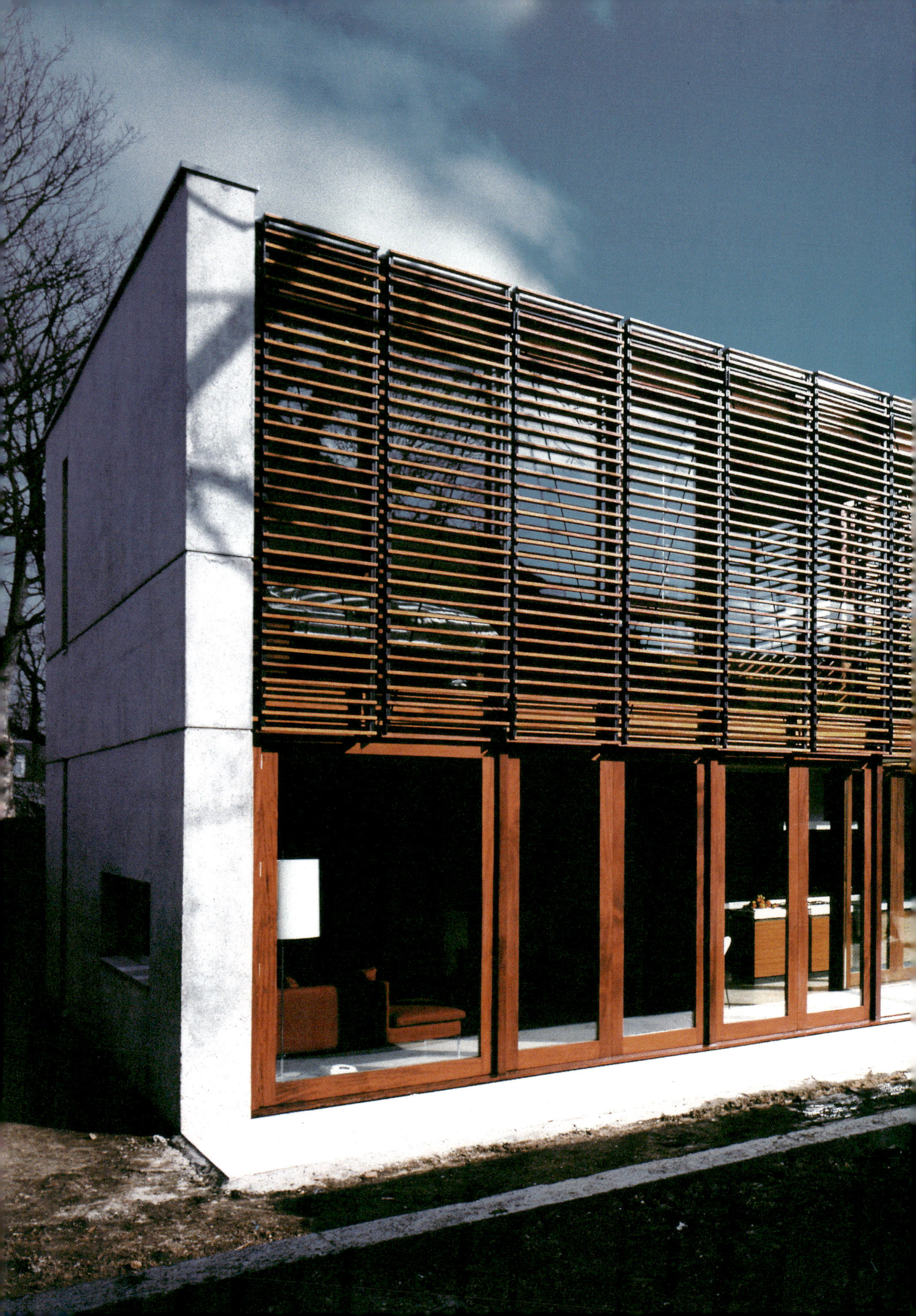

PROJECT TEAM

Architect: Azman Owens Architects
Contractor: Varbud Construction
Structural engineer: Brian Eckersley
Services engineer: Mendick Waring
Completed: 2003
Project value: £500,000

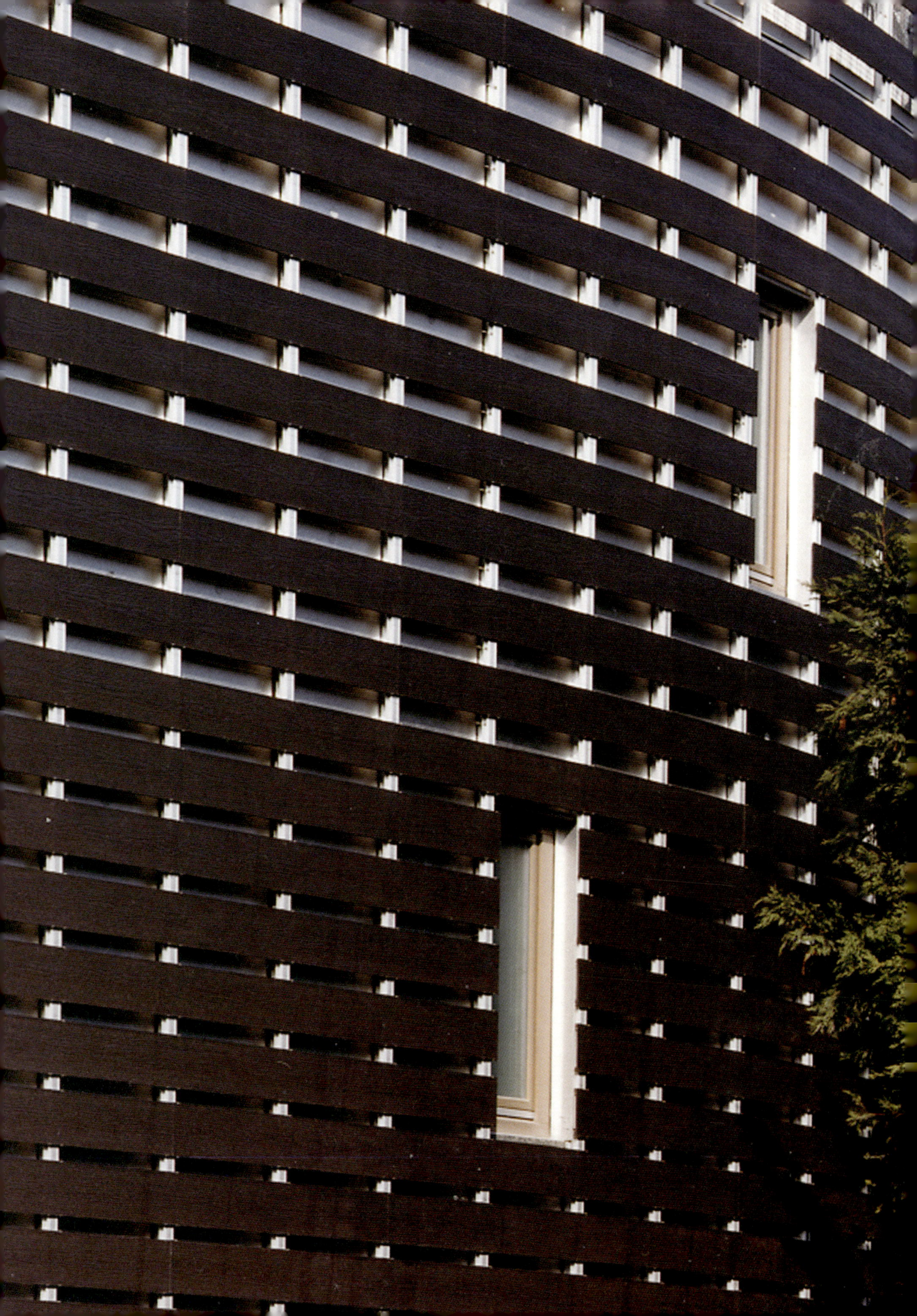

ONE CENTAUR STREET
LONDON

Architect: de Rijke Marsh Morgan

Location

The building is close to Lambeth North tube station on the Bakerloo Line; take the first right along Hercules Road and One Centaur Street is there beside the viaduct. Its unmistakable modern freshness, cut glass crispness and ribbed rainscreen façade give a much needed lift to its otherwise dull surroundings.

Introduction

Centaur Street was conceived as an 'inside-out' building. Internally, the walls are of textured concrete, and externally they are overclad with a chocolate-brown concrete graduated 'timber' rainscreen. The building consists of four apartments, each enjoying a dynamic interior organised as a large, open, double height living space, interpenetrated by adjacent enclosed bedrooms and stairs which form a buffer to the viaduct.

The four-storey block contains four large apartments arranged over multiple floors, all featuring double height spaces and each with their own design idiosyncrasies and features. The block and its shared planted garden sit within their own landscaped environment.

'One of the best new apartment blocks in Britain, a minor masterpiece that ... points the way to the kind of compact housing our cities need but which so many of our architects and house builders find difficult to achieve.'

Jonathan Glancy, *The Guardian*

The developer's vision
Roger Zogolovitch, Lake Estates

December 13, 1999 was a lucky day. It was the day I successfully bid for the site at auction. The site was a piece of waste ground that was adjacent to a monumental railway viaduct and was being used as a car park. To some people it looked uninviting and derelict, but to me it was an opportunity. I worked on the assumption that I wanted to create living spaces which were protected from the brutality of the landscape that encompassed the site – the huge brick arches of the railway viaduct, an overpowering office building and the untidy backyards of town houses. I have always been interested in the way the terraced house provides a vertical 'half landing life', which leads to the formal scale of the first floor space and connects with the rooms to the rear. I wanted to recreate this experience as the idea behind Centaur Street and proposed that the volume was divided into three zones – the cocoon zone, the function zone and the living zone. To give more emphasis and priority to the living space I arranged it on more than one level, which created a double height space.

As a young architect earning a £300 fee per flat, designing terraced house conversions in the 1970s has been of immense value. In those days of hand-drawn plans, the work gave me a detailed understanding of the manipulation of internal space as my career progressed from architecture to development and then to Centaur Street.

When writing the brief for this project I decided that the appropriate starting point of the terrace house could be met by the three zones and that the architecture should capture it. Reinforced concrete was quite logically the material of choice and I became very emotionally attached to it. I confess that I had never built with it before – it had been bricks, render and stud work. It's a material that sculpturally suits my ideas for volume emerging from the site context.

Ground floor duplex apartment

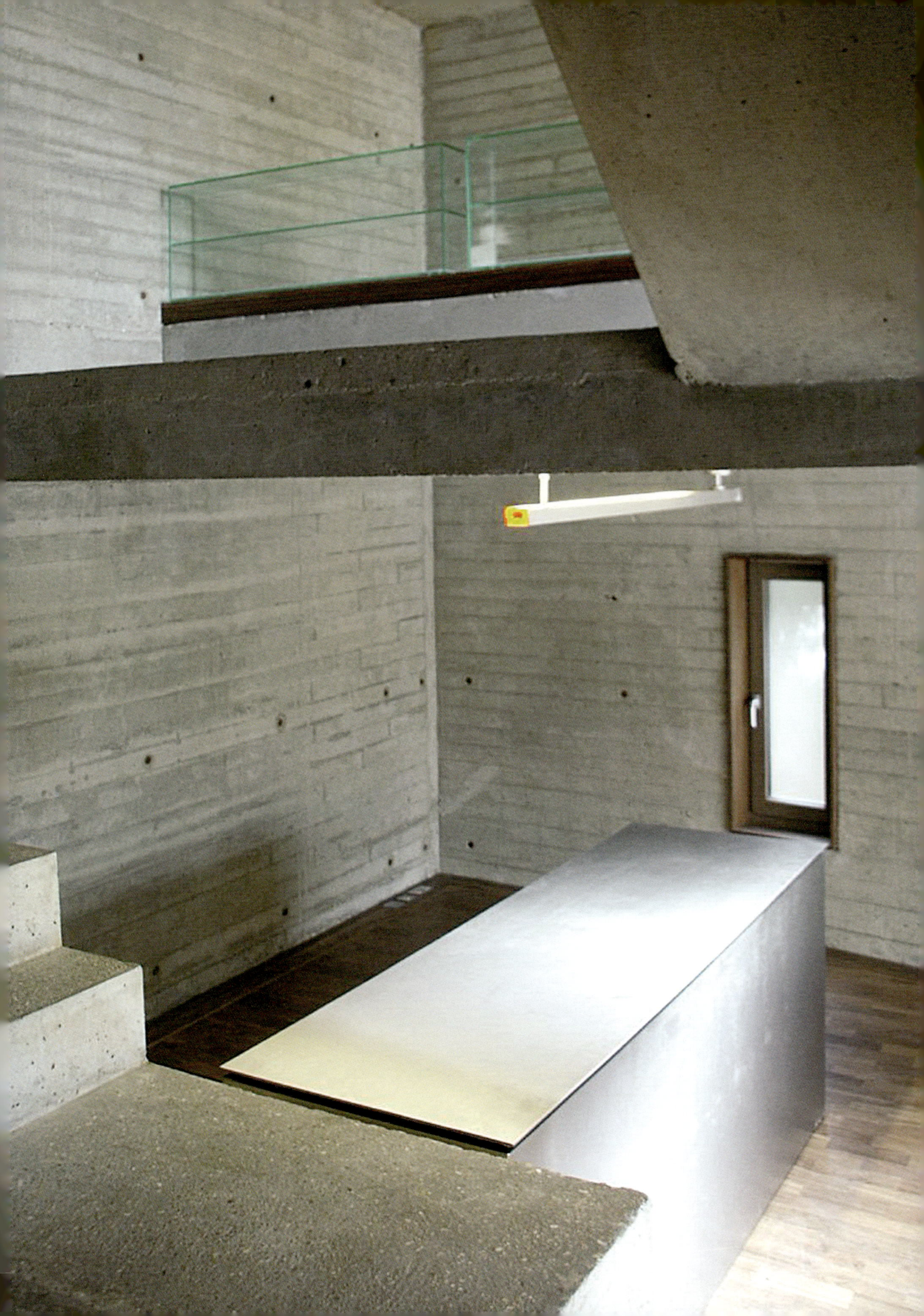

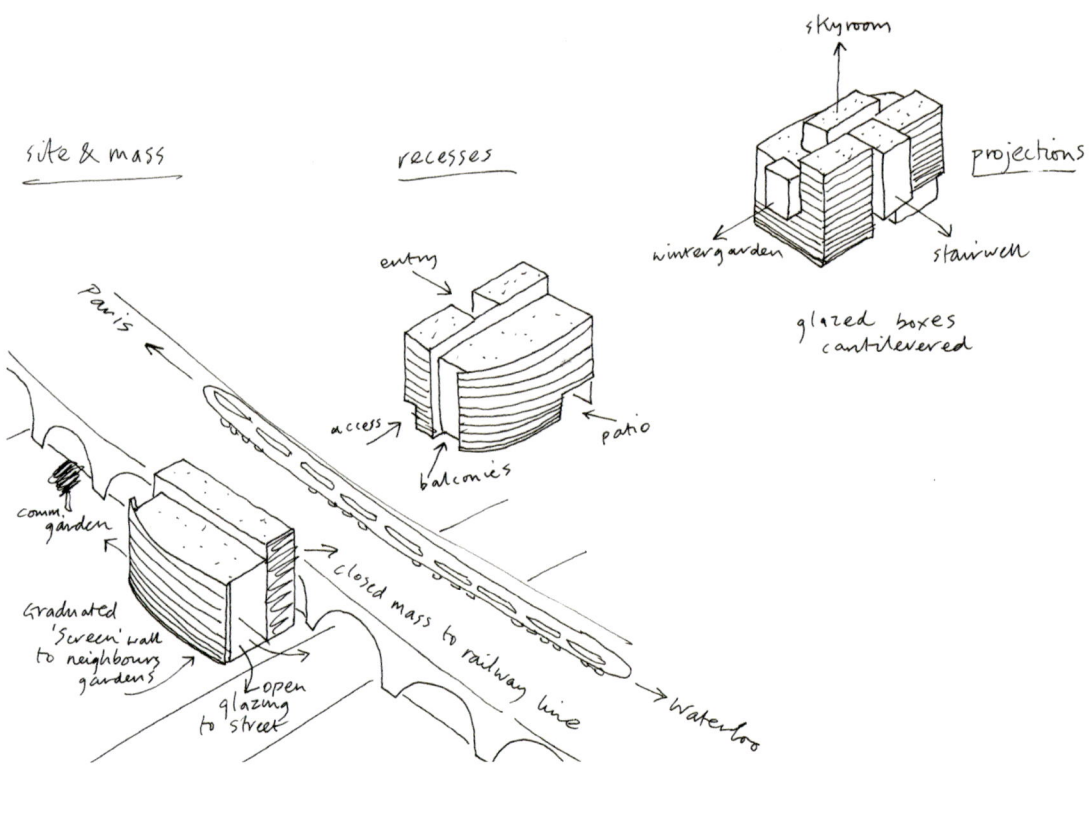

top: South elevation
bottom: Concept sketches of design development

130

ONE CENTAUR STREET

South Elevation

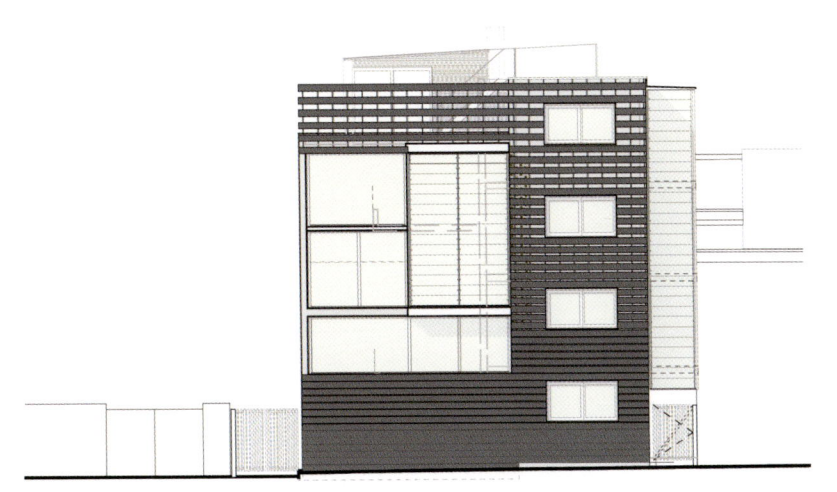

North Elevation

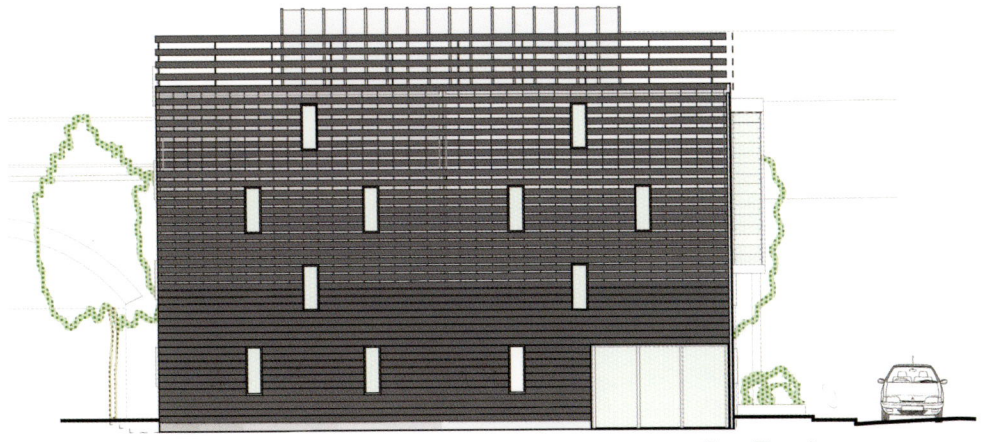

East Elevation

131

The surprising thing about concrete is the beauty of its surface texture as it is struck from the mould. The scheme as it was originally conceived contained four units to be sold as raw shells with walls of concrete. The concrete needed to have a patina, a texture like old brick. The board-marked finish with its patina and depth of colour was a revelation. Nobody builds load-bearing brickwork any more in the tradition of the Victorians. A thin brick veneer with no structural capacity has no appeal to me. The material of today is concrete and I wished to create a typology with it which is as adaptable as the terraced house once was.

City-dwellers suffer from noise pollution and concrete gives a sense of security to the internal space. It is stable and has good thermal mass, which helps to reduce internal temperature gradients and therefore heating costs. The perception of scale and generosity of the split levels was remarkable. People would walk into the apartments and get lost in an area of just 1,500 ft^2. I believe that this was due to the changing eye level as they moved up through the space from room to room. This altered perspective gives the impression of being in a much larger apartment.

The brief was explicit, I wanted influence in the design process. It was a collaborative effort by everyone involved and gave me the freedom to be as innovative or as experimental as I wanted to be.

I am pleased that the project turned out in a way that totally satisfied my intentions and that is a tribute to the architect and all the other participants who helped in its realisation. Too often we look just at the finished building without the proper recognition of the human resource and effort that has gone into the project's making. Good buildings need more than delivery programmes, they need the human commitment of a team of dedicated people who work on it. Centaur Street was lucky to gain that support from all who worked on it

Architecture discussion
Alex de Rijke

The brief was to create an experimental housing type which examined and exploited the potential of density on small gap sites. The Eurostar viaduct was an unlikely genus loci, but its close proximity generated a clear zoning of the plan; a powerful entry route, with stairs and services, and finally living space, all arranged as parallel strips. The accommodation is varied, intimate and prioritises space and light.

We are interested in all building materials and the art of construction, working entirely with standardised materials that we find in catalogues. Yet we try to create non-standard architecture by carefully assembling the components. The client came to us with a proposal to design an entirely insitu concrete building, which was quite a challenge as we had never designed with concrete before. It was to be exposed internally and clad externally which is odd because there seems to be a precedent for it to be the other way round in the UK. This project was more about texture and surface quality than composition or colour. It was achieved internally by the process of casting and forming the walls, the floors and the roof. We had a site into which we simply poured concrete right up to the boundaries and to the rooftops. Had we had more time to reflect and freedom to consider our options we might have elected to go with precast concrete for the staircases. But we had a client who was completely in love with insitu concrete 'the grey gold' as he would affectionately brand it. He wanted a jointless, seamless building and I quite understand why. The contractor was not completely equipped or experienced to form it to the standard we had expected, but they made an honest and diligent effort and after the surface was lightly sand blasted we were very satisfied with the resulting cave-like quality. There is something wonderful about this material – it is the structure and architecture; it is not stitched together, bolted or nailed. We can design double height load-bearing walls, overhanging cantilever slabs and neo-Georgian steps in concrete.

Our working relationship with the client, who was an architect, was both open and exacting and a strange, intense process. It was the paradox of having an architect commission you to design his own building, but he trusted us enough to let us bring our ideas forward and infuse them into the design. We had board marking on some walls, making it in softwood, and flat finish in the bedrooms using phenolic resin overlaid plywood panels. We enjoyed the illusion of wood that the internal concrete conveys. On the outside we had fibre-cement board-marked panels, pretending to be wood very effectively and maintenance-free. It was an Eternit product that was developed for weather boarding on seaside chalets but not on a building. Since we adapted it for use as a rainscreen on our building, we see it copied everywhere.

This building is didactic, the cladding does not cover up the weathertight insulation behind it, and the insulation does not hide the load-bearing structure nor the structure hide the finished architecture. It's a layered composite design like the shirt, waistcoat and jacket of clothing that you can see at the edges. I think building architecture goes downhill once the main structure

top: Stairwell and terrazzo landing
bottom left: Board-marked internal wall
bottom right: Cantilever mezzanine floor

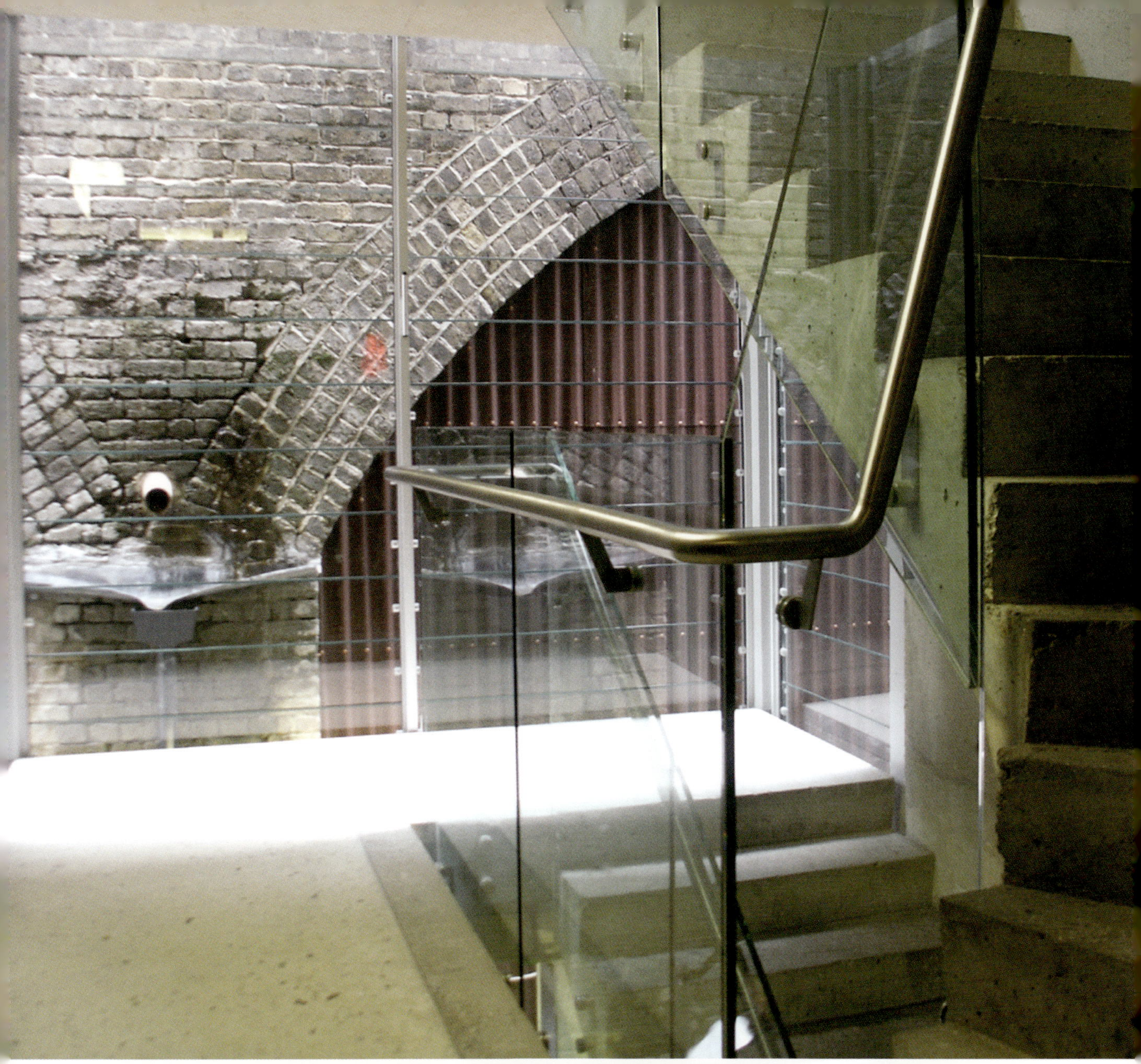

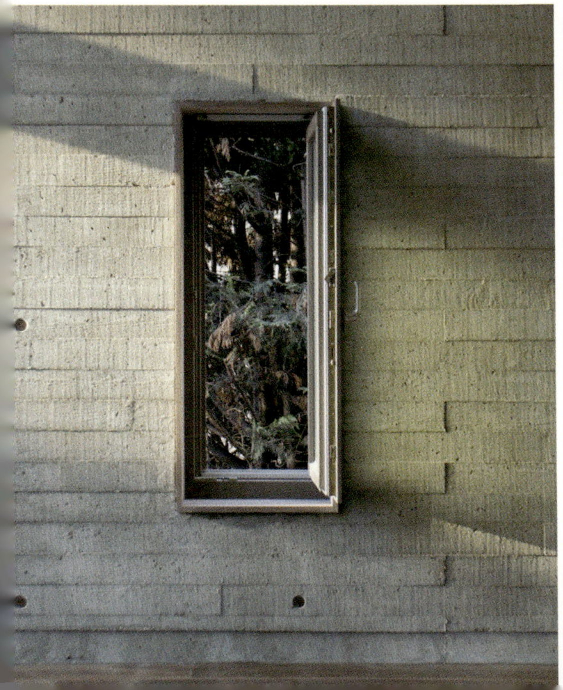

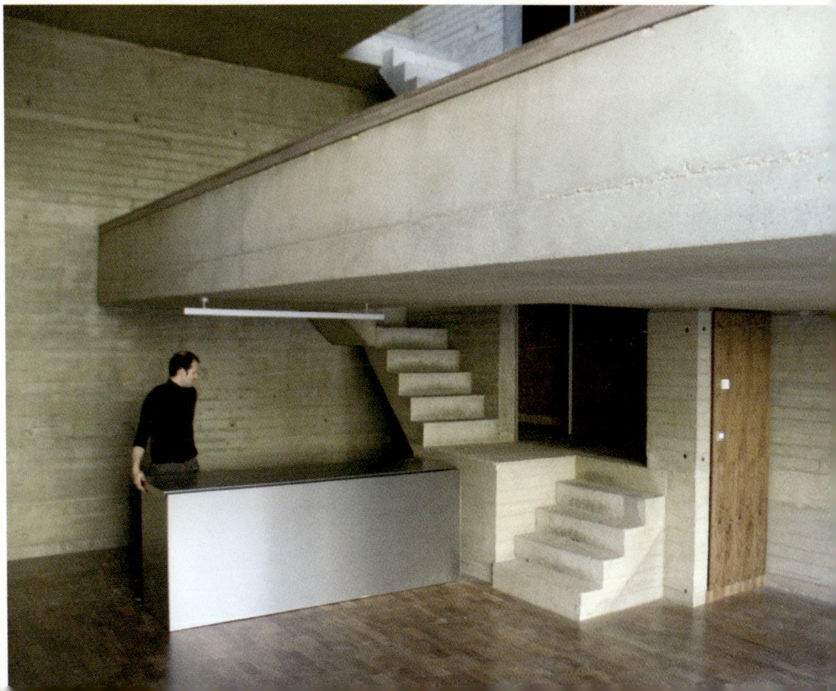

is covered up like the upholstery of a furniture frame. I hate the notion that you need five trades to make a wall – stud frame, plasterboard, insulation, plaster and paint. Concrete is the perfect medium for house construction – it has character, it can support itself, it is self-finished and is one process.

We introduced recycled materials from the construction into the furniture and flooring of our new office in the ground floor apartment. The ply, which had a phenolic resin overlay and which was used for the floor slabs, we used for the bookshelves. The rough boarded timber laths that formed the board-marked wall, we used to make into a sliding door. We tried to limit the palette of materials by using the formwork that cast the concrete. The glass parquet on the ground floor was cut into 300 mm squares with bevelled edges. However, to lay the glass parquet on a fire-retarded, Styrofoam substrate has proved a technical nightmare. A self-levelling screed was poured over the existing floor which was not flat enough for the Styrofoam. The Styrofoam has routed ducts for running the computer cables and for sleeving the underfloor heating pipes and once these were in position we placed an acrylic sheet over the support ridges to distribute the load and to bed the parquet into the foam substrate. The glass panels are held down by double-sided adhesive tape and the joint edges sealed with glass glue. We found that the glue was too brittle and de-bonded with the slightest movement, so we replaced it with silicone that seems to work well.

The spatial organisation of the interior is based on the Raumplan principle pioneered by Adolf Loos. The double height central space has a tight mezzanine floor slotted into it with smaller self-contained rooms leading off to the sides. This was not to be a static event with rooms assembled in rows around a central staircase, like a terraced Victorian house. Here, the smaller rooms wrapped around the large void and were interconnected by a series of discrete stairs. It is a hybrid of the open plan European apartment and the English vertical terraced house and it works well.

Concrete works
David Bennett

The board-marked panels were nailed to the backing ply and were made a feature of the internal surface finish. The boards were varnished before applying a high-performance chemical release agent. This ensured at least three reuses of the timber. For the plain concrete surfaces in the bedrooms, a medium density phenolic resin (MDO) overlay board supplied by WISA was specified. Everything was screw fixed from the back of the panels to keep the contact face of the MDO free of potential blemishes or splits. In addition, the strong backs and walings that support the inner wall forms were designed with minimum tie bolts over the body of the formwork.

Most of the board-marked concrete wall surfaces were lightly sand blasted to remove surface stains and grout runs that occurred when casting the floors above them.

To achieve a consistent mid-grey colour, the OP cement from Blue Circle's Northfleet works was maintained at 350 kg/m^3 and the water/cement ratio held at 0.47, with the help of a water-reducing admixture. The coarse aggregate in the concrete mix supplied by Tarmac Topmix was proportioned to reduce the 5 mm stone content for better control of workability and to reduce the risk of aggregate bridging. The mix constituents were kept constant from one batch to the next, particularly the cement and the 63 micron sand content, which dominate the finished colour. But the dry batched concrete gave poor consistence in workability and the slump would sometimes vary from 80 mm to 175 mm for pours on the same day. The contractor did a sterling job working with such variable material and it was inevitable that there were going to be a few hiccups. I feel it is important to communicate the concrete requirements to every truck mixer driver, batcher, dispatcher and operations manager in the set up in order to achieve the level of control that is needed.

On the main staircase the landings are an insitu terrazzo topping of white cement, buff Redditch sand and 15 mm pearly quartzite aggregate obtained from CED in Thurrock. These create visual links between the natural grey of the staircase flights. Externally, the off-white concrete of the hardstanding was cast with site batched GGBS cement mixed with OP, buff Redditch sand and Thames Valley aggregate, and given a light sand blast for texture and to even out surface reflection. The bench seat has been cast with the same site batched concrete.

PROJECT TEAM

Client: Solid Space
Architect: de Rijke Marsh Morgan
Structural engineer: Adams Kara Taylor
QS: Measur
Main contractor: Parkway Construction
Completed: 2003
Project value: £1 million

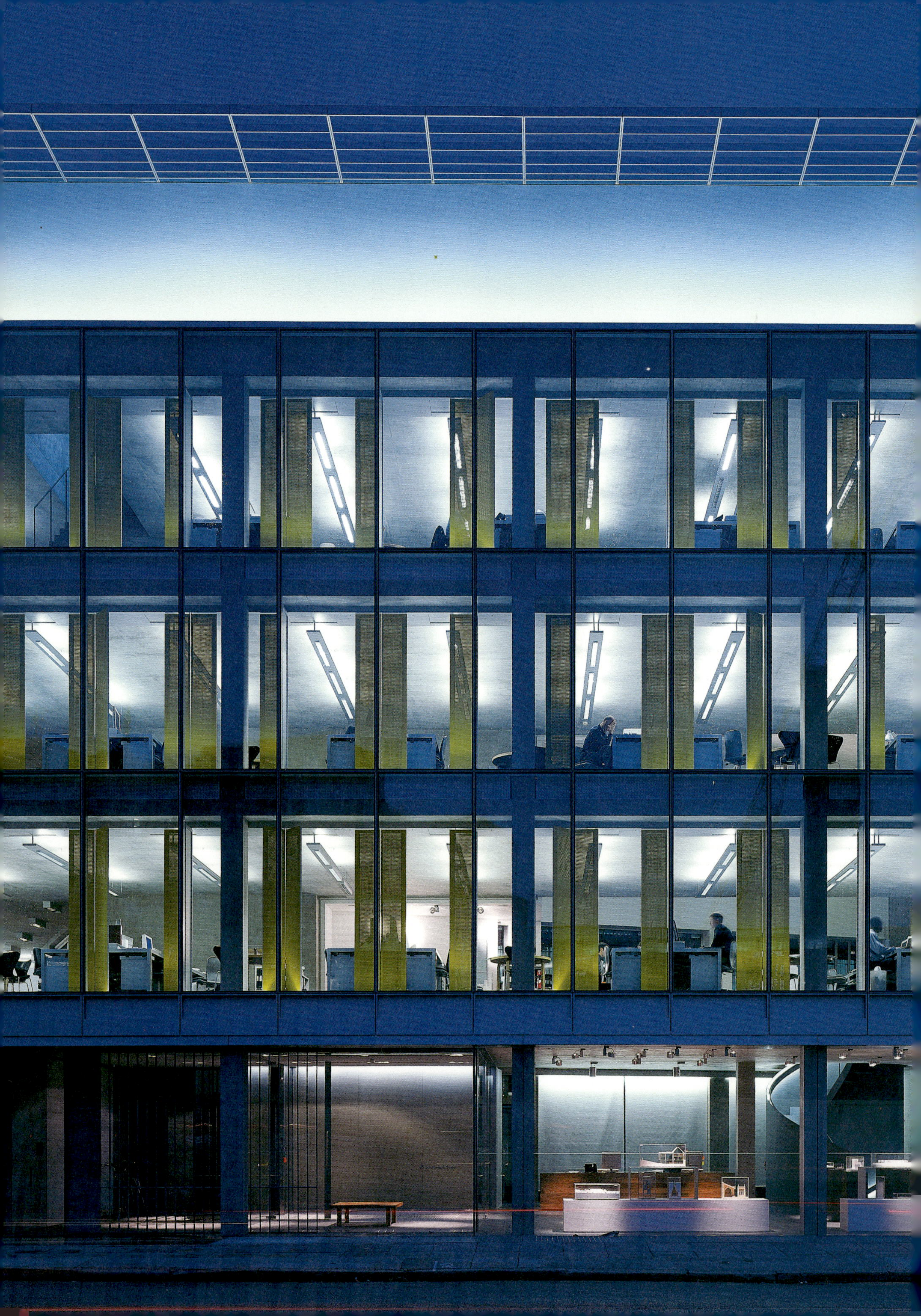

85 SOUTHWARK STREET
LONDON

Architect: Allies and Morrison

Location

The building sits midway along Southwark Street on the south side of the pavement, within the Bankside district and close to Tate Modern. From London Bridge station it's a 15-minute walk; a pleasant stroll if you wish to circumnavigate Southwark Cathedral and wander through Borough Market. The building is bright, luminous and cool, wrapped in glass with a chic ground floor café that serves good coffee.

Introduction

The building on Southwark Street has been designed by the practice for its own occupancy. The upper three floors house the studios which are linked by a stepped atrium. The ground floor contains the reception and café, while the basement contains the library, printing and model making facilities. At ground level a new public route has been created through the building connecting Southwark Street to Farnham Place. The two façades of the building – on Southwark Street and Farnham Place – are radically different from each other. The elevation to Southwark Street is almost entirely flat and glazed, revealing the internal concrete structure and vertical yellow louver screens which protect the interior against glare. The elevation to Farnham Place is rendered with relatively small window openings and folds and bends in response to the complex geometries of the site and rights of light of the adjacent buildings.

Architecture discussion
Alex Wraight

The site was bombed during the Blitz and was latterly used as a petrol station and car park. Its complex shape, onerous rights of light constraints and soil contamination made it an unlikely contender for commercial redevelopment but these constraints and its location appealed to us. The contrast in scale and orientation between the front and rear of the site was the starting point for the way in which the building was ordered and each façade developed.

To the rear, the medieval street pattern of blocks and alleys creates a complex and fractured footprint, an intimate scale and various rights of light issues. In response, simply detailed openings are cut into the largely rendered surfaces with precision, not only to control the ingress of daylight from the south but also to allow specific views from the interior spaces. The complex geometry created internal spaces more suited to servicing space than open plan studios; consequently, this is where we located the circulation and toilet cores. Additionally, in response to the rights of light envelope, the building steps down from the roof to form a series of planted terraces.

In contrast, Southwark Street cuts straight along the north of the site, a busy major road running parallel with the Thames. Here the elevation is fully glazed in order to maximise the amount of daylight reaching the interior, and relates to the scale of the major street in its simple tripartite composition – a frameless glazed shop front at pavement level allowing views through the building, a large and

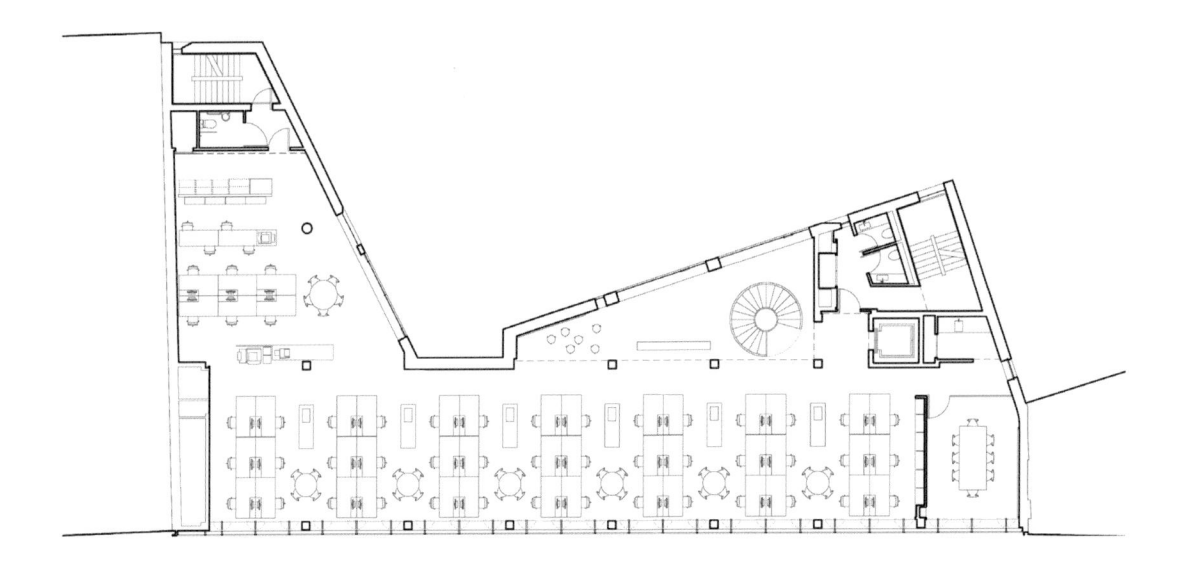

First floor plan

regular grid of glass curtain walling to all three floors of studio space and, at the top, setting back to produce a long reflector with an aluminium canopy.

This glazed skin reveals and makes explicit the build-up of layers within the building – behind the regular 1.5 m glazing module, the larger 4.5 m structural grid of the concrete frame is clearly visible, together with the horizontal extent of each floor slab, the raised floor zone (demarked by perforated anodised aluminium spandrel panels) and the inner layer of vertical coloured pivoting aluminium fins.

The intention of the new building was to gather all employees into one location south of the river, instead of being scattered in disparate groups accommodated in four or five buildings in the West End. We were keen to generate a relaxed, studio-like atmosphere so the office floors are open plan and connected by an open atrium, itself connected to the ground floor reception and basement via a sculptural, spiral staircase.

Early site studies revealed a road network in Bankside whose geometry was formed by the parallel lines of the terraced pike gardens constructed in Tudor times on the south bank of the Thames. The parallel roads were linked perpendicularly by narrow lanes, evidence of which can still be seen around Clink Street. This grain was subsequently lost when the Victorians built Southwark Street and when Bankside Power Station and the St Christopher House development to its south were constructed. Our proposals for the site of St Christopher House for Land Securities (now called Bankside 123) and our own site, sought to reinstate the perpendicular pedestrian route through the sites. Ultimately, this will cut through the Turbine Hall of the Tate Modern linking Bankside to the Millennium Bridge and the City beyond. The public route through our building divides the ground floor space into the office reception area and a retail unit, which we run as a café.

At ground floor level, the fully glazed north elevation forms a shop front with unobstructed views into the generous office reception.

Ground floor entrance and reception

ARCHITECTURAL INSITU CONCRETE - CASE STUDIES

Our aim was to engage with the street and display a large selection of our models in the window for the public to see. The reception is additionally animated by informal meetings, CPD lectures and exhibitions.

There is a large conference room and roof terrace at fourth floor level as well as a plant corral housing the boiler room, chillers and AHU concealed below coping level.

Concrete and materials

The location of the building provided a starting point for our selection of materials and our detailing strategy. Bankside, until very recently, was an industrial area with a very different architectural vernacular to the City. The warehouses and factories were simply detailed, of modest materials with very little decoration. We sought to continue this approach and selected and used materials in their natural state – aluminium has a natural

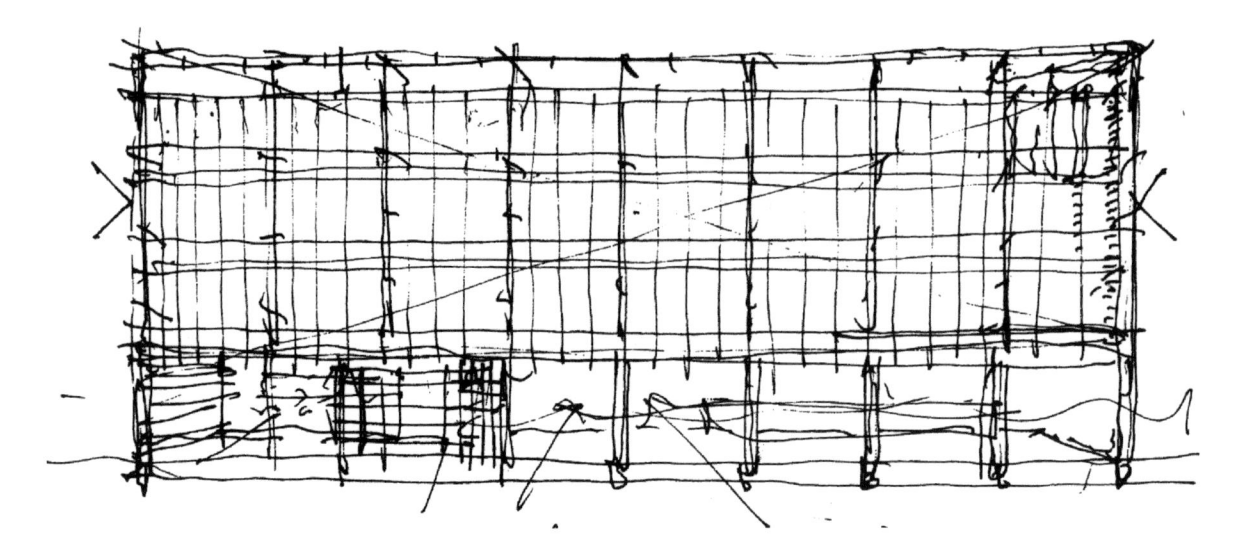

Conceptual sketch of main elevation

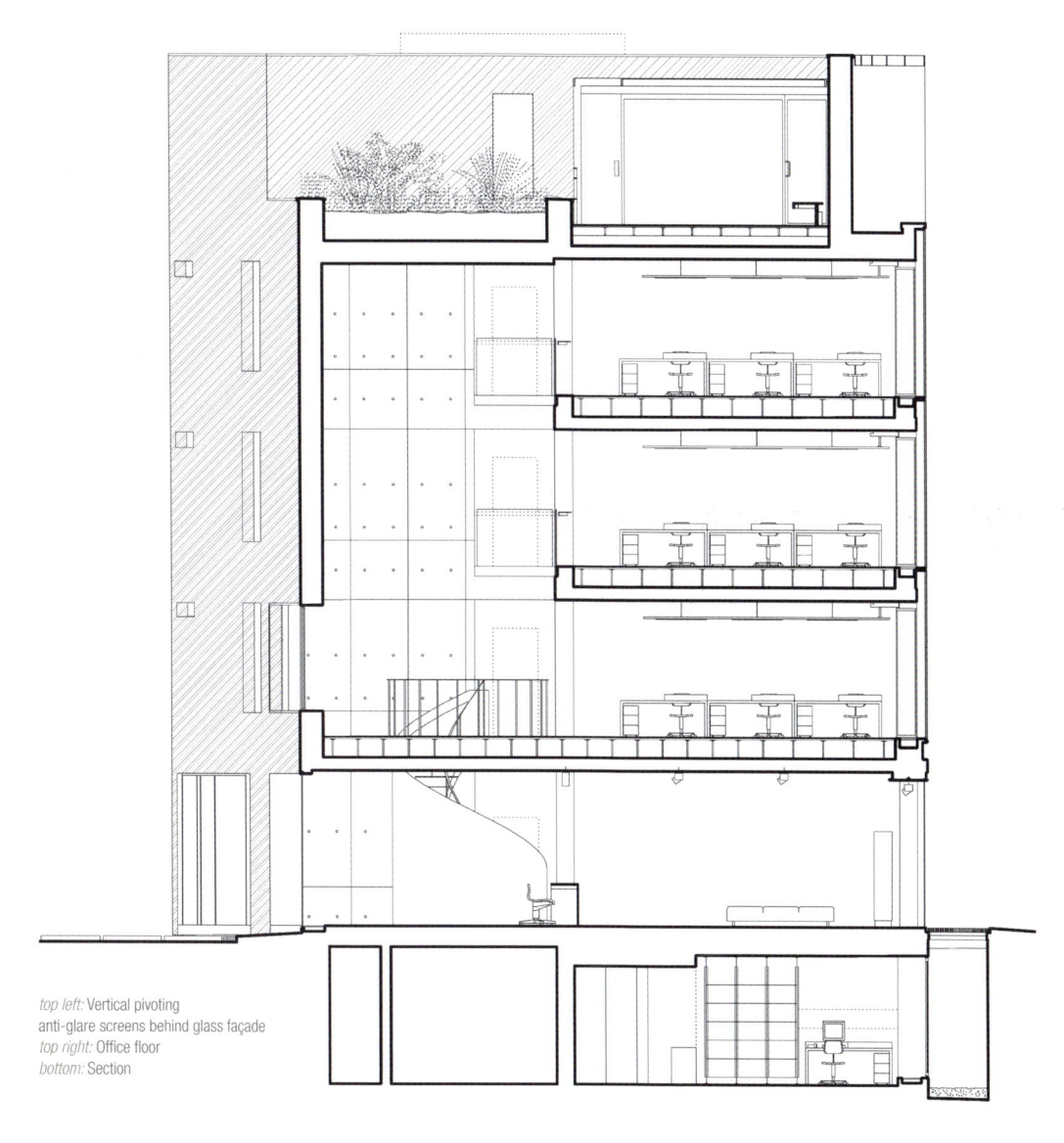

top left: Vertical pivoting anti-glare screens behind glass façade
top right: Office floor
bottom: Section

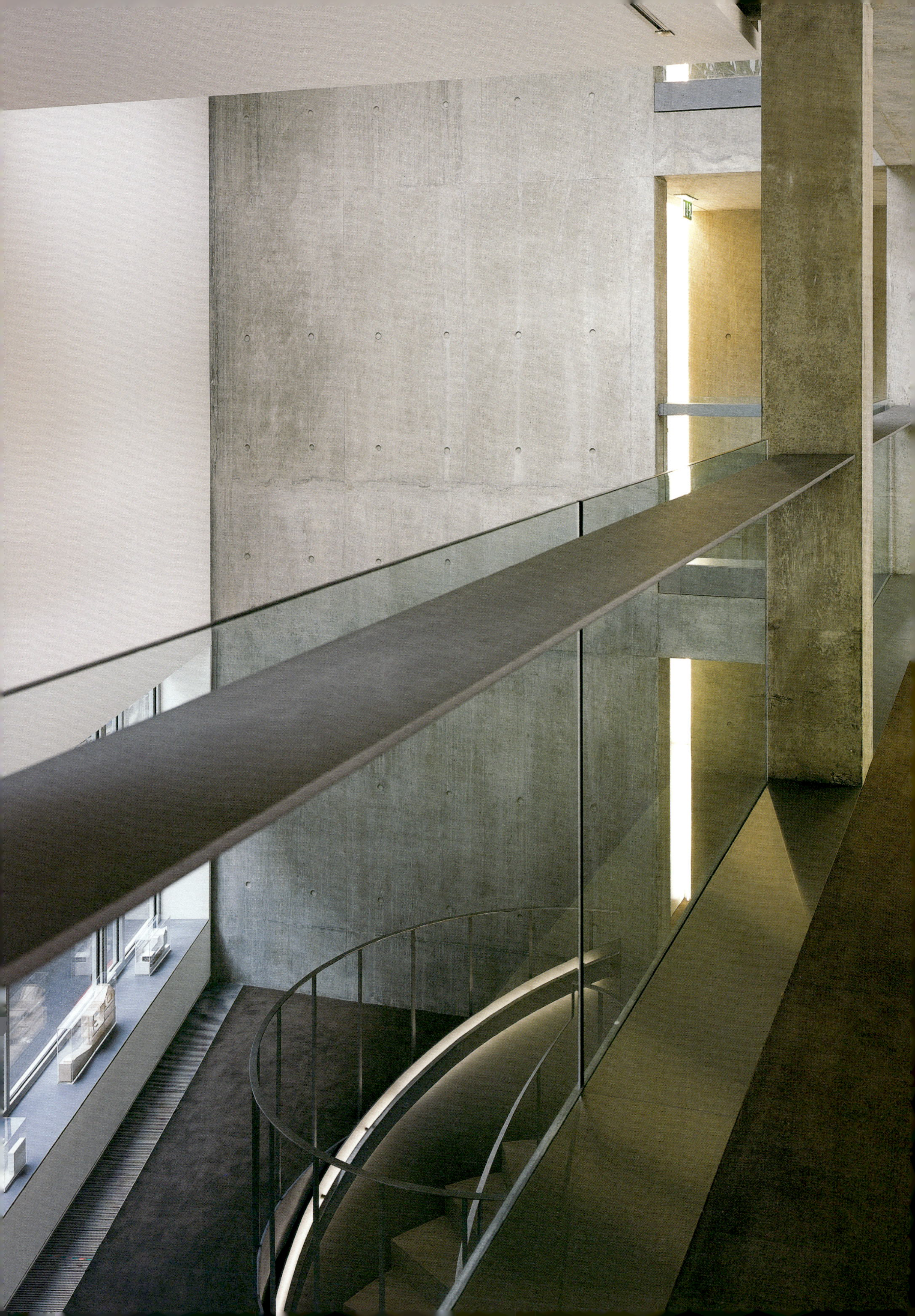

anodised finish, concrete is left fair faced and exposed. In this way, the building structure and cladding remain constant while the more ephemeral items of the fit-out can be changed and replaced as they wear out.

In order to accentuate and express each of the materials individually, the interface between them is detailed so that they do not touch. Sometimes the gap between materials is expressed and sometimes it is concealed.

The selection of exposed, insitu cast concrete for the structural elements was made for a series of reasons. First, it provides a massive solidity – when you knock on the cladding to a column in a steel framed building it sounds hollow. In contrast, in our building, the concrete is dense and solid and cool to the touch. It feels like an immovable piece of structure and its autonomy is accentuated by the separated detailing of the adjacent materials.

The surface and appearance of insitu cast concrete was a further factor in our choice. The uncertainty of the cast finish, affected by the weather on the day of the pour, the number of times the formwork has been used, the exact consistency of the concrete and the thoroughness of compaction, appealed to us. The variety of imperfection, albeit bound within a precise setting-out regime, exposes the true nature of concrete and is not something that can be artificially reproduced.

Constructing our own office building gave us the opportunity to explore the use and appropriateness of fair faced concrete, particularly soffits, within a commercial environment to a greater extent than would normally be permitted to us by the letting agents advising our clients. However, we have had fantastic feedback from clients who have visited the building so we hope we will be able to use this approach again. The exposed flat slab soffit provides very flexible floor space, both for an open-plan and subdivided configuration, and the lack of ceiling void can reduce the overall height of the building. The basic column grid is 4.5 m along the slab edge by 6.9 m into the depth of the plan. This grid is further subdivided into a 1.5 m cladding module. The flat slabs are 250 mm deep. They were not designed as two-way spanning floors as continuity was in one direction. There are stability cores and shear walls to provide lateral restraint to the frame. The 4.5 m grid gave us a reasonably dense grid on the façade which is exposed on the half module of the glazing. This enables the cladding and concrete frame to be read as separate layers, with a third layer formed by the internal pivoting shutters.

The office environment is conditioned by a displacement ventilation system within the raised floor. Air is ducted from the air handling unit, located at roof level at the eastern end of the plan, and emitted via swirl diffusers located throughout the office floors. Exhaust air is extracted by a riser at the eastern end of each office floor to complete the air circulation loop. The raised floor zones are expressed on the façade by perforated anodised aluminium spandrel panels set behind the glass. The top of the glazing transoms are aligned with the top of the raised floors and the concrete slab edge is clearly visible below the spandrel behind the glazing.

The pivoting aluminium fins located behind the north façade glazing counteract the effects of glare at computer screens and provide some privacy for the building's occupants, as well as animation of the façade from the street. The design was developed with Art Andersen of Copenhagen: each fin is made from aluminium plate, perforated to reduce the contrast between dark and light as the eye adjusts to look out through the façade from within, and bonded to a slender, solid aluminium frame with an overall thickness of only 20 mm. The fins are paired up to fold like butterfly wings in line with each curtain walling mullion. The colour is only revealed as a fin is swung open.

Concrete contractor

For tender purposes, we opted to use the structural engineer's concrete specification and included a statement on architectural intent for the finishes. We highlighted three classes of finish, A, B and C. Class C, which was the highest quality, referred only to the curved wall that enclosed the spiral staircase. Most of the concrete was classified as class B. This described the maximum permissible size of blowholes, while allowing some discoloration and alignment tolerances. Class A was a basic concrete finish intended to be covered up with lining finishes.

Whelan and Grant was the selected concrete frame subcontractor. Our experience of exposed concrete finishes has taught us that, when trying to convey a desired quality of finish to a subcontractor, there is no substitute to viewing precedents together and constructing benchmark samples. Often, contractors assume that the architect is seeking a perfect, blemish-free finish while we were actually very keen to maintain the character of struck concrete without extensive finishing work. We were successful in setting up a dialogue with Whelan and Grant, through site visits, design

Lift wall and staircase

ARCHITECTURAL INSITU CONCRETE - CASE STUDIES

team meetings and samples, which enabled us to understand the limitations of the material and develop a meaningful and realistic expectation of the finish they could achieve.

Our discussions included how they would assemble the formwork and build the walls and floors, the tie bolt hole positions, the panel arrangements, the joint lines and other important features. The support system they proposed was designed by Peri. A double layer of ply was used to ensure that the contact face was flat, smooth and true. The facing panel in contact with the concrete was a WISA MDO paper faced ply which gave the concrete a matt finish. This facing ply was cut down to a 1,125 × 2,275 mm module to subdivide the structural grid while avoiding the risk of misaligning the panel joints to the glazing mullions set at 1,500 mm centres. The nail heads to the facing ply are quite visible on the cast surface, so we agreed in advance a regime for setting these out consistently from sheet to sheet. It was Whelan and Grant who suggested a neat method of filling in the tie bolt holes. They proposed to cast a plug, using the same concrete batch as was used for the adjacent wall, to fill the tie bolt holes once the formwork was struck. This proved a very successful detail with the visible, cast surface of the plug exactly matching, in colour and texture, the adjacent walls.

To confirm our discussions, we transferred the advice Whelan and Grant provided onto our working drawings. We were able to confidently and accurately elevate every surface, setting out panel joints, tie bolt holes, kickers and construction joints which Whelan and Grant followed to the letter. It was at this stage, on hearing of problems encountered on a previous project, that we decided to omit the cast-in ceiling pockets into which we were intending to recess smoke detectors.

The frame construction was immaculate in terms of line and level. In programme terms, the construction was slightly slower than expected (approximately 3 weeks per floor) although it must be remembered that there are no following finishing trades, since the concrete is the finished face. We were very fortunate that Whelan and Grant employed a sculptor on the project. He was keen to learn how to work in concrete and worked as the finisher, repairing excessive blemishes. Again, we developed a good dialogue with him to establish the absolute minimum he needed to do in order to achieve our required finish. For the most part he gently rubbed down the concrete with hessian sacking to remove staining and became very skilled at toning patch repairs where we had honeycombing caused by excessive grout leakage.

Lift wall and atrium

PROJECT TEAM

Architect: Allies and Morrison
Structural engineer: whitbybird
Services engineer: WSP
QS: Barrie Tankel and Davis Langdon
Main contractor: Mansell
Concrete contractor: Whelan and Grant
Completed: 2003
Project value: undisclosed

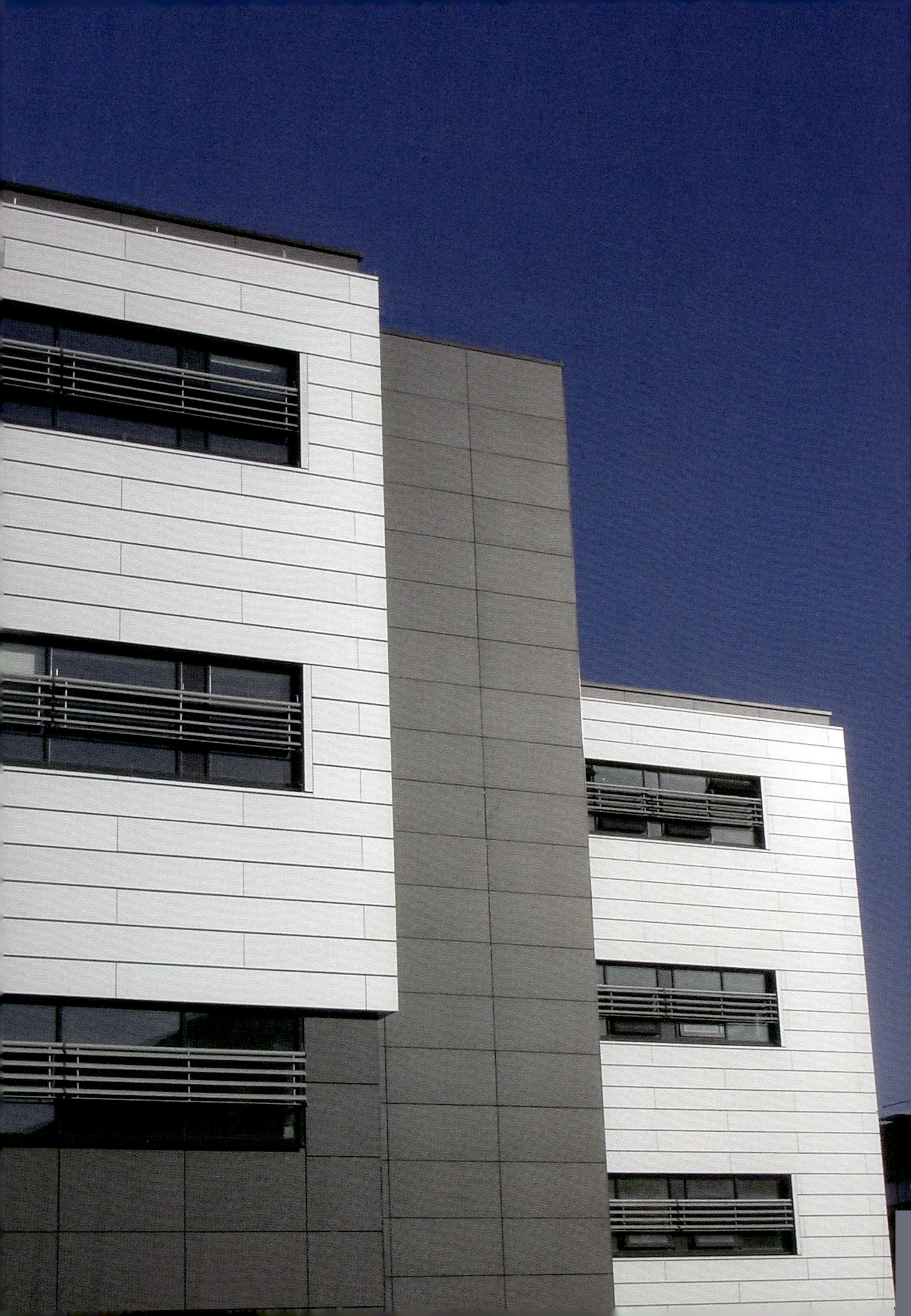

THE BANNERMAN CENTRE
BRUNEL UNIVERSITY, UXBRIDGE

Architect: Rivington Street Studio Architects

Location

The new extension forms the east wing of The Bannerman Centre, in the centre of the Uxbridge campus just off Cleveland Road, the main road through the campus. From the town centre it's a 5-minute bus ride on the U1 or U2 line, stopping at Cleveland Road. It's a good 20-minute walk from Uxbridge underground station which is served by the Metropolitan Line and the Piccadilly Line at peak times.

Introduction

Rivington Street Studio was appointed in January 2002 to design the library extension at the Brunel University campus. The new building is not only the focal point of the campus but also the flagship of an ambitious masterplan to rationalise accommodation across four sites.

The recently completed extension doubles the space available to the library, providing for a larger book and journal collection, additional study areas, more PCs, an Assistive Technology Centre for disabled users and a café with outdoor seating. Student services, such as the Job Shop, Placement and Careers Centre, and finance offices, are now assembled under one roof to supply a more integrated support system and improve the university experience.

The project brief called for an image and accessibility improvement, an increase in the number of study spaces and a substantial increase in the book stock capacity. Great emphasis was placed on the building being adaptable in the future for more computer-based research and the development of its e-library. The plan is that as the library becomes less book-focused, more of the space will be given over to IT and computer rooms.

The four-storey building was conceived as two large offset blocks separated by a core and void area. The existing library building is symmetrical about both axes, and the offset blocks of the new building help to articulate the elevations as well as signalling the new combined entrance. The structure was designed to accommodate library book stack loading throughout the building and provide column-free spaces. A fair faced insitu concrete frame was chosen for its economy and thermal performance.

The building is naturally ventilated for the office and study areas. Louvers at the top of the atrium draw air through the building by a vertical stack effect. Automatic opening windows are controlled by the building management system and open during the summer night-time in order to remove the stored heat in the concrete slabs and absorb cooler air. A shallow raised floor system is utilised throughout the building.

A shiny aluminium cladding system and a dark-grey, matt fibre cement board were specified for the external wall. These contrast with and complement the surrounding concrete faced buildings which are the predominant material used on the Uxbridge campus. Deep reveals and aluminium brise soleil shade the east and south windows from the summer sun.

Internally, maple glazed screens are used to provide views through the building while providing enclosure to the silent study and group study rooms.

Architecture discussion
David Tucker

The building is 5,600 m² over four floors and is organised into two rectangular blocks that are 10.8 m wide. The grid of the new building is a 6.84 m module at right angles to the existing building and a 10.8 m module in the other direction. The floor-to-floor height is the same as the existing building which it ties into and it comprises the two rectangular boxes with an atrium and service core in the middle. The atrium space is used to ventilate parts of the building by stack effect. We wanted the building to be as naturally ventilated as possible but in the library we had to provide mechanical ventilation due to concerns with opening windows at night as this might encourage the theft of books. The offices and classrooms on the east are all naturally ventilated.

The ground floor of the new extension is the main entrance for both the new building and the existing one and it's where you will find the student services and a café. The Library Building is the centre of the campus and students congregate in this area to use the cash office, welfare facilities and student union offices.

We wanted to design a concrete building. The campus is one huge statement of expressed concrete (although not all of it very pleasing). We like insitu concrete and wanted to expose it but first we had to convince the client that it would look graceful and not turn into a drab, overbearing structure in the way that many older buildings have. Our case was also helped by our environmental argument about thermal mass and natural ventilation. It was difficult to quantify any hard savings in long-term running costs but, for no extra construction cost, this was the bonus.

To start with we were asked to design a much smaller building, then the student services and café were added on and the building enlarged, but the budget stayed very tight. We were working to

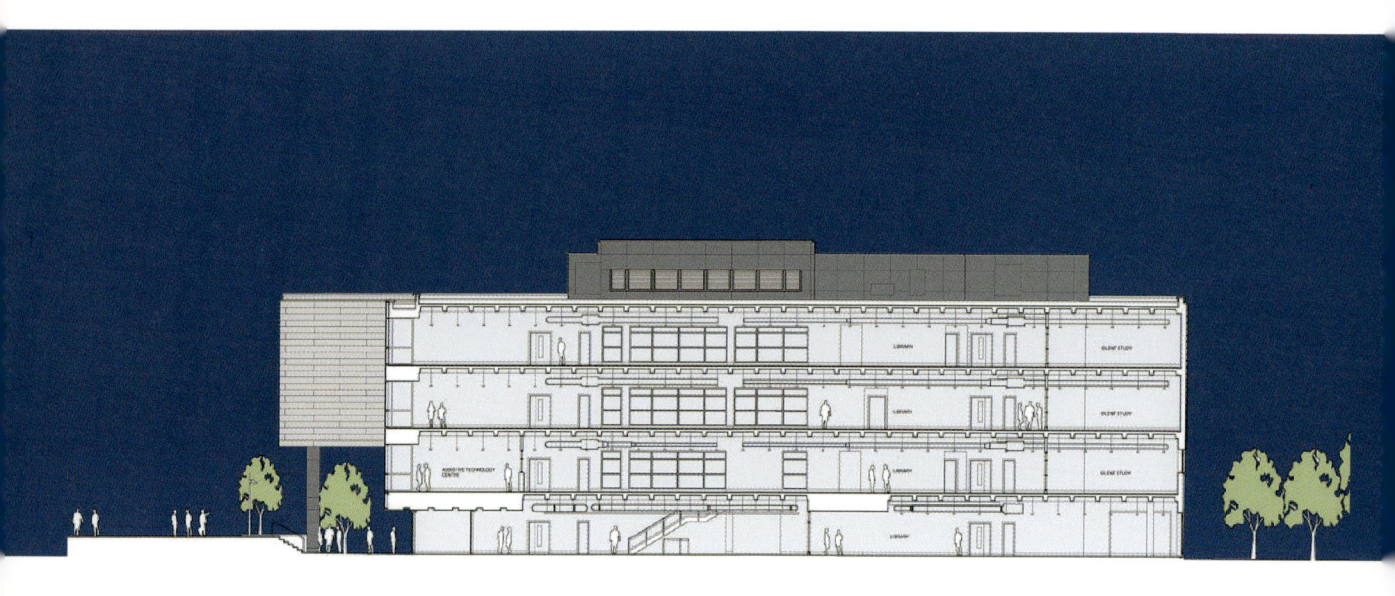

Longitudinal section

top: Atrium
bottom: South elevation

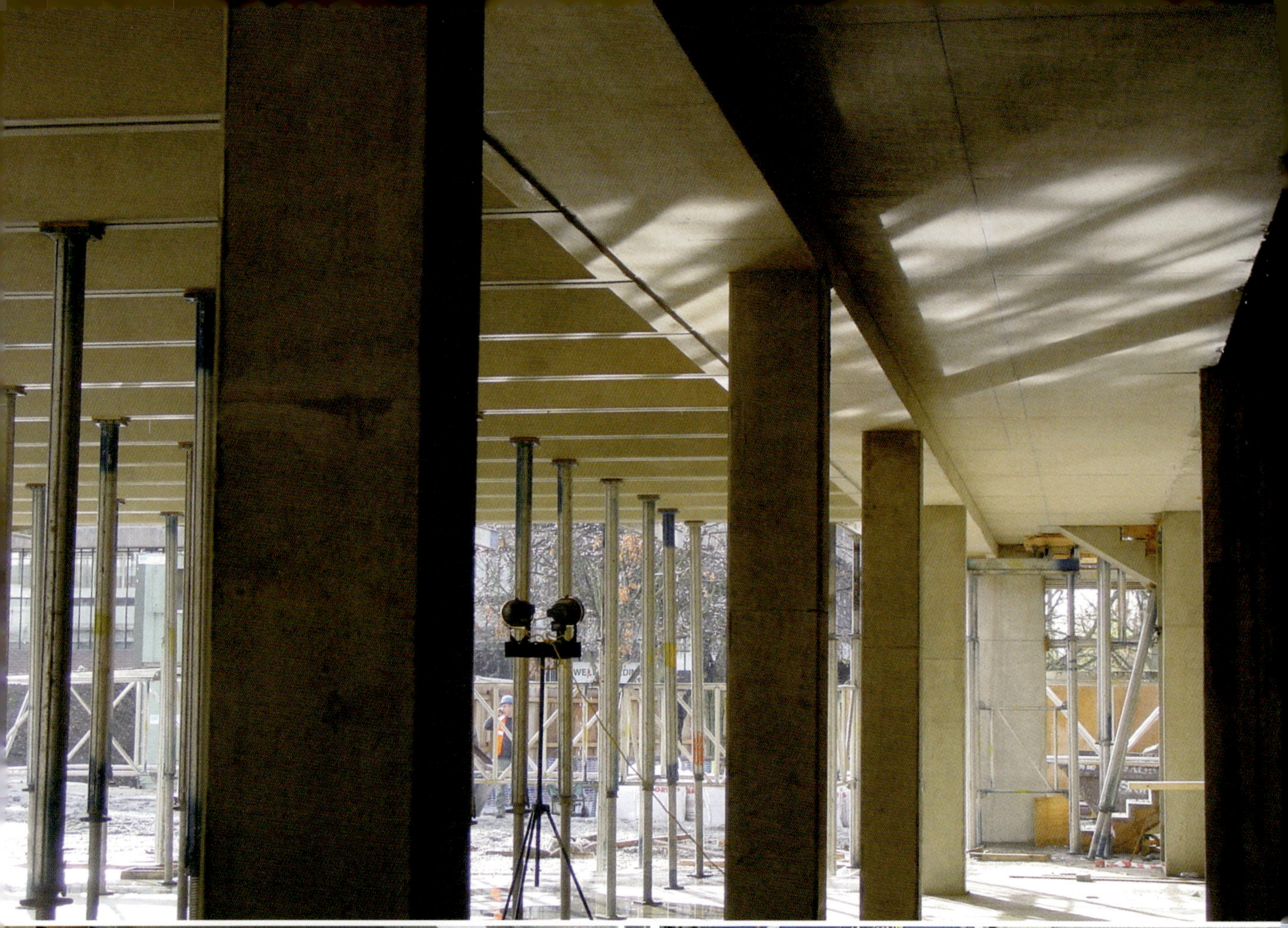

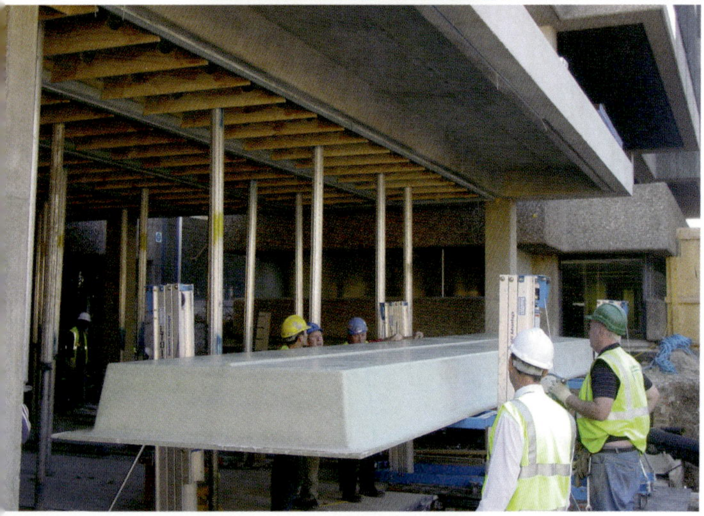

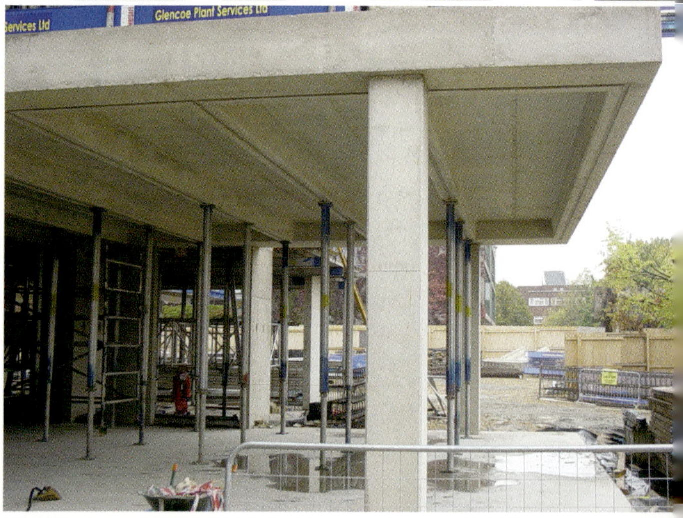
Services Ltd
Glencoe Plant Services Ltd

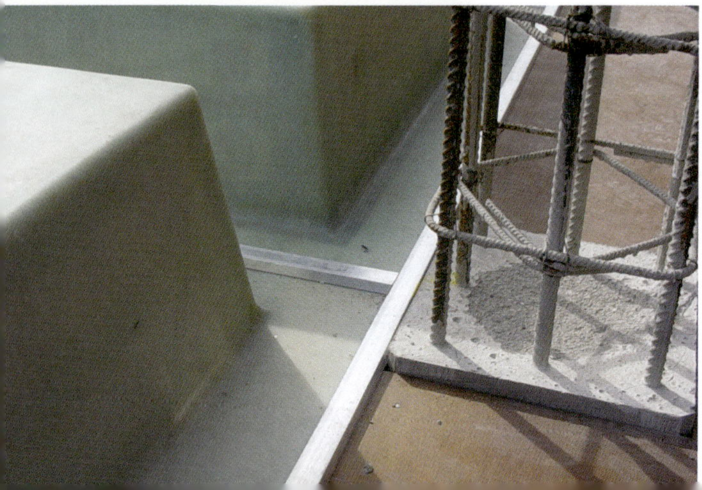

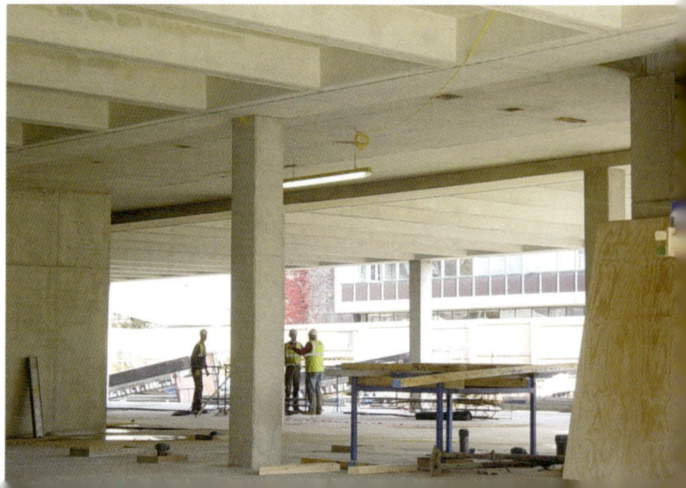

£1,200/m² for construction. The structural engineers did look at the steel option but with the heavy floor loading from the book stacks and the large spans that we required, it was more efficient to design the frame in concrete. There were lots of underground services running across the site, and a long span solution was vital to minimise the piling penetration. We actually designed a very lightweight floor using a ribbed slab 450 mm deep with a topping of just 125 mm. The ribs were set at 1.7 m intervals and span 9.6 m. The top of the concrete was power floated and power trowelled to give a smooth, dense surface finish for the raised floor. The tolerances required for the shallow raised floor system we specified dictated a very tight level tolerance to the floor. The columns were 400 mm square and set at 10.8 m centres with lateral stability carried by the lift shaft box.

The exterior of the new building had to be anything but insitu concrete as the university board did not like the dirt-encrusted concrete façades on the older campus buildings. We went with an aluminium tray cladding system and aluminium windows. The cladding panels are composite units with insulation which are clipped onto a supporting frame. The silver panels reflect light and give a sparkle that was missing with the old concrete façades. There is also a curtain wall system with double glazing, which is fixed by drilling the mullions to the face of the concrete.

The structure is concrete and, as architects, we wanted to leave it as an exposed finish but we were advised by the cost consultants that this would be expensive and would exceed the budget. Pressure was put on the design team to reduce the fine finish to the concrete and remove the timber cladding on the interior walls but we fought hard to retain the curved concrete soffits that were exposed on the underside of the floors. If we had not been persistent and ambitious, this building might have ended up as an anonymous box with false ceilings and walls covered in render.

We looked at a number of buildings for our concrete reference.

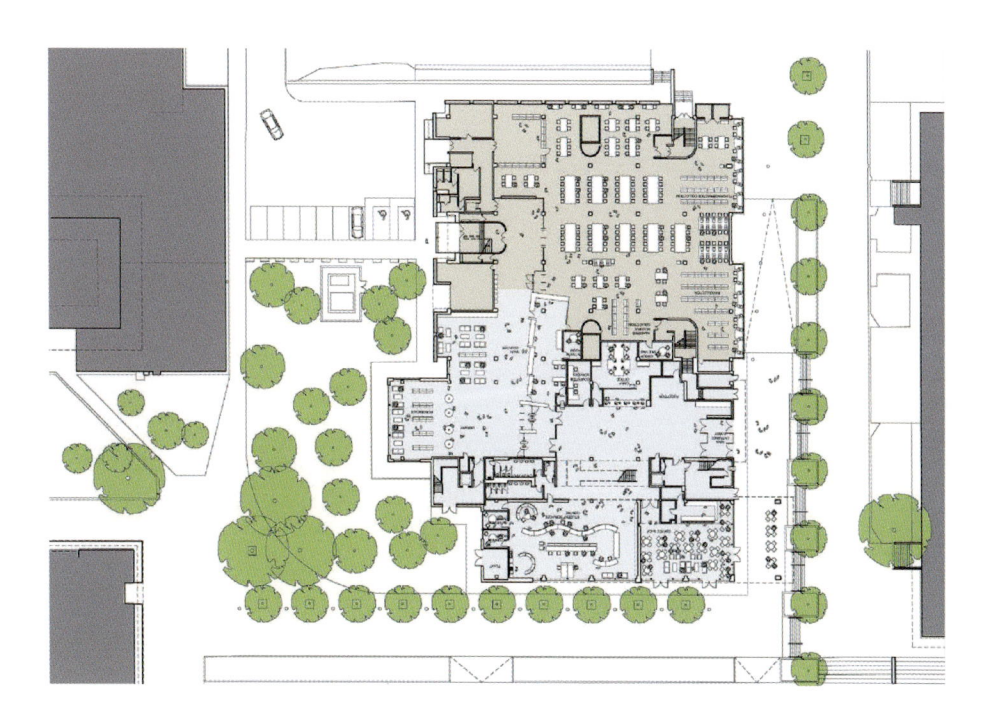

Site plan and building layout of ground floor

opposite:
top: First floor soffit with props
middle left: GRP moulds
middle right: First floor slab – café area
bottom left: Close up of GRP mould
bottom right: Finished ribbed soffit

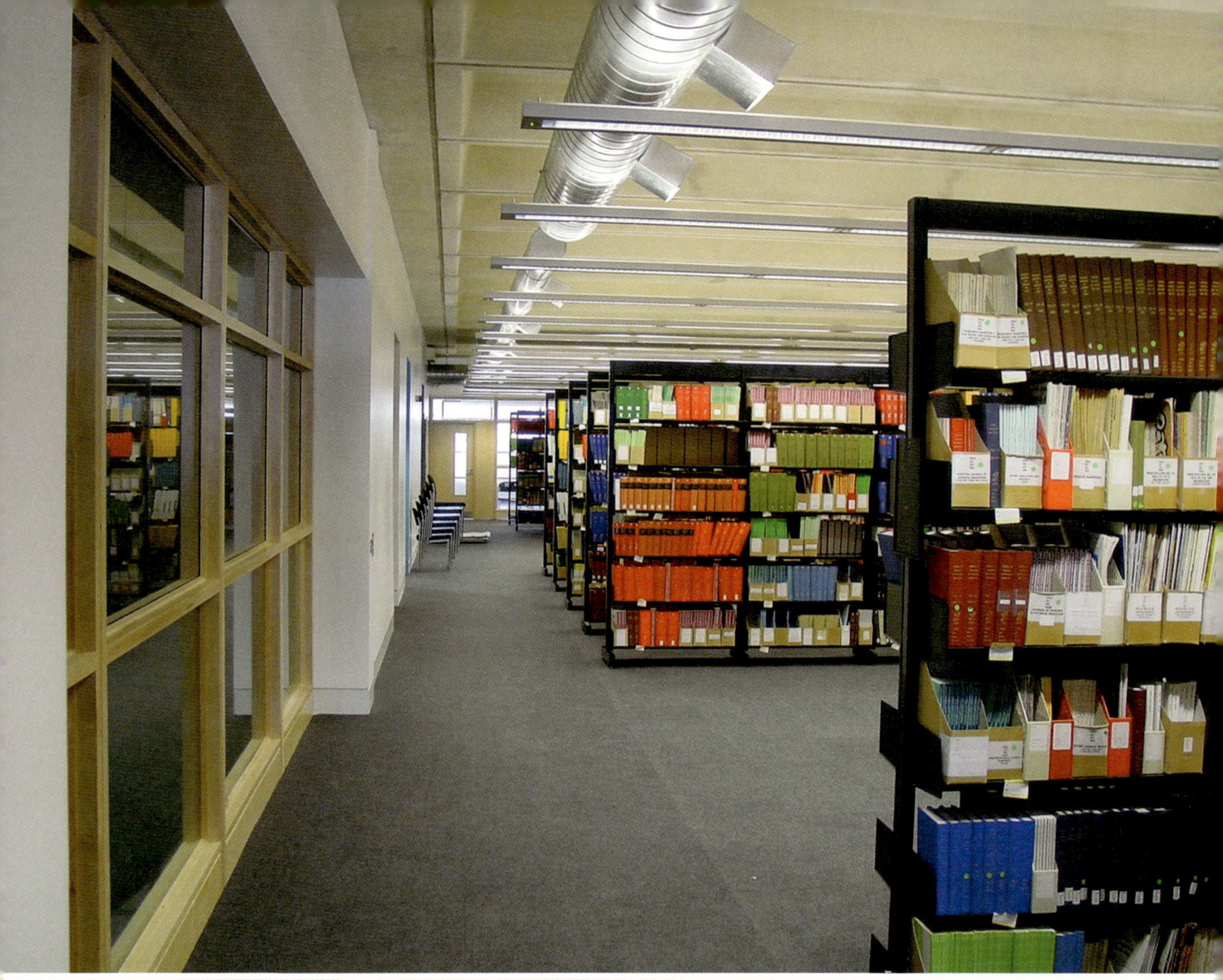

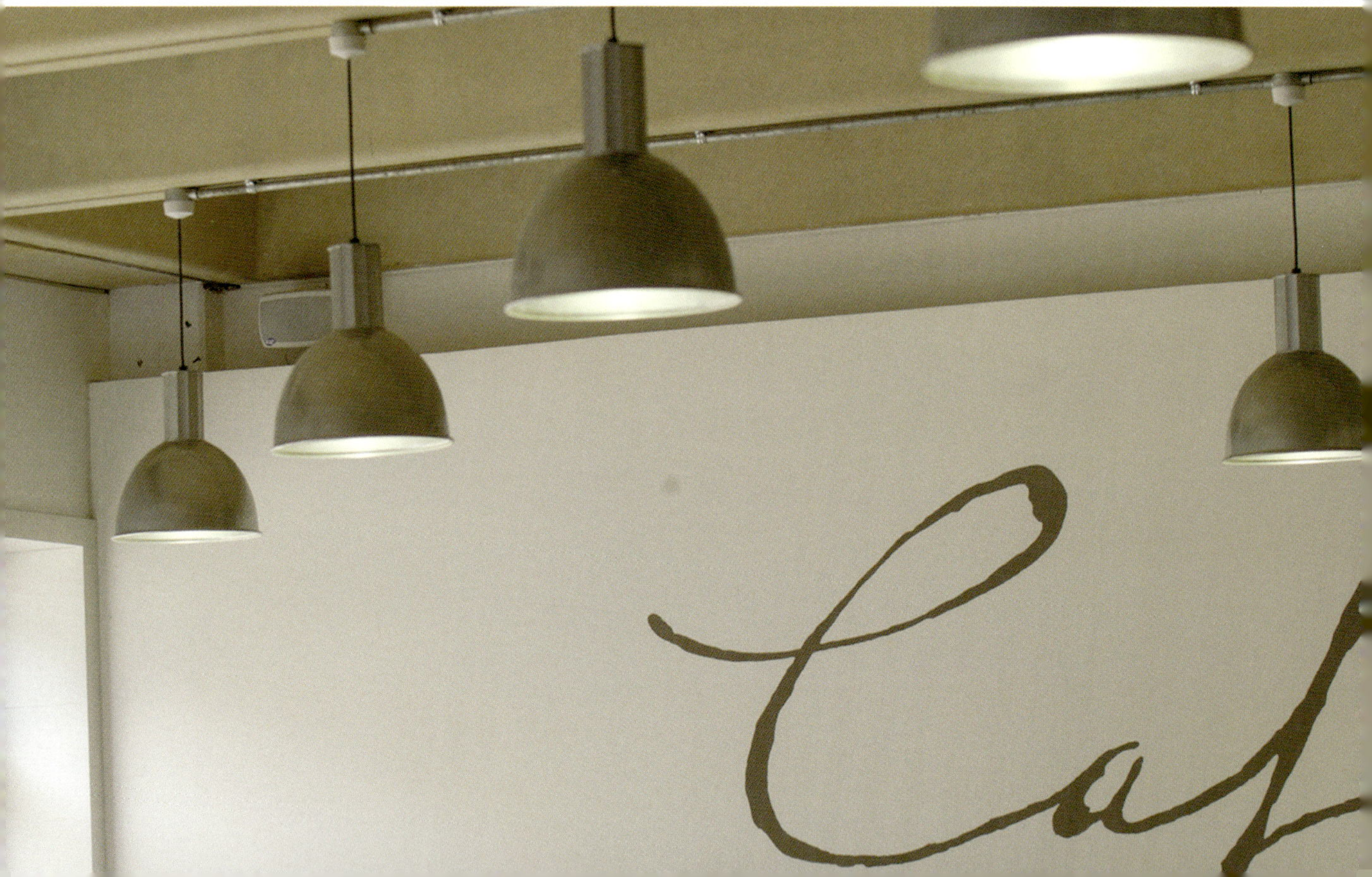
Ca

The Wessex Water HQ by Bennetts Associates had a fine precast soffit which we could not afford but its thermal mass and energy efficiency was something we wanted to use. Davidson House by Lifschutz Davidson in Covent Garden, London, was another source of inspiration; a high-spec office building with fine insitu soffits. We won the client over with the thermal mass argument and the reduced running costs and they finally agreed to the exposed concrete soffits but not to the beams or columns, those were to be painted.

At the value engineering stage everything was reviewed, from the floor light boxes, the electrical trunking, the door types and even the carpet specification. All decisions were cost driven, although much of the data used for comparison purposes was anecdotal and derived from other projects.

In reality, the insitu concrete frame turned out to be a fine exposed concrete structure. The beams, the columns and vaulted soffits were well formed and had a good surface finish. The only minor problem was the grout loss at the kicker joints of the columns, which were just above finished floor level. But then the columns were to be painted, so it was not an issue. In December 2003 the concrete frame was topped out and it looked great.

We've got a standard university fit-out in terms of plasterboard partitions, glazed screens where we can afford it, exposed ductwork providing all the mechanical ventilation, vinyl on the staircases and a simple overhead lighting system. This is a T5 fluorescent lighting system which gives two-thirds downlighting and one-third uplighting. It is located between the bookshelves directly above the aisles. It gives a very even spread of light and we were also keen to uplight the curved ceiling. We never envisaged the lighting being set within the depth of the curved ceiling so that it was hidden from view. We did discuss the option of using the ribs as the location from which to hang the lighting, but we were concerned that the rib would cast a shadow over the underside of the soffits.

CONSTRUCTION NOTES

The exposed concrete specification was based on a smooth finish that was described in the Concrete Society Technical Report 52. The guidance notes were used in the structural concrete specification. The ready mixed supplier batched the concrete using the same cement colour and maintaining the same water/cement ratio. We had sample panels made by the concrete subcontractor Getjar for the vaulted soffit using glass reinforced plastic (GRP). The concrete colour and surface finish were checked before we went ahead. The decision to use GRP was proposed by the engineers as we wanted a smooth surface and good corner and joint line definition. The GRP soffit liners were supplied by Patterns and Moulds. We ordered a whole floor of 56 GRP moulds. Each mould was numbered and taken up floor by floor and cast in exactly the same position.

A recess was cast on the underside of the ribs to hide the joint between the two GRP curved moulds. The recesses were made of aluminium, which proved difficult to remove. They had to be levered out with some force to release the GRP moulds, which came away very easily. The other problem we had was nuisance from purple bird droppings produced by pigeons when the elderberries were ripe. The GRP had to be jet washed before any concrete was poured as the stain would appear on the finish.

We tested plastic star-shaped cover spacers for the floor soffit, but these showed up on the surface. A precast concrete spacer standing on its narrow end was the better solution as it was invisible on the surface. The only occasions on which it was seen was when the spacer had been accidentally pushed onto its flatter side.

The contractors cast 550 m^2 of floor slab in one pour – that's one-half of the floor area. The concrete mix for the floor included 50 per cent GGBS to lighten the PC greyness. The columns were a standard grey mix with no GGBS. It did not matter that the columns were a different colour as they were going to be painted. The floors were all power trowelled to get a smooth, level surface to tight level tolerances for an inexpensive pedestal raised floor system 60 mm high which had very little level tolerance.

The decision to paint all the exposed concrete surfaces below the soffit line should perhaps have been reviewed as the exposed concrete looked great and leaving it unpainted would have saved money and time. It was a shame not to have shown off such good workmanship.

top: Library
bottom: Café soffit and ceiling lights

PROJECT TEAM

Client: Brunel University
Architect: Rivington Street Studio
Project manager: Mace
Structural engineer: whitbybird
Services engineer: Brian Warwicker Partnership
QS: Hand Deere & Cox
Main contractor: Bluestone
Concrete subcontractor: Getjar Ltd
Completed: December 2004
Project value: £6.3 million

Cash Office

POSITION CLOSED   POSITION CLOSED   POSITION CLOSED

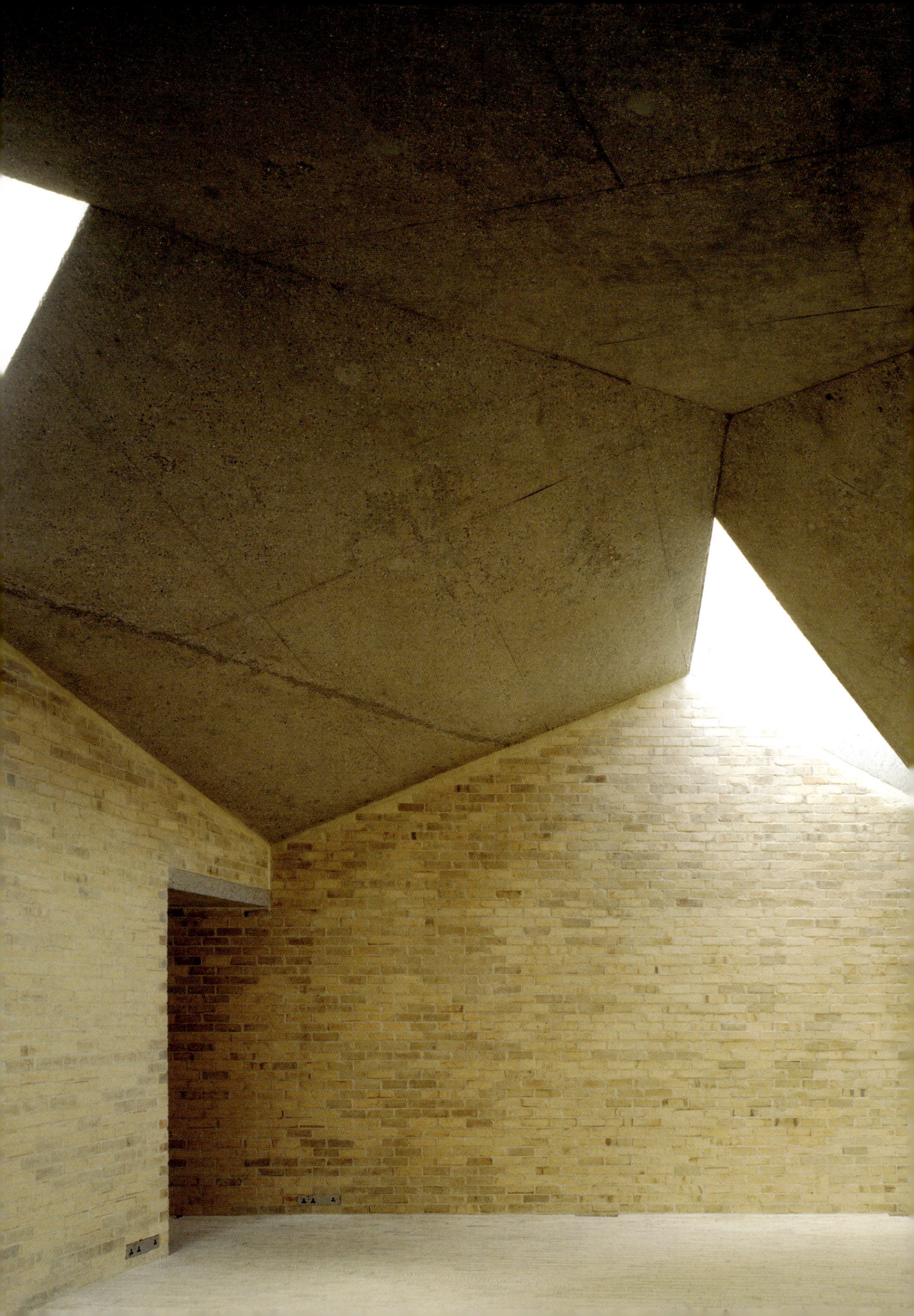

THE BRICK HOUSE
LONDON

Architect: Caruso St John

Location

It is the clients' wish that the location of the house remains confidential.

Introduction

This family house stands among dense residential streets in a busy part of West London. The land is shaped like a horse's head, surrounded by three taller buildings and can only be reached by a carriageway through the façade of an adjacent Victorian terrace.

The wildly spatial shape of the site has been used to form the living spaces. The interior plan is completely separate from the typologies of the London townhouse or the inner-city loft, while still retaining a strong sense of dwelling at the heart of the city. Walking around the house takes you across broad spaces, to corners with windows overlooking small gardens, to intimate rooms deep inside. The exterior form of the house that is generated by this varied arrangement is incomprehensible from within it. Instead, the form appears unbound and soft, as if an internal force is pressing the walls and roof out against the buildings around it.

The floors and walls of the house are built of brick, inside and out. The use of one material binds the whole building into an enveloping body, emphasising a skin-like character over any tectonic expression. The arrangement of the bricks within the mortar shifts as surfaces stretch, bend and twist, making them appear elastic. The ceiling of the upper floor is cast concrete and adopts different levels to make particular spaces within the overall deep plan. A flat ceiling appears to press down over the dining table, and a domed profile forms the high ceiling over the main living space.

ARCHITECTURAL INSITU CONCRETE - CASE STUDIES

Architecture discussion
Adam Caruso

We met the clients quite a few years ago when a recommendation led them to approach us to design their new house. Having lived for years in a five-storey Victorian terraced house, they wanted a change. They wanted to live on a single level so they could spend more time together as a family, rather than avoiding each other in a vertically stratified house, living behind closed doors. They also wanted to be in a 24 hr part of town where they could always get a coffee or snack, whatever the time of day.

An estate agent found a site where we could build 500 m^2 of space on two floors – it was a garage with a one-storey shed that covered a backyard approached by navigating between existing buildings from the main road. The site had planning consent for a courtyard house but very quickly we rejected that idea because we saw the conflict between wanting a large courtyard and providing a large living space on one level.

We had the idea to flood the site with accommodation and to enter the site through a narrow passageway which was 3 m wide and 3 m high. Instead of going straight into the building we thought of ramping up the entrance and building the main floor as a piano nobile, from which you would then step down into the bedrooms. The idea for the bedroom floor was to have three courtyards at the points of the site. Above it was one big area for the living space, with one window next to the dining area looking north to the back of the neighbouring terrace and a south-facing window looking out from the study, and we also introduced three roof lights. If you are going to create a big space that has no aspect you need to get natural light into it – and when you are in the space you have to feel like you are somewhere.

We were thinking of the roof construction and the walls and the way the materials can give a real sense of presence. We had the idea of using brick very early on, inspired by Lewerentz and his great churches, and the tradition of London brick. This house

First floor dining area

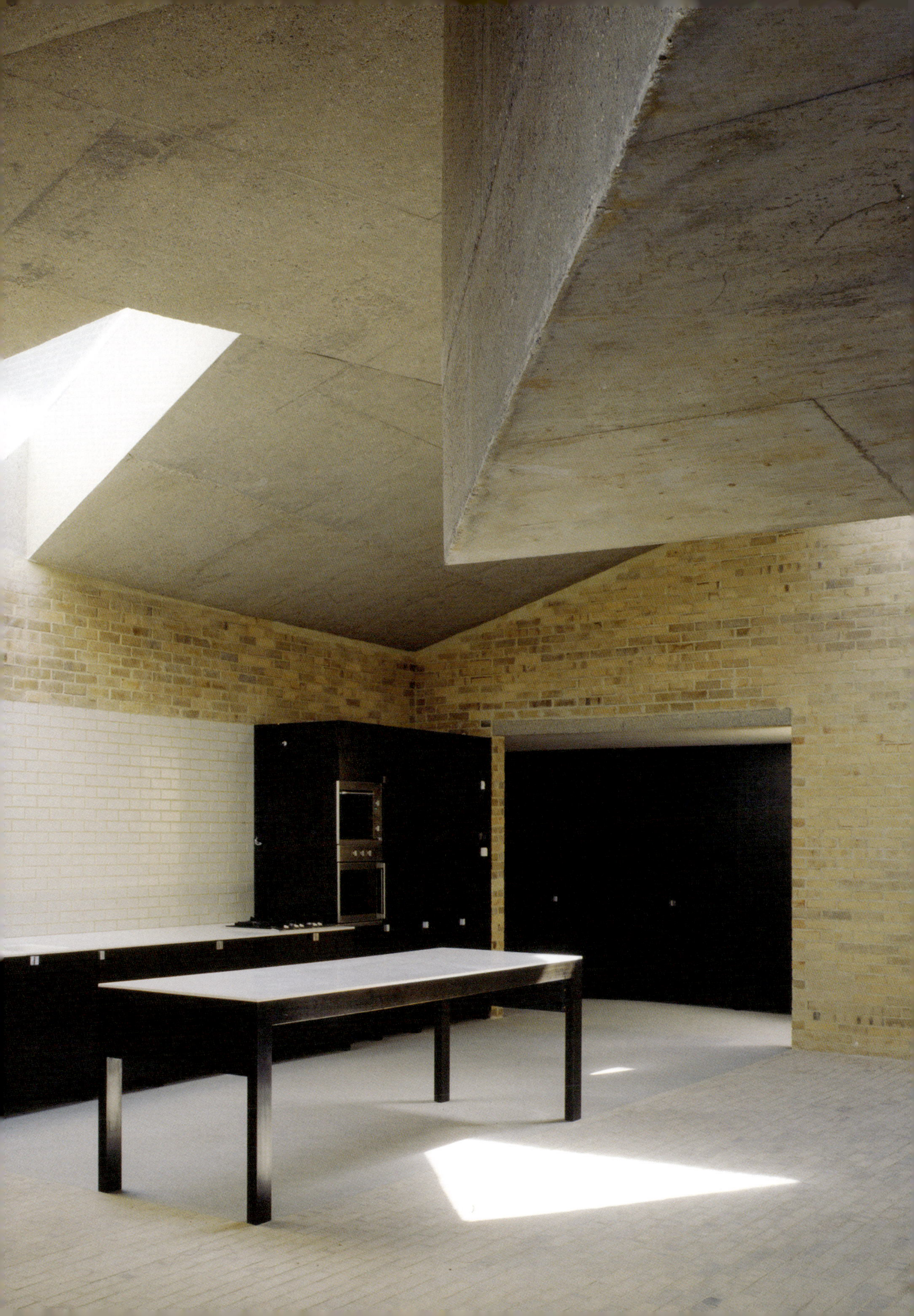

ARCHITECTURAL INSITU CONCRETE - CASE STUDIES

top: First floor plan
bottom: North elevation

is surrounded externally by every variety of London brick. We wanted a brick interior that was rough, handmade and primordial, something like the beauty of ruins from Roman architecture which is so powerful. We were also influenced by 1960s architecture and brutalism and wanted to try to revalidate concrete by introducing a formal exuberance rather than a structural didactic. The bricks form load-bearing diaphragm walls, but you can't tell that from the interior, they are just a running stretcher bond, very understated and simple. The floors and the underside of the roof construction were always concrete.

We used brick slips over the concrete surface in the living room floor, to give the feeling of living in a brick volume. We then explored the different treatments you can apply to concrete and the post-texturing of the surface – smooth trowelled finish, grit blasted finish and terrazzo. The reveals of the roof lights have been accentuated by introducing concrete upstands. The concrete pitched roof is only 450 mm thick, but the upstands make it feel like it is one gigantic metre thick.

The concrete box insulates you from the bustling noise of the city. The clients say it is like living in Rome, walking down an alleyway, turning left, going through a door and finding yourself in an abandoned chapel. You feel you have arrived somewhere that is deep in the fabric of the city.

The roof has angular slopes and geometric shapes with raised levels and lower levels like vaults and this is reflected in the floor whose spaces have different settings. We had the idea that even though the living area was one big space, by manipulating the height of the roof soffits and the slopes – it is lower over the dining area and higher over the seating area – we could create different moods. It is very like arts and craft houses which have volumetric elaboration that is very explicit – the bay window, the inglenook and the gracious centre of the living room. We were conscious of these influences but were mindful that we were creating a house and not a chapel, so too much formalisation would not be good.

To develop and define the roof shapes and long room spaces, we made models at 1:20 scale which we photographed, introducing lighting and even miniature furniture to give us a real sense of the interior. It's a play between volumes and the material surfaces and how they interact with the space. If we had plastered and painted the walls it would have looked and felt totally different, and looked

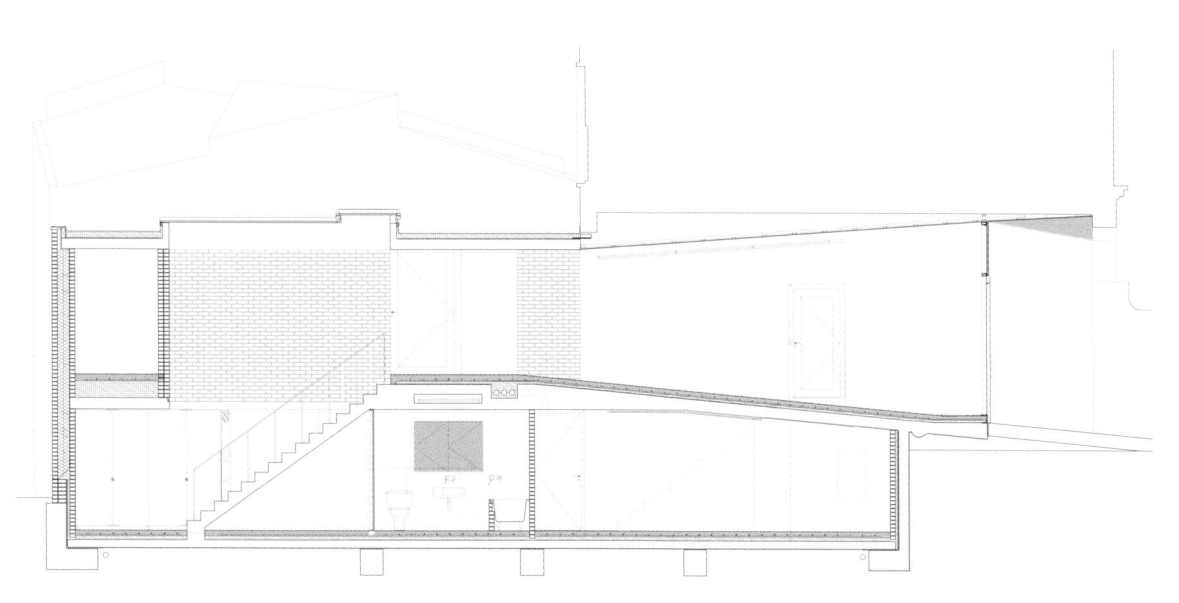

Section

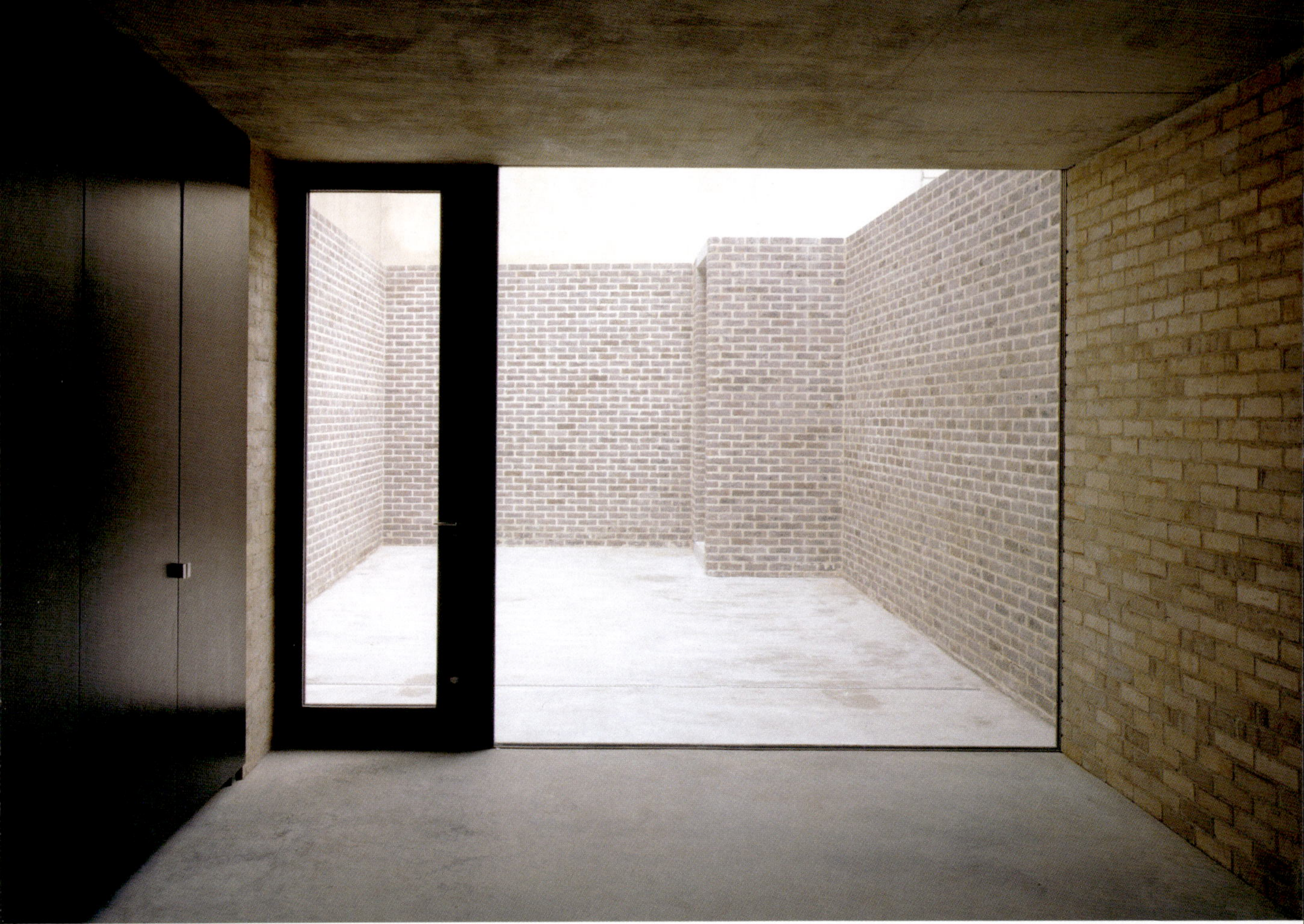

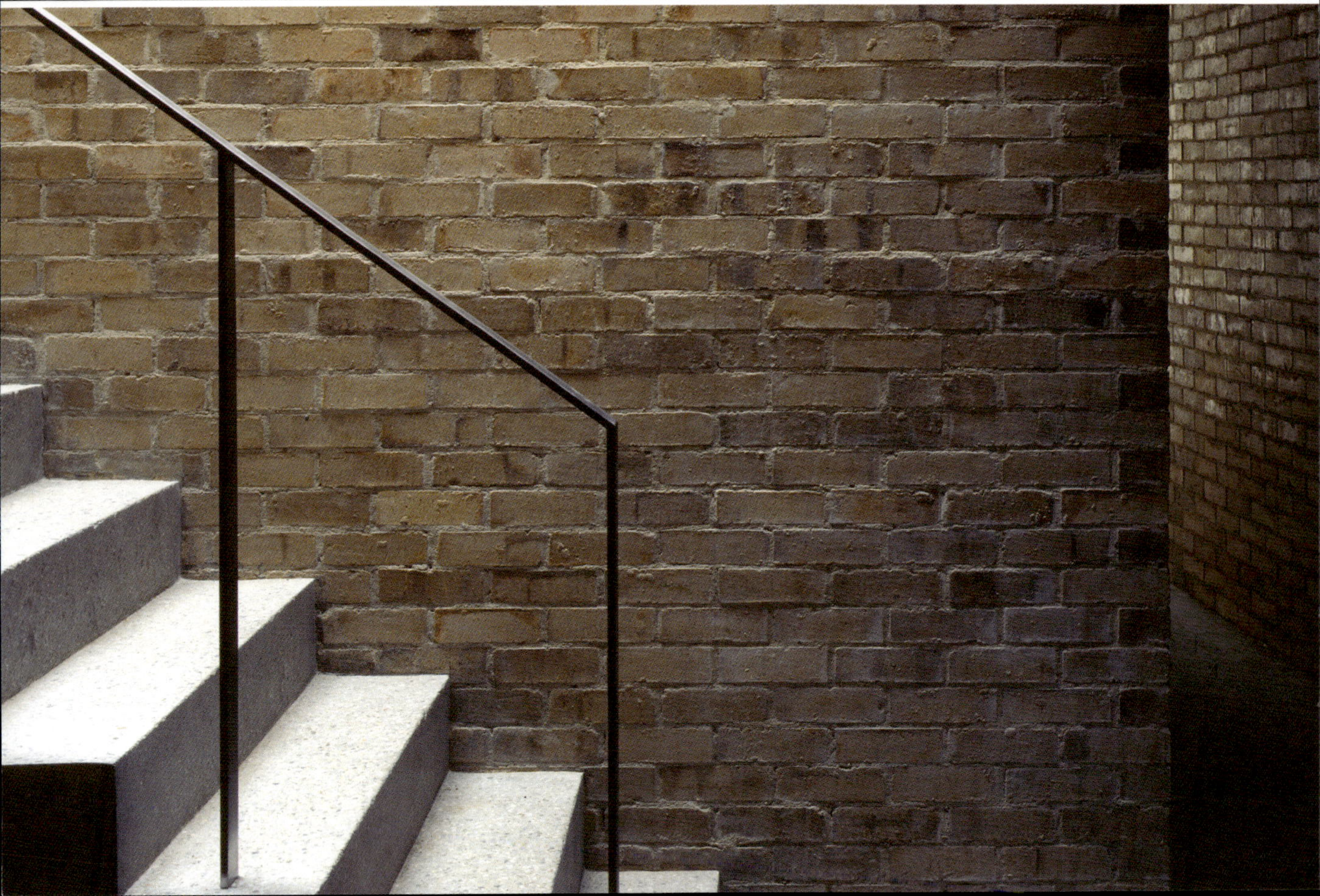

dead compared to brick and concrete.

One major concern for us was the inability of UK contractors and small builders to form and cast these shapes in concrete. The brickwork selection was quite tortuous as well. Originally, we were going to use one brick type for the whole building. It was a bit overburnt and crumbly on the face, the client was concerned that the brick might fall apart and perhaps it was too dark a colour. So the bricks used for the internal face are lighter than those used outside and have a harder surface. Lime mortar was used throughout, which was almost the same tone and colour as the internal bricks. Hydraulic lime, which was easier to handle than lime putty, was specified and we had a lot of help from the Hydraulic Lime Association.

We did not like the standard brick cavity wall, because the cavity is too small and often fills with debris. The diaphragm walls have a large cavity of 350 mm and at set intervals they are braced by brick returns to create a series of boxes which are the diaphragms. The air gap of 350 mm is not quite enough insulation by itself so it was filled with cellulose-type insulation. We did not bond the diaphragm bricks to the inner skin as we wanted to prevent moisture ingress. We did use a damp-proof membrane (DPM) liner and connected the diaphragm bricks with steel ties. It is a very stable construction, even though most of the brickwork on the exterior skin had to be laid overhand as the site was so inaccessible. There is just a small gap between the exterior bricks and the party walls so it will not be seen. As long as it was grout-tight it would weather well.

We had the idea of using insitu concrete walls but this proved difficult to build and form against the party wall. The brick construction worked well. On the inside skin there are no cut bricks to create the sloping wall lines to receive the roof slab. They are whole bricks that are stopped as castellated steps and the make up to the underside of the roof floor was of concrete which was recessed 20 mm from the brick line. The concrete was later rendered with lime mortar to lie flush with the bricks to overcome the detail. The floor slab and roof slab sit on the inner leaf of the diaphragm wall on dry pack with a DPM. The concrete to the lower floors was all flat slabs.

There were 18 party wall conditions; the first drawings we did were to show all these interfaces so that they became part of the party wall agreement. We prepared 22 drawings of wall sections that formed the party wall agreements. The structure was supported on footings, there was a lot of belt and braces waterproofing involved because we would not be able to resolve problems if any repairs were required at a later stage. We had a drainage layer and a geotextile fabric with sumps on the external side to eliminate any risk of leakage. The sumps allow you to pump away any groundwater behind the wall if a problem should arise. There was insulation and underfloor heating to the base slab with the usual DPM layer; the insulation was 18 mm thick. On top of that was 100 mm of reinforced concrete, which was power trowelled to the ground floor accommodation. There were saw-cut joints made on the 100 mm floating slab and movement joints formed on the perimeter. In the bathroom areas, the floor was ground to create a terrazzo finish.

Upstairs, the structural slab is a two-way spanning floor 185 mm thick, the soffit of which was expressed. Insulation was placed on top of it, then 90 mm of concrete topping, encapsulating underfloor heating pipes. The surface was covered with brick slips and lime mortar. The bricks were later sealed with a waterproof coating. For the first floor soffit, which was expressed in the ground floor, we used a high-quality ply shutter laid out to a joint and sheet plan.

We specified a standard concrete mix for all the exposed walls and floor soffits and roof, which was supplied by Cemex from their Kings Cross depot. As we wanted to have a grit blasted surface and a terrazzo surface, they incorporated a 10 mm limestone aggregate to produce a better exposed aggregate finish. The sloping roof has the same mix, every slope and angle and sheet size was drawn out but, regrettably, the contractor did not follow the plan. There were day-work joints which we had carefully detailed and positioned so that they would not be unsightly. Fortunately, as we were carrying out sand blasting we could get away with some repairs and patching of the surface. There were more imperfections than we would have liked, but overall it has an amazing quality and presence.

The grit blast finish for the house started off with a light impact but in the end we preferred a much deeper abrasion to fully express the aggregates. However, if you do too heavy a grit blast you get surface pitting, large blowholes and aggregates falling out. We did some preliminary testing on the plant room ceilings. The flat bedroom soffits were to remain as struck. On the stairs, the bathrooms and the ramped walkway to the front door we used diamond grinding.

top: Courtyard at ground floor
bottom: Staircase to ground floor

CONSTRUCTION NOTES
Damien Collins, Harris Calnan Ltd

We demolished the existing buildings on the site and excavated the courtyard space down to a depth of 2 m to create the new ground floor. The existing boundary wall was underpinned through the entire perimeter of the site before excavations commenced. The initial dig and underpinning took 3 months to complete. Our approach was to do an overall reduced level dig to 1 m, then carry out the underpinning in sections along the perimeter before completing the excavation to 2 m. Restricted access through the narrow archway between the adjoining buildings slowed our progress. We blinded the surface with concrete and then placed a damp-proof lining. We didn't have any groundwater problems.

The structure is of brick diaphragm walls supporting concrete floors and a concrete slopping roof. Two skins of brickwork make up the diaphragm wall with cross walls to connect the two every 900 mm. We built the exterior skin first with the cross walls protruding as fins, then placed insulation and then built the inner skin. The floor loads only sat on the inner brick wall.

We had to build the wall this way because we were hard up against the party wall with virtually no room for access. The exterior wall was built overhand but in a couple of places where we could approach the work from the outside, it was built normally. The brickwork was supported on a structural slab on the lower ground level which had edge beams. The bricks were laid on lime mortar with no movement joints as the lime mortar can accommodate moisture movements, unlike dense cement mortar. To bed the concrete floors on the brick we used neoprene rubber. There was some settlement in small areas due to the lime mortar joints being squeezed by the concrete floor load.

For the terrazzo in the bathroom, staircases and the ramp areas we used a hand-held 9" angle grinder. We did not even change the angle grinder heads that we started with and we achieved a smooth, scratch-free finish.

Limestone filler was added to the concrete mix to create a warm surface colour. All the soffits to the pitched roof slab are grit blasted heavily to express the aggregates. The roof was difficult to build as it was 450 mm thick; it had five different pitches and two apexes.

The ends of the roof slope had to have thrust blocks to resist any movement of the concrete during casting. There were problems with the day-work joints due to poor construction detailing and our concrete subcontractor under-estimating the complexity of the work. For the pitched roof with its complex geometry we built a series of trusses off a flat, level decking. The trusses then supported timber and metal beam supports which carried the ply soffit. We built the complete roof in wood with the roof trusses, cast the concrete and when the slab had hardened we disposed of all the timber. No one from the design team had ever built a concrete pitched roof before and neither had the contractor, so our approach was to be cautious and conservative with a very steep learning curve.

The paper film ply that we selected caused a lot of brown staining on the concrete, resulting in a poor surface finish due to the discolouration. When we formed the pitch roof, plain Douglas fir was used which warped and became unusable. The steepest roof pitch was 35° and the shallowest was 10°. We needed to create a good finish to the top of the roof so we had to use a top shutter to lay on insulation and thin grey sheeting which was matched to the concrete colour. The biggest problems were the thrust blocks which moved on occasion causing concrete to leak through and create very poor joints that we patched and repaired. We had to pump the concrete through mouse holes left in the top shutter and then close them as we carried on further up the slopes. The underside of the concrete was critical for the finish; it appeared that the larger aggregates seemed to settle towards the lower part of the roof.

We worked through winter with the concrete and hydraulic lime, monitoring the air temperatures the whole time, and did not cast concrete or lay bricks below the minimum recommended temperatures. All the bricks were lined with hessian to keep the frost away. Our major problem was the variable quality of the fair faced bricks for the internal walls – in both dimension and colour consistency. There were a lot of bricks that were rejected. At the start of the project the small brick company that was producing the bricks could not supply the volume we needed. We ordered 16,000 and received only 4,000. It was a further 6 weeks before the next batch came out of the kiln. When it did arrive on site a very high percentage were too burnt and had to be rejected. We also had a yellow staining exuding from the bricks; this was caused by pigment leaching from the clay. In the end we used a mild hydrochloric acid to wash it off and it was a slow, labour-intensive operation.

Staircase and light well

We did not power trowel the concrete where it had to be ground, we just hand trowelled it. If it had been power trowelled it would have pushed the aggregates deeper into the concrete, making them difficult to expose without an excessive amount of grinding. On the first floor the concrete screed was covered with brick slips and lime mortar. The brick slips were cut from bricks used for the internal wall and these were stuck to the floor with tile adhesive. We did have an amount of surface cracking in the basement (lower ground) floor slab which contained the underfloor heating pipes. We used a resin seal to fill them and in the bathroom where micro-cracking occurred we used cement grout, colour matched, which has worked well. The first floor topping slab did not crack.

PROJECT TEAM

Architect: Caruso St John Architects

Structural engineer: Price & Myers

Services engineer: Mendick Waring

QS: Jackson Coles

Main contractor: Harris Calnan

Brickwork contractor: Liberty Brickwork

Completed: April 2005

Contract value: confidential

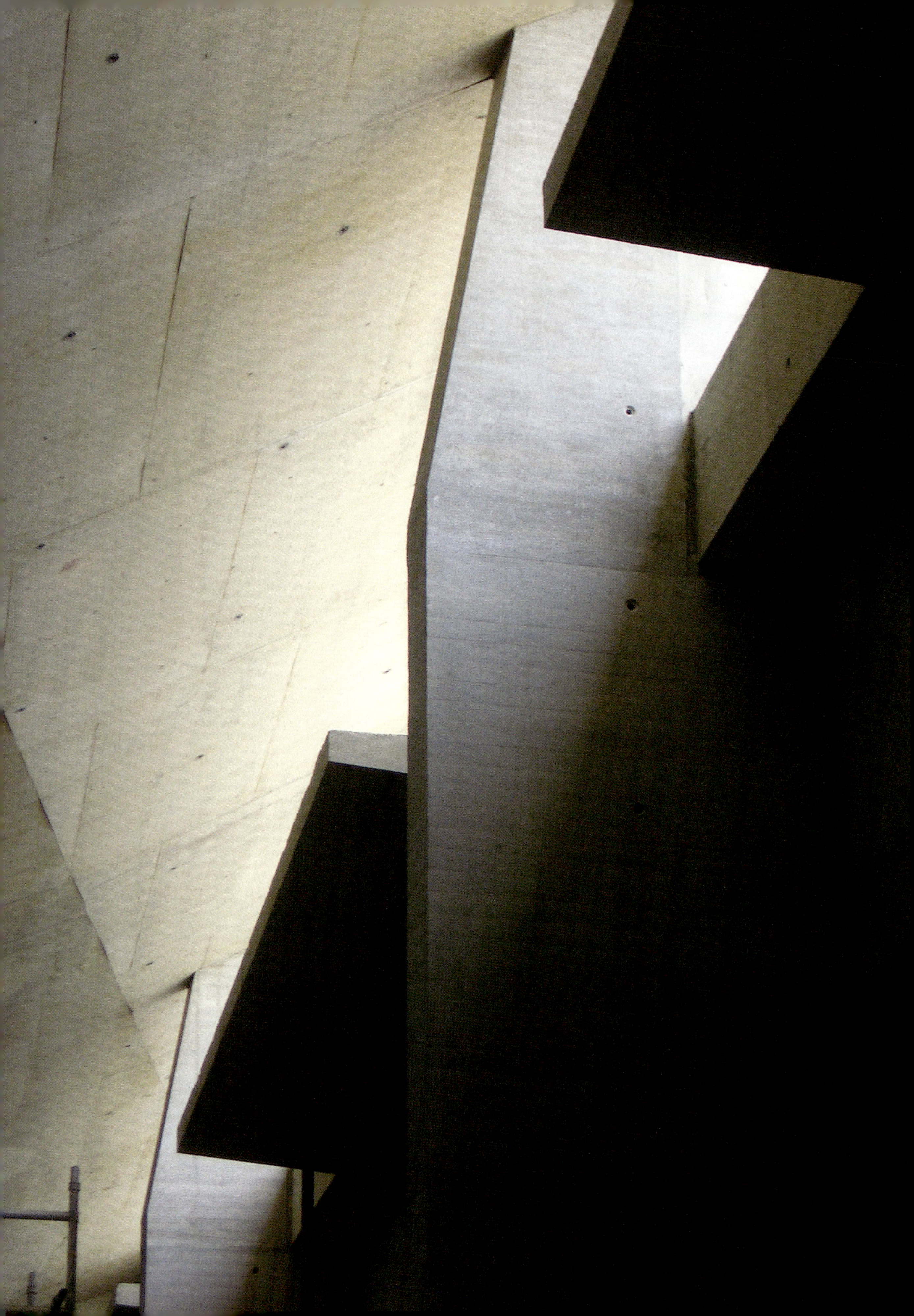

THE COLLECTION
CITY AND COUNTY MUSEUM, LINCOLN

Architect: Panter Hudspith

Location

The museum is located in Danes Terrace, off Danesgate which is a pleasant 20-minute walk uphill along Broadgate heading north towards the cathedral after leaving the levels of the drab railway station. It's a 5-minute taxi ride from the station for the unadventurous.

The museum is only a 30-minute drive by car from the A1 and an hour from Hull, Nottingham, Peterborough and Sheffield. When visiting, also make time to watch the exciting theatre presentations in the auditorium before browsing in the shop and enjoying a visit to the courtyard Café Bar.

Introduction

The City and County Museum project was won in competitive interview in 2000. The project involved master planning the Flaxengate area and the design of a new museum to house a substantial collection of archaeological and historical artefacts. The design of the museum creates connections through the site and improves links with the Usher Gallery and adjoining Temple Gardens.

The form of the museum is a direct response to the character and scale of medieval Lincoln. The museum reads as a group of buildings that work together rather than as iconic objects. Conceived as five large blocks of stone that have an informal relationship, the rooms being either 'carved out' or left as inhabited spaces in between. The largest of these creates a new public square to the east that opens out onto Temple Gardens and will engage directly with the entrance of the Usher Gallery once the Curtois extension has been demolished, allowing both buildings to work together.

The external courtyard was designed to allow activities within the museum to spill outside in the summer months, while also creating a new public space for Lincoln. The forms of the building fracture as they follow the contours up the hill, towards the cathedral, opening up routes and views through the site. The external walls are angled planes of rough-faced limestone; the internal finishes are cast-in-place, smooth, board-marked concrete. The windows, external doors and canopies are built of bronze and the windows are lined internally with Forest Stewardship Council (FSC) European oak, which articulates the internal doors, acoustic walling and handrails.

On the museum's west side, a route wraps itself around the curved form of the Audio–Visual Theatre to form a passageway into the courtyard. This passage through the heart of the museum site establishes a link between the Lindum Hillside and the High Street.

Architecture discussion
Simon Hudspith

When we first traveled to Lincoln we spent some time understanding the locality and the site. For us, the monumental rampart walls of the Bishops' Palace which are set into the hillside and support formal gardens within them, and the unique space of the Cathedral Green which governs how the full elevation of the cathedral is viewed, were extremely influential local features. Although we were designing a new building it was important that it related to its medieval surroundings and to the topography of the site. We were interested in designing a modern building that could re-express some of the history of the city.

Having won the competition there was an impetus to put in a Heritage Lottery Fund (HLF) application to secure the matched funding for the project. In response, the HLF asked us to think very carefully about the site and what was happening around the new building. They suggested that we produce a master plan for the area as they wanted the new building to be the catalyst for regeneration. We then spent 3 months working on the master plan. That gave us time to really get to know the city, the pedestrian movement, transport networks and local amenities.

The one thing that had unsettled us was the client's request that they would like the building to have a 'wow' factor. In truth, this was something we wanted to aspire to, but felt uneasy about as it as it is such a subjective notion. We felt strongly that the design should not try to be iconic or attempt to compete with the spectacular cathedral. We wished to produce something that was much gentler, that sat well into the landscape, did not shout out at you as you approached it, but intrigued you, and was understated. The site, which contained a multi-storey car park, measures 50 × 45 m, is on a 1:8 slope, and sits next to the Temple Gardens, the park which contains the Palladian-style Usher Gallery that houses Lincoln's art collection. We thought the building in its park setting was a very important landmark, so we schemed the new museum buildings to be much lower and not as imposing in scale as the monumental proportions of the gallery. One of the key design ideas was to create a courtyard that would become the centre of the life of the new building. We noticed that there was a diagonal route from De Montfort University's art department at the far side of Temple Gardens that leads directly to the High Street via the site and we decided to make the museum courtyard part of the route. Those who use this route will always engage with the museum, even though they may not wish to step inside.

We showed the client a series of drawings to explain the thrust of our arguments but we did not reveal the full extent of our ideas until a model was made at 1:200 scale. I started making models as a student at the University of Pennsylvania and then at SCI-Arc in Los Angeles. Models were the main design tool for a number of architects in LA, such as Frank Gehry and Morphosis, as three-dimensional computer programmes were still in their infancy.

The good thing about making models is that everyone can engage with them and relate to them more easily than computer images. You can see round a model, touch it and appreciate various angles in an instant. It's a very important part of the design process, and helps the client understand the building. We start with simple block models to establish the massing of the building within its context. Each time we change the scale (from 1:200 to 1:100) more of the design is developed. The final models are at 1:50, where we work with 1:5 and 1:2 detail drawings to establish how the building will be built and make decisions on the subtleties of materials and junctions.

At 1:50 you can also experiment with how daylight enters the building. And we spent considerable time exploring this, especially how light enters into the gallery spaces.

We knew we wanted to have a gallery that was as efficient as possible in terms of its flexibility, usage and energy management.

Limestone façade and
external courtyard

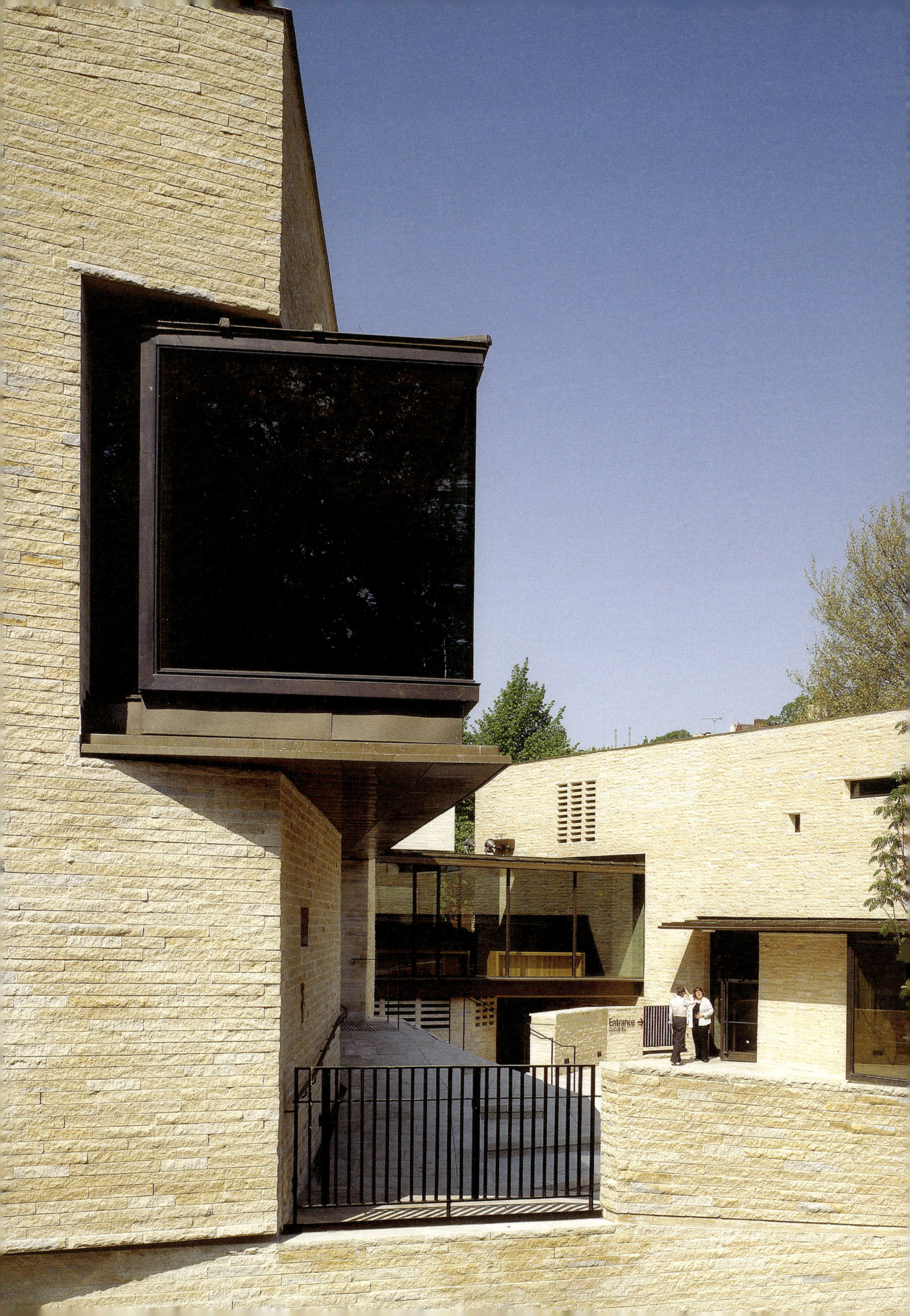
Entrance

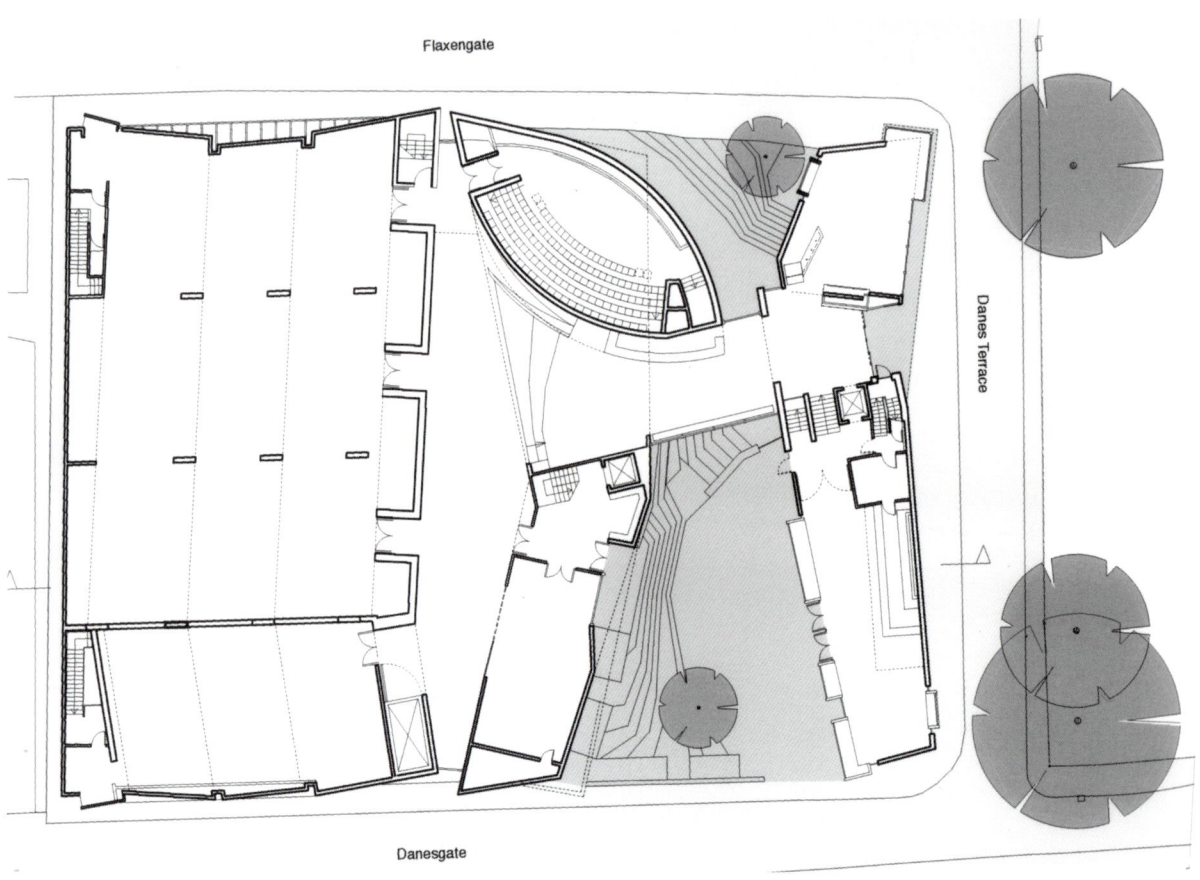

top: Constructing the board-marked walls
bottom: Building layout and ground floor plan

170

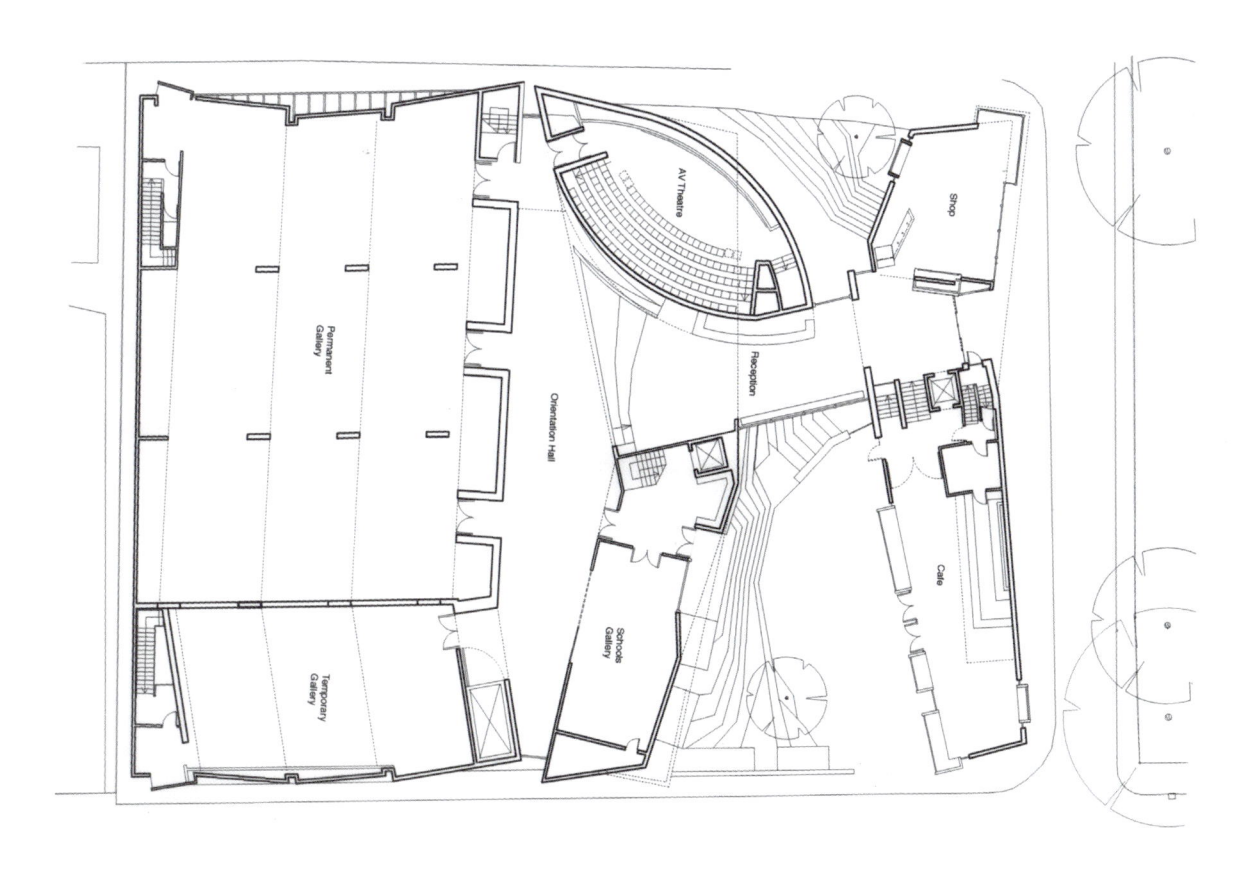

top: Constructing the sloping roof
bottom: Upper floor plan

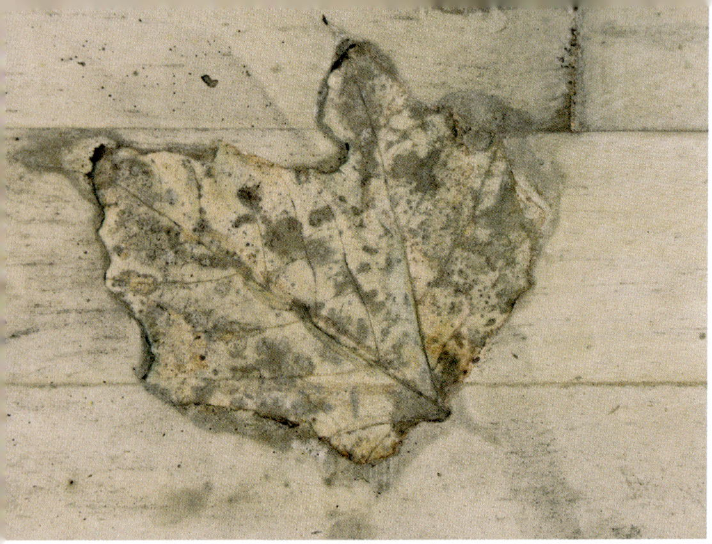

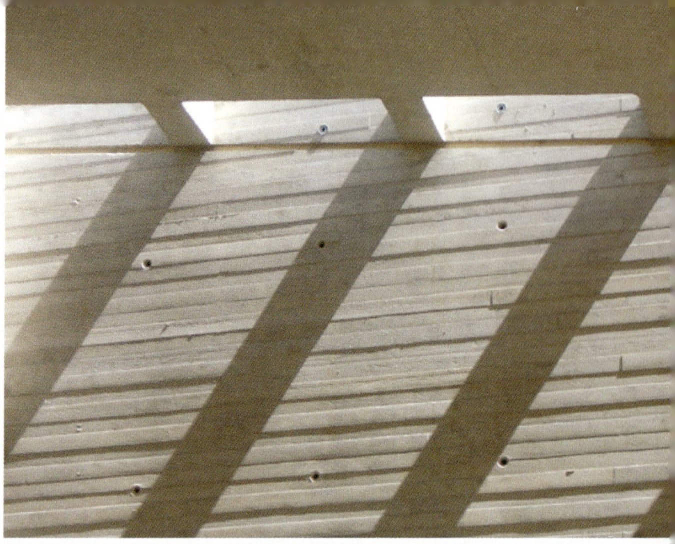

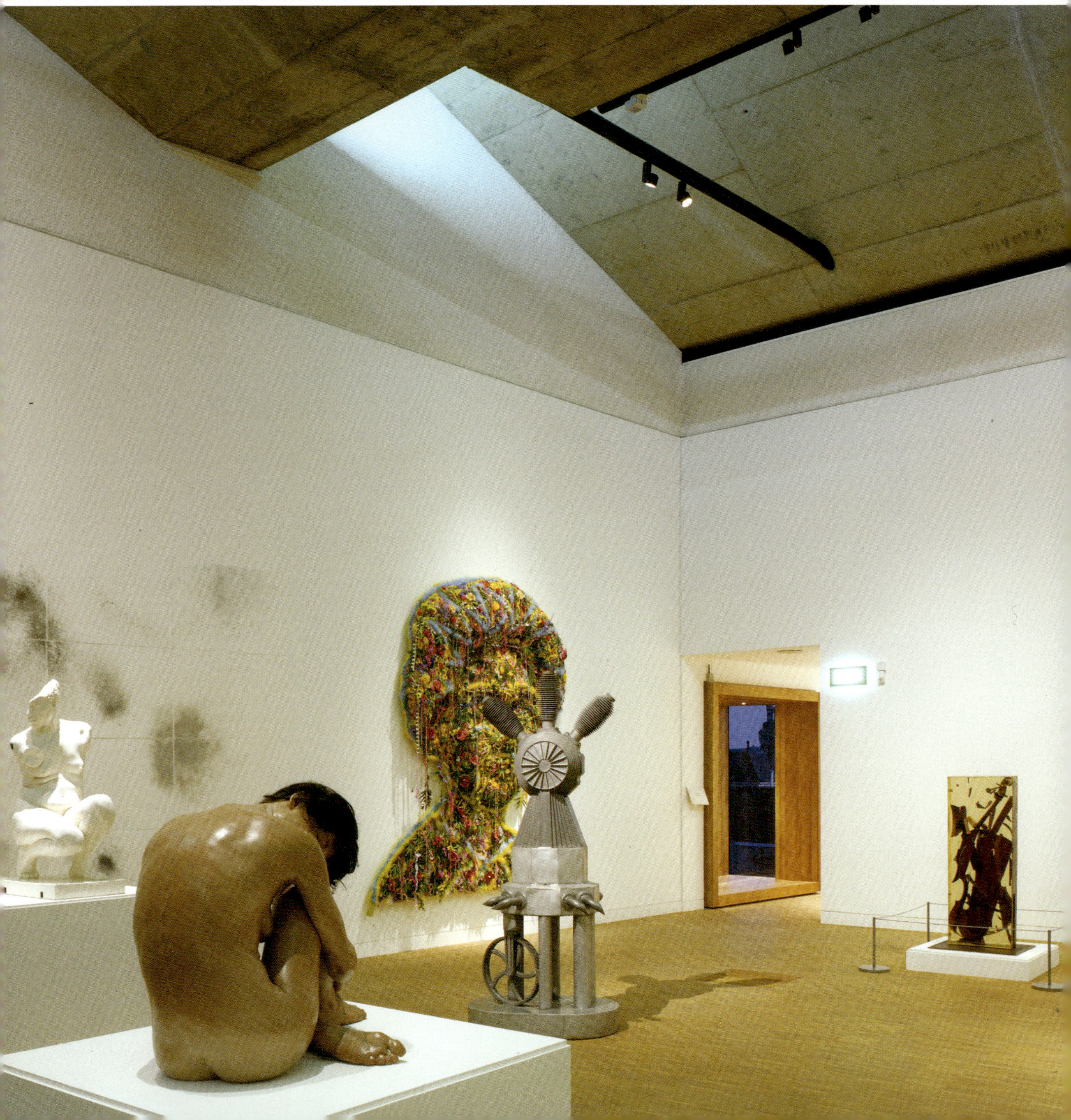

As a result the configuration of the gallery needed to be simple yet flexible. There are four bays of three modules and each can be, if required, divided into twelve individual rooms. Currently there are two rooms: three bays housing the archaeological collection and one bay containing the temporary gallery for contemporary art.

From the start of the project the stability of the thermal environment was central to how we designed and constructed the building. Concrete proved to be the ideal material due to its high thermal mass and compatibility in providing structural support for the stone façades and having similar movement characteristics so there were fewer expansion joints.

The external envelope of the building is 100 mm of Lincolnshire limestone coursed at 70 mm with a 5 mm bed joint. The limestone is of varying lengths and 'split faced' to give a rugged appearance that was sympathetic to our idea of the building being a series of stratified stone blocks.

Internally the self-compacting concrete was shuttered using planned pine boards that matched the 75 mm coursing of the external stone. This gave a much smoother finish, almost crystalline, as if the limestone has been metamorphosed into a much harder material. The boards were planed to three different thicknesses, which emphasised this stratification, and the self-compacting concrete picked up the subtle variations in the grain pattern and knots.

To continue this idea we placed sycamore leaves inside the shutter prior to pouring of the concrete, which then appeared as fossils within the walls.

On a practical level the benefits of self-compacting concrete were discussed at length with our structural engineers, Price and Myers. We were naturally concerned about the finish and getting the right people to supply it. We were advised that self-compacting concrete would give an even colour, that it would have no blowholes and it would be the best material to cast the steep, variable pitches and twisted planes of the gallery roof. A standard concrete mix for the roof would be impractical. Whatever we chose for the roof would be specified for the rest of the vertical concrete work. We had samples produced of various different grey and white standard concrete mixes to find the right colour. In the end we much preferred the self-compacting concrete just for its natural colour, as it had a warm tone from the limestone fillers. We spent a lot of time preparing layout drawings of the internal surfaces, trying to tie the board marking to the 70 mm coursing of the external stonework. The drawings would be sent to the concrete contractor/formwork design specialist, Doka, and they would comment back about preferred tie bolt locations, construction joints details, etc., and the working drawings were amended until we were happy with the solution. It was very much a team effort. The self-compacting concrete gave a porcelain finish tongue and groove board-marked walls that truly exploited the potential of the material and was perfect for the spirit of the building.

Self-compacting concrete
Mark Thomas, Lafarge Ready Mixed Concrete

Flowing, self-levelling, self-compacting concrete was supplied by our ready mixed plant in Lincoln. The self-compacting concrete which is branded 'Agilia' has 400 kg of Portland Cement from Rugby's South Ferraby works, a nominal 10 mm graded limestone coarse aggregate, Trent Valley sand and a percentage of fine limestone filler. The early trials showed the surface finish to have too many air pockets, which was attributed to the angular-shaped limestone aggregates. That was changed to a smooth, rounded single sized gravel and the problem was cured. To obtain the self-compacting properties, the carefully blended material was mixed with a cement/aggregate ratio of 1:3, a water/cement ratio of around 0.6 and a suite of high-range water-reducing admixtures and surfactants to achieve the cohesion and flow. No viscosity agent was used. The concrete achieved a flow of 50 cm and the fluidity of the mix was maintained for 2 hr before the concrete started to set. The concrete was placed by skip using a flat hose attached to the discharge port. The hose was lowered down to the bottom of the formwork from one central point on the top of the wall panel. The concrete flowed down the hose to find its own level and fill the formwork and the hose gradually rose with the rising level of concrete. For the sloping roof, the concrete was placed through openings made in the top shutter near the base of the slope, and then from the top of the slope using a flat hose. The concrete mix was ideal for vertical pours but was not suitable for flat slabs where a power floated finish was required. The formwork pressure can be much higher for this type of concrete, so the full hydrostatic head was assumed in the design of the formwork support system with a minimum movement permitted at tie bolt positions. That ensured no grout loss, sharp corners and neat day-joints. Properties of the hardened concrete do not differ significantly from those of ordinary concrete with similar cement contents. For compressive strength the standard concrete cube test is used. The extra cost of the higher cement content and special admixtures can be justified by savings in labour and the blemish-free finish that can be achieved.

top left: Leaf imprint
top right: Board-marked finish
bottom: Permanent gallery

CONSTRUCTION NOTES

Seamus Reagan, Northfields Construction

We mainly tender for reinforced concrete structures and but don't capture many jobs with a standard frame and simple construction approach. We win the more difficult bespoke designs. We were approached by the structural engineers Price and Myers to tender for the job, having worked with them before. We attended design meetings before the tender drawings were detailed, assisting the architects and engineers with construction information on how to form and pour the sloping roof, detail awkward corners and arrange the board-marked panels. We proposed the use of self-compacting concrete to pour the sloping, tapering roof section. We were quietly confident that this would achieve the finish the architects were looking for. We gave our advice freely, knowing that there was no guarantee of winning the tender.

By being proactive in this way and making a contribution at the design stage we felt we were in a better position to price the work correctly, which we won by competitive tender.

The sloping roof was at such a steep angle that we thought it was impossible to cast without a top shutter. We could manage an angle of 22° but not a 28° pitch. We had cast a sloping roof of 18° using a fairly low slump concrete and no top shutter in Nottingham and that worked well. One problem with a top shutter is being unable to clean all the debris from within the shutter and not leave sawdust and lose tie wire behind. We had an industrial vacuum cleaner with a long nozzle which we pushed through an opening on the top shutter to remove all the debris, but it was not perfect. We had some areas where the debris was not completely removed and this caused an unsightly line at the junction of the roof and the wall.

To test the quality of the finish of the board-marked walls, we made a section of it on the back of a plain fair face wall that we were constructing in Loughborough using the same self-compacting mix.

The finish was good and it gave the architects the confidence they needed, so they stuck with it.

The architects provided idealised layout drawings of board-marked concrete and tie bolt positions with the builder work package. We elected to use a Doka Framax formwork system which came in squares of varying sizes, faced with a rough ply surface. The units come with fixed tie bolt centres. Due to the high pressure of the self-compacting concrete, the tie bolts were increased to 20 mm DyWidag rods over the lower half of the wall section. Doka designed the formwork system to withstand the pressure of the self-compacting generated by the fast placing rate.

The formwork panels were assembled on the ground, face down and then lifted and held in vertical position for the 75 mm tongue and groove board-marked panels to be nail-gunned to the ply face on the Doka system. The nail gun fires very tiny metal pins which you can barely see on the surface. For making the floor connections to the wall we used continuity strips to leave the wall a free standing structure to be poured in one. We would cast walls around 7 m high and up to 9 m long and as much as 300 mm thick in one pour.

The board-marked strips were used only once and then discarded. There were a few places where they were used twice but that was exceptional. The pour sizes were non-standard, as the walls had a sloping top line or were angled and, with not much repetition, for practical reasons the boards had to be removed to dismantle the Doka system and then rebuild it for the next pour. It was a choice between having more than one support system on site and using one set which could be adapted for every wall pour. We chose the latter which meant removing the boards each time. It was quicker to use new boards than have to cut the existing ones to suit. There was lot of wastage as a result.

With the board marking we had to work to a set datum and to match up with the horizontal lines of the previously cast wall. To overcome the timber tolerances and creep of the wood, we ordered three different widths – 72 mm, 74 mm and 76 mm – to accommodate any variations. The pieces had to be hand picked before they were

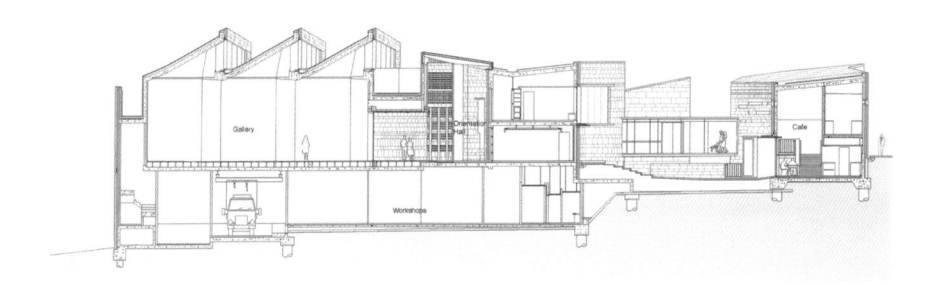

above: Long section
right: Shop

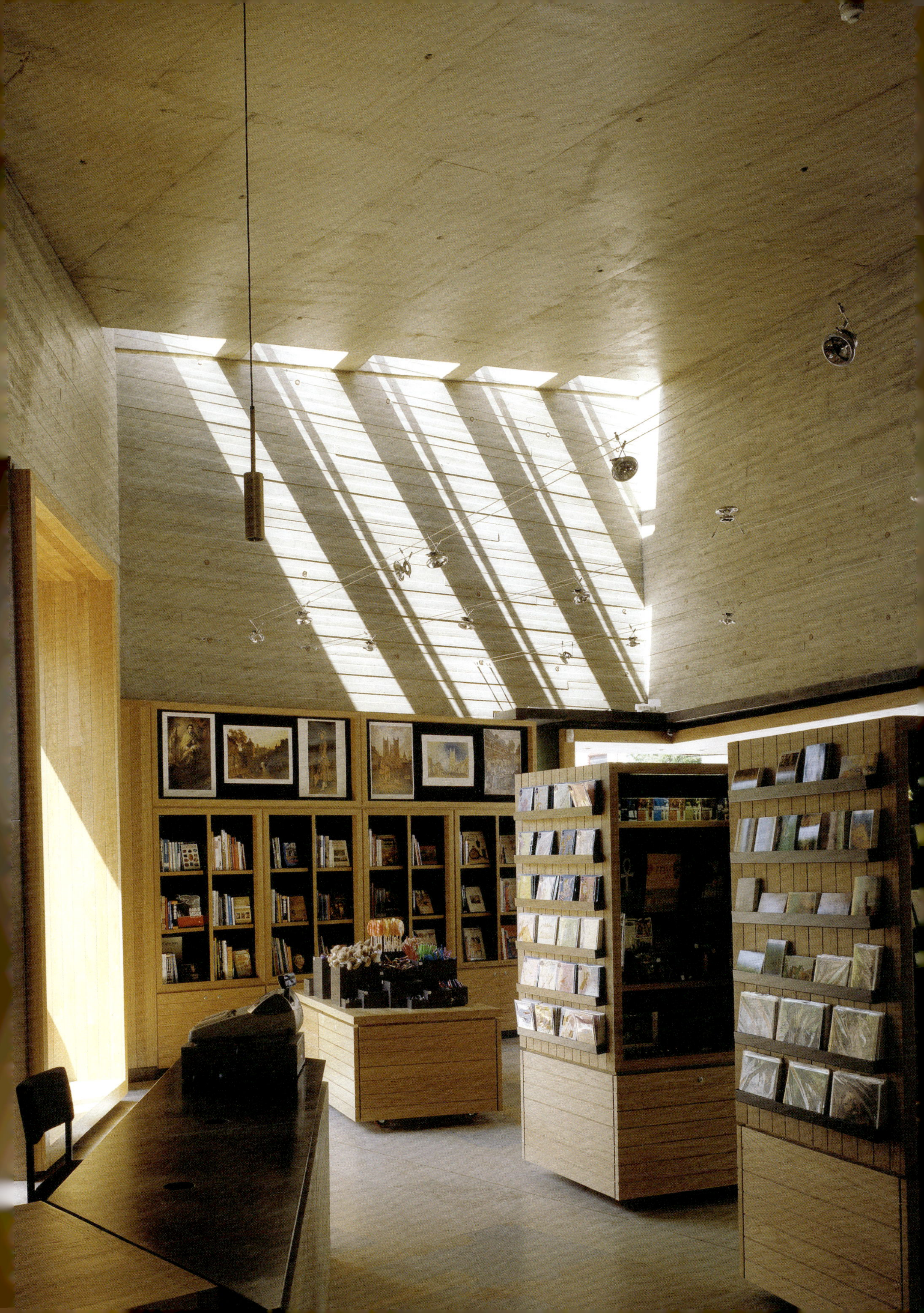

fixed. To get the best match with the previously cast wall the Doka panels were made up on the ground, bolted together and then erected in place and held vertical. An access scaffold was built to allow the board-marked pieces to be pinned to the Doka ply using nail guns, working from the bottom up.

The timber was a redwood with a very uniform grain that was easy to work with. The thickness of the timber was varied from 18 mm to 22 mm to create stratification lines. The pieces were cut in random lengths and they fitted well together because they were tongue and grooved. It would not have given such a good finish if they had just butted up.

It took two days to complete each fair face formwork wall, a further day to fix the rebar and by the end of the week we would be pouring the wall section. The back face of the formwork, which was insulated, was just an ordinary ply. For the high wall we lined the exposed Doka ply with Douglas fir backing sheets to give an even surface to fix the boards to and to stiffen the ply section to prevent any warping or movement under hydrostatic pressure. We applied a wax release agent to the untreated boards, called ADO wax. We put it on with a roller, and any excess had to be wiped off with a cloth otherwise it would retard the concrete and cause the surface to dust.

The formwork was removed after 48 hr. If we removed it earlier we tended to spoil the arrises and corners, which were weaker. Later, the concrete surface was sprayed with a dust sealer called 'Stop dust' which gave the surface a waxy sheen. We employed a team of eight joiners and a foreman to carry out all the wall formwork.

The self-compacting concrete was placed in the wall forms using a 'lay flat' hose. The lay flat hose was 5–6 m long and was connected to a hopper at the top. The hose reached down into the vertical formwork to within 0.5 m from the bottom. The hopper was positioned midway along the wall and supported by the crane. The fluidity of the self-compacting concrete would allow it to flow to all parts of the formwork and fill up evenly and smoothly through the flat hose, without entrapping air. To ensure that there was no air trapped in the hose after each cubic metre was poured down the hose, the hopper exit port was closed as the lay flat hose closed.

For the sloping walls we omitted the lay flat hose over the lower half of the faceted sloping roof as it was impossible to lay it down the slope without entangling it with the rebar and tie bolts. We poured the self-compacting concrete through an opening part-way up the slope through the back shutter. When that section was filled we moved the hopper to the top of the forms, closed the opening and finished the pour without the flat hose again. Where the concrete had entered the mid-section it created pale flow lines like a waterfall that had marked the formwork and hardened on the concrete surface. It highlighted the pouring points all along the roof slope. It seems that the shutter face was being coated with fine limestone and cement particles as the self-compacting concrete flowed down to the base of the shutters. This may have partially set as it clung to the surface, which was sticky with release agent. The bulk of the concrete then reached the pour point but did not fully intermix. With vibration this might have been avoided but if we had done so the self-compacting concrete would have segregated. We had no idea this was happening. All the nine pours for the roof had to remain shuttered and propped until the roof structure was fully built and structurally connected. Had we anticipated the problem or been able to spot it after the first pour, assuming we had been allowed to remove the shutters, things might have been different.

The most difficult element of the concrete work was forming the sloping faceted roof slab with an enclosed shutter which had to resist the concrete pressure and the uplift forces. The tapering faceted roof profile meant that everything was handmade, cut out of timber. No two sections of the roof pour were the same so there was little opportunity to recycle material.

Despite the difficulties we did not suffer any grout loss or deflection during all the concreting operations. The surface definition was of exceptional quality, particularly the board-marked walls. Considering that the formwork and temporary works supports were being designed for the first time using assumed concrete pressures generated from the self-compacting concrete, it was a fine achievement.

PROJECT TEAM

Architect: Panter Hudspith
Structural engineer: Price and Myers
Services engineer: Arups
Main contractor: Caddick Construction
Concrete subcontractor: Northfields Construction
Budget: £7.8 million
Completed: 2005

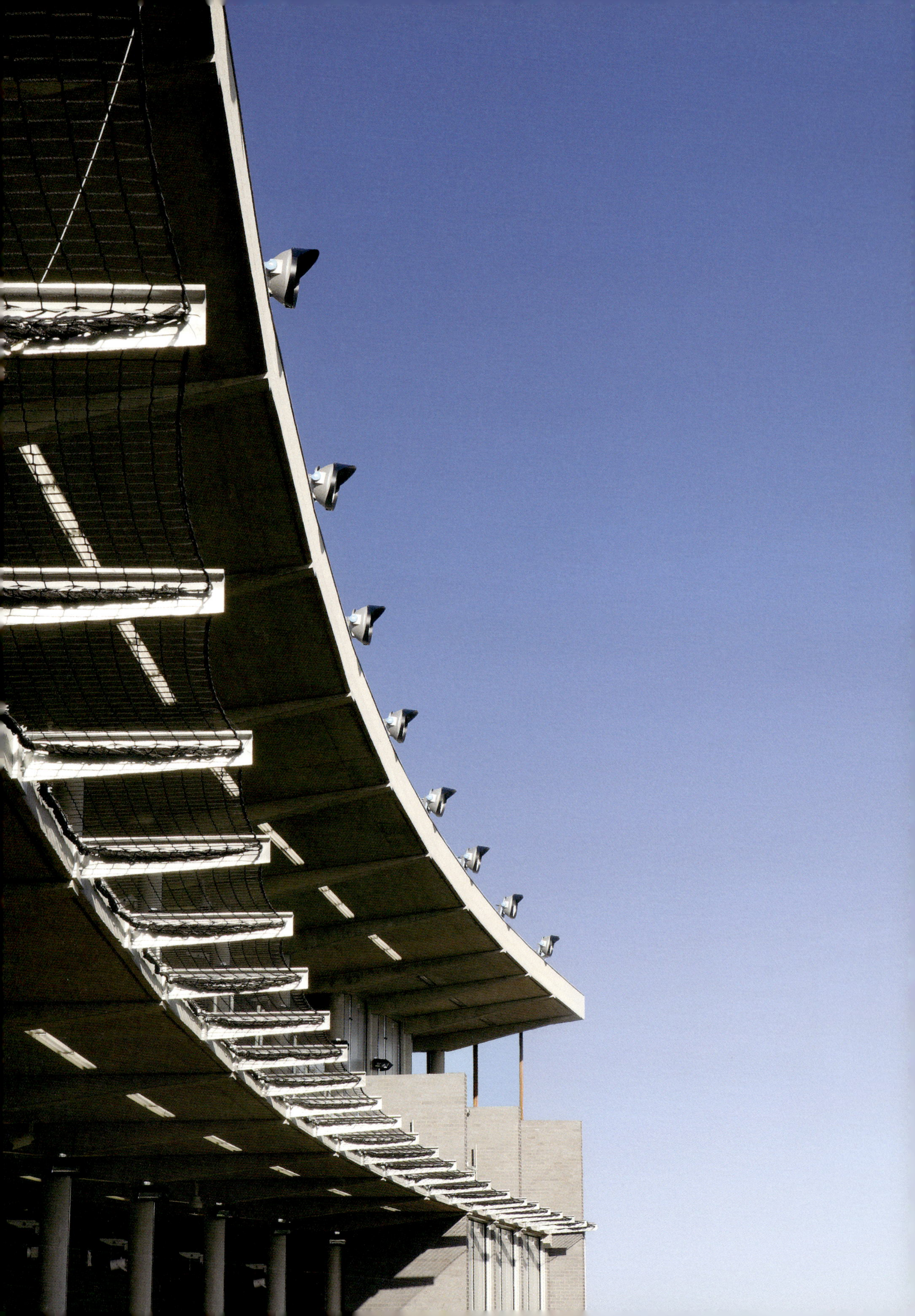

PLAYGOLF
NORTHWICK PARK, HARROW ON THE HILL

Architect: Charles Mador Architects

Location

The golf centre is on the A404 Watford Road past Northwick Park Hospital as you leave Harrow on the Hill. It's on the left-hand side and clearly signposted. The nearest tube stations are Harrow on the Hill and Northwick Park on the Metropolitan Line. With a full set of golf clubs I would suggest the car journey, but if it's the ambience and après golf that appeals then a 5-minute bus ride or 20-minute walk from the station is well worth the effort.

Introduction

The site was owned by the London Borough of Brent and is part of a public park. They had been trying unsuccessfully to develop this parcel of land into a golf course but it was believed to be contaminated with toxic fill. When Playgolf were introduced to the council some years ago they had soil samples taken which proved that the fill was not badly contaminated and was suitable for laying out a short golf course.

Northwick Park's six hole pay-and-play golf course, which you can get round in an hour, was designed by Peter McEvoy OBE with meticulous attention to detail, and to high-quality construction standards. Each hole has been crafted to replicate – as accurately as possible – the features of a portfolio of legendary holes, all key holes at major championships, including the Postage Stamp at Royal Troon with its infamous 'coffin' bunker, Augusta's beautiful but treacherous 12th and 16th holes, the Belfry's challenging par four 9th and Royal Birkdale's classic short 6th hole.

As you approach the golf complex, the impressive sweep of the concrete framed club house and driving range gives you that feel-good factor – expectation, promise and adventure. This is a state-of-the-art golf course despite its few holes, which is matched by a stunning, sleek concrete structure that sets new standards and aspirations in modern golf course design.

ARCHITECTURAL INSITU CONCRETE - CASE STUDIES

Architecture discussion
Charles Mador

The building is a golf centre with a two-tiered driving bay, café, shower rooms and golf shop and it functions as the club house for the golf course. It is 120 m long by 20 m wide and has two floors. The elevation facing the golf course is open and contains the driving booths on two levels where golfers can hit practice balls towards the various distance targets on the golf range. The first 8 m of the building is open to the elements because of the driving booths; the remaining 12 m is an enclosed heated space.

We have built a number of these golf centres before and they have all been steel frame structures until now. The problems we have had are to do with corrosion of steelwork exposed to the atmosphere. The open golf bays trap the morning air, which condenses on the steelwork and causes a build-up of corrosion on the stanchion bases. The other concern we had was that a steel frame ends up with a very messy architectural appearance, as the services that are required are clipped onto the frame and produce a cluttered effect.

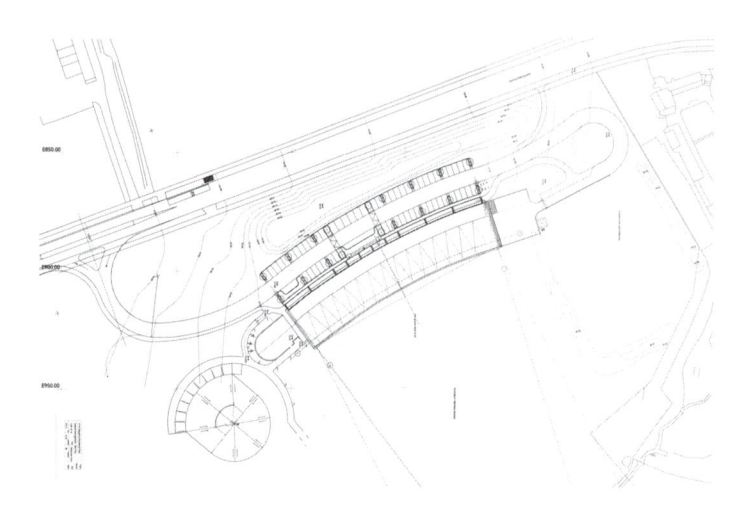

Site plan of main building and car park

At Northwick Park we had planning issues that made us think again about our choice of construction material – the lower level of the building was to be set into the landscape due to the sloping site which meant constructing a concrete retaining wall. The other point was that we were obliged to incorporate a green roof to soften the impact of the building on its park environment. The original project managers for this scheme, who were later fired, were pushing for a steel-framed building. They argued that the programme benefits would override all other criteria. How misinformed they were. In my view the extra complications and cost of finishing and fireproofing steelwork make it an uneconomic choice.

After further deliberation and value engineering, the design team and client were convinced that the structure should be in insitu concrete. This was the logic behind the decision – as soon as the concrete structure was finished it would provide a finished soffit, a structure to take the green roof and a power-floated floor finish that eliminated the need for screeds and following trades. There were no services to be hung from the ceiling as they were cast in with the slab to remove such visual clutter.

The lower level floor was a slab on grade with a waterproof concrete retaining rear wall. We did not have the funds to do anything 'posh' to the concrete. We accepted its industrial grade finish but we controlled the arrangement of the formwork.

The panel proposed by the contractor was a rigid construction, and had an absorbent film faced surface that gave a matt finish. In the end, the soffit finish was not as good as we had hoped as the paper faced panel produced brownish staining on the concrete and a patchy appearance on reuse. It was a cheap material which did not give a high reuse factor, which in my view is a false economy. But it was true to level, there were no grout losses or poor construction joints and once the concrete had dried out and carbonated it was more acceptable.

In every other way the construction and programme time we had hoped for was achieved successfully. Had we employed the concrete subcontractor as the main contractor we could have saved substantial cost with just the work packages for brickwork, windows and services to secure. Due to financial planning and the funding conditions, this simple scheme had to have a project manager and a management contractor, who would do the job for a fixed price. It cost more as the management contractor took a sizable fee for administrating the contract. Large 'blue chip'

top: Entrance
bottom: Lower level driving bays

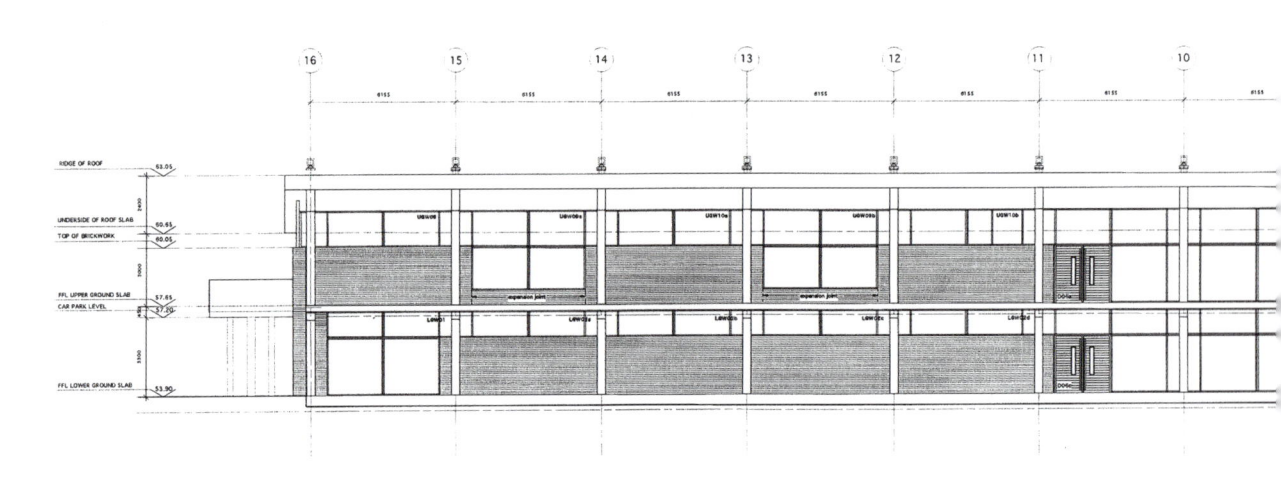

top: Construction of concrete frame
bottom: Driving bay elevation

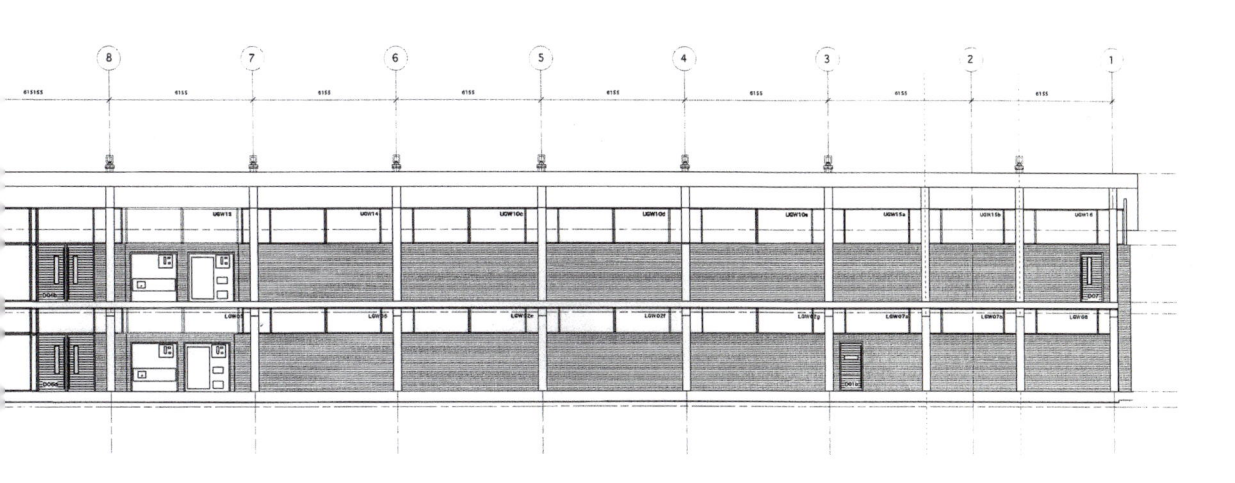

top: Casting the curved roof slabs

contractors with next to no experience of building with exposed concrete put high risk and cost against it quite unfairly.

As architects, we try not to make the concrete too precious or jewellery-like, as it can be cast with standard materials and ordinary concrete and still retain a high quality by drafting a sensible and achievable exposed concrete specification. If you look at the work of Le Corbusier or Louis Kahn, they achieved quite a basic finish but the joinery was so precise.

The casting of the short upstand edge to the roof produced a problem. We wanted a crisp leading edge to the roof, which was 500 mm high, made up of 250 mm roof slab and 250 mm upstand. It was to be cast monolithically with the roof slab pour, but it was actually cast in two stages. When the roof slab had hardened the upstand fascia panel was fixed and linked to the existing formwork of the slab edge, which had been left in place mistakenly. The roof slab had contracted, leaving a small, continuous gap from the formwork that had been left in place. When the 250 mm upstand was cast on top of the slab, grout oozed into the gap and caused an ugly line and irregular finish at the joints. The upstand edge was also proud of the roof slab edge so there was no crisp leading edge.

After trying a number of remedial solutions the face was bush hammered, which removed the discontinuity, masked the day-joint line and evened out the surface. A saw cut was scribed along the lower slab edge to maintain a smooth arris and to provide a guide line for the bush hammering.

Around the reception and café area the higher level of electrical services necessitated a false ceiling. This was created quite simply by off-setting from the concrete using 25 mm battens for running wiring. The ceiling was faced with birch ply, which we had hoped to salvage from the birch ply we had proposed for the floor shuttering. To have reused it this way would have saved money and made it aesthetically pleasing.

An interesting aspect of the concrete detailing was making provision to prevent cold bridging. We wanted the roof slab to project beyond the external wall of the enclosed area on the west elevation and for the leading edge on the opposite elevation to cantilever over the driving bays. We had a simple solution for the roof slab projecting beyond the wall line using insulation that was covered with birch ply and running it 1.5 m into the building. On the driving bay elevation the 8 m long projection of the upper floor slab is carried on a series of tapered beams pinned back to the main structure and supported on a central column. Where the tapered beam is pinned to the main structure there is a thermal break along the floor slab. It was a robust detail that worked well.

The most dramatic part of the building architecturally is the driving bays and the cantilevered slabs overhead. The columns supporting the tapered beams are circular and set away from any enclosing walls. The columns were cast using proprietary cardboard column liners. The finish at the junctions of the column and floor slabs was so good that we left them as struck. The special feature of the cantilever slab over the driving bays was the tapered beams, which commenced at 350 mm deep, reducing to hardly any thickness at the tip of the cantilever. Given the repetitive nature of the construction, the concrete contractor was able to recycle the formwork and cast these elements in a way that worked very efficiently.

The brickwork was designed to reflect the durability and strength of the hardwearing surface of the concrete structure. A dark engineering brick was selected for this reason. We fixed the glazing in warm iroko wood frames, which contrast well with the grey concrete and birch ply soffit and internal panelling. Besides the smoothness of the power trowelled floors, in the entrance area and reception the concrete was diamond polished to create terrazzo, which came out beautifully. It was a dusty process using dry grinding but it was worth it.

Often a main contractor will find all sorts of excuses and reasons to make an issue out of simple routines in concrete operations. For example, we wanted to make provision for services in the slabs by forming recesses and inserts in the soffit and in the circular columns for the lighting conduit to run down a preformed recess. We had sorted all the construction logistics of it before the project was tendered. The benefits of this arrangement, despite the increased cost of the inserts, is that the client cannot introduce changes at a later stage as everything is fixed. It saves the client a lot of money by being forced to stick with the original plan. Once on site, the concrete contractor had no difficulty in placing the inserts and forming the recesses in the correct location. They did not have to fall back on remedial work as it worked like clockwork.

The reinforcement to the slab appeared to be quite light – it was far less material visually than we had ever seen before. This was

top left: Window close up
top right: The first hole and lake
bottom left: Main staircase
bottom right: Staircase detail

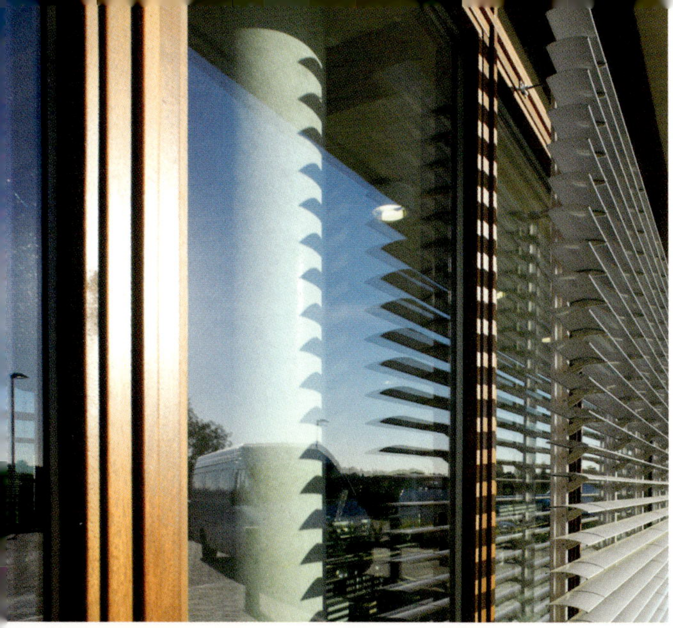
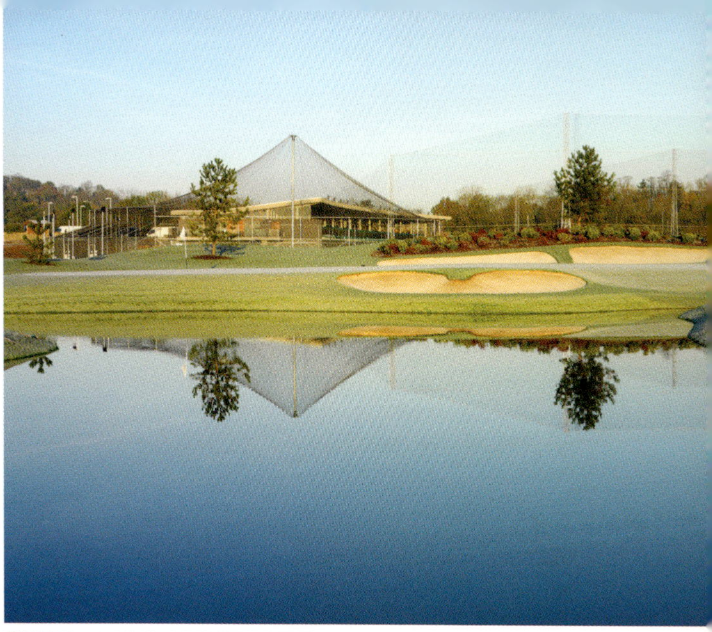
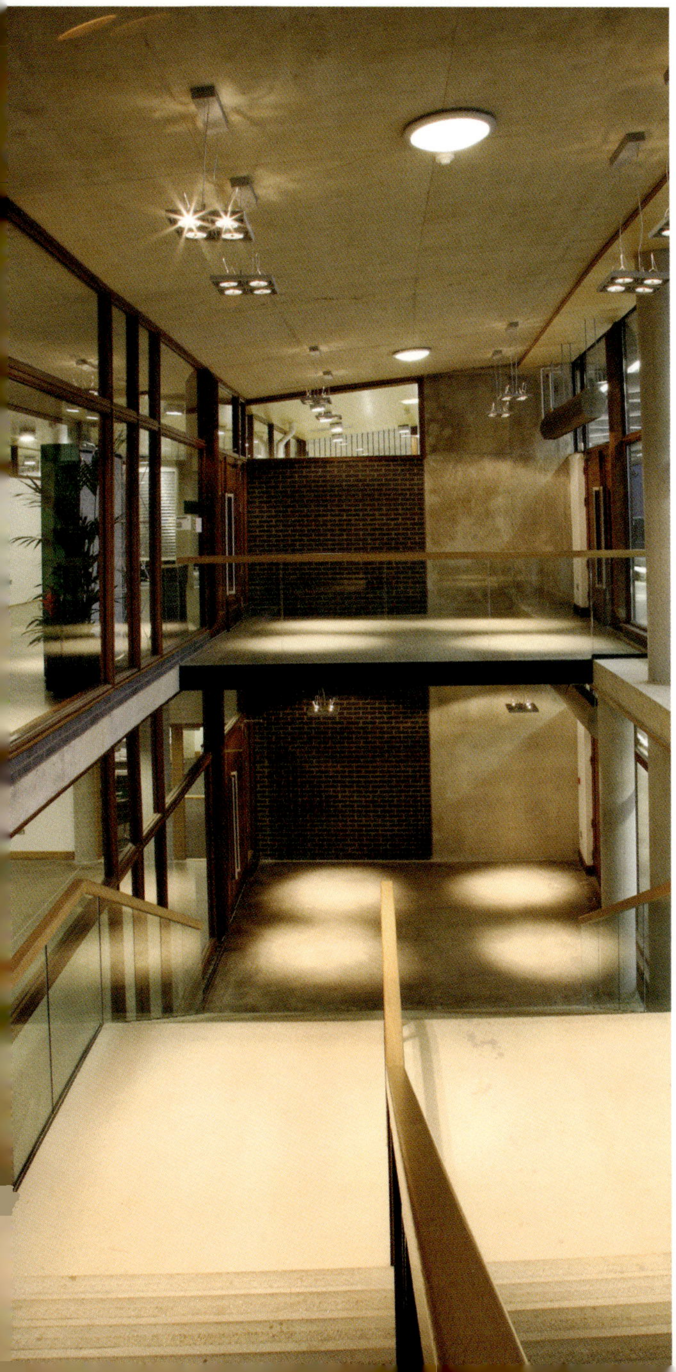
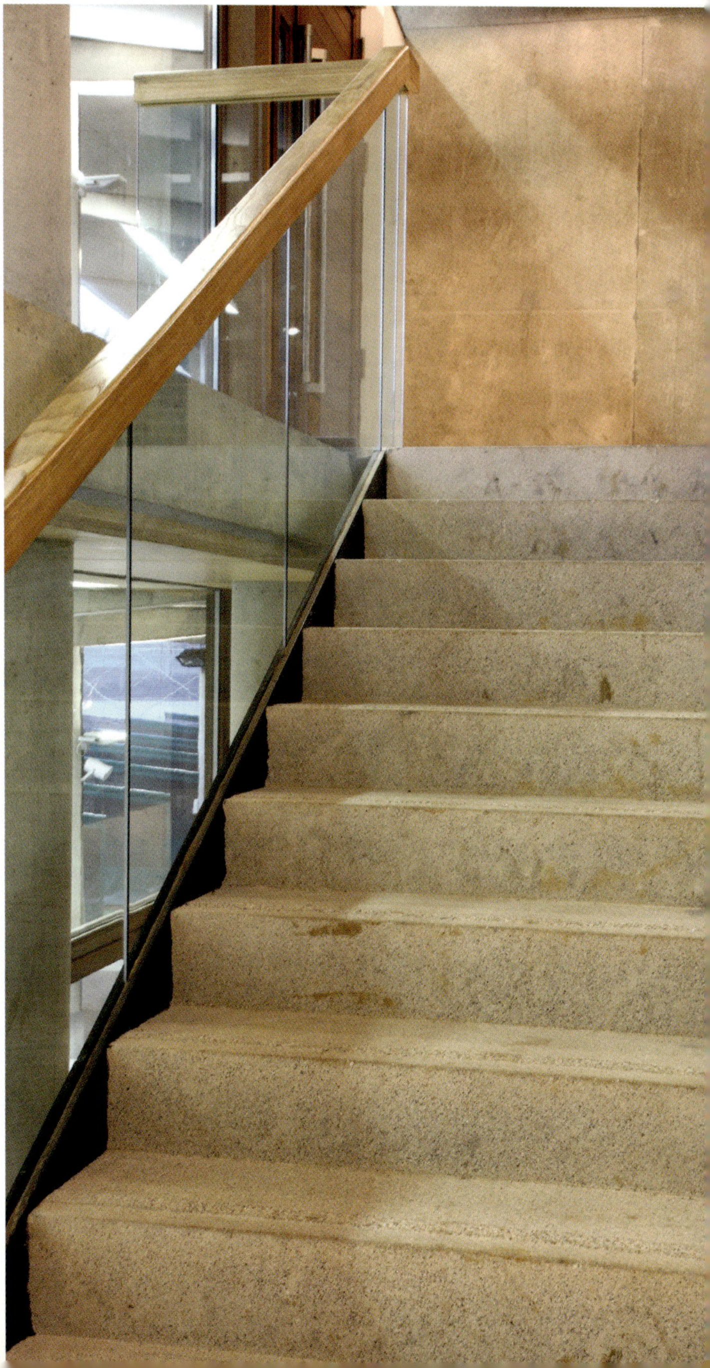

the result of the slab design by yield line theory, a special skill of our structural consultant Powell Tolner and their now retired partner Gerrard Kennedy, whose design work and structural expertise is widely admired in the concrete industry. This approach meant that it was so much easier to position inserts and form box-outs for the services. Yield line theory is a well-founded approach but often a neglected method of designing reinforced concrete slabs. Yield line theory investigates failure mechanisms at the ultimate limit state of a slab. The resulting slabs are thinner and have low amounts of reinforcement in very regular arrangements. They are therefore easy to detail and to fix, and are quicker to construct.

The robustness of the concrete is standing up well to public usage compared to the steel composite structure we have used in the past. This building is also weathering well. The building was completed for a budget of £800/m^2 inclusive of infrastructure costs, roads, car parking and incoming services.

We had a big feature entrance staircase which was 3.5 m wide that drops 4 m from first to ground floor with a half landing. The staircase was precast by a local firm and has an anti-slip surface on the nosing, just like those on the London Underground steps. There is also a precast spiral staircase at one end of the driving bays for the professional to access from the golf shop.

There is a similar Playgolf centre in Manchester but you would only call that club house a modern tin shed by comparison. This is a proper building which jumps into a new class of golf centre and says much more about its architecture. It's a great breakthrough for us as a practice and this is how we wish to progress in the future. As architects, we can become separated from the trade contractor when negotiating and dealing only with middlemen, and that is a concern. We believe we can only improve the quality of our design if we communicate and negotiate directly with the contractors who actually do the work.

PROJECT TEAM

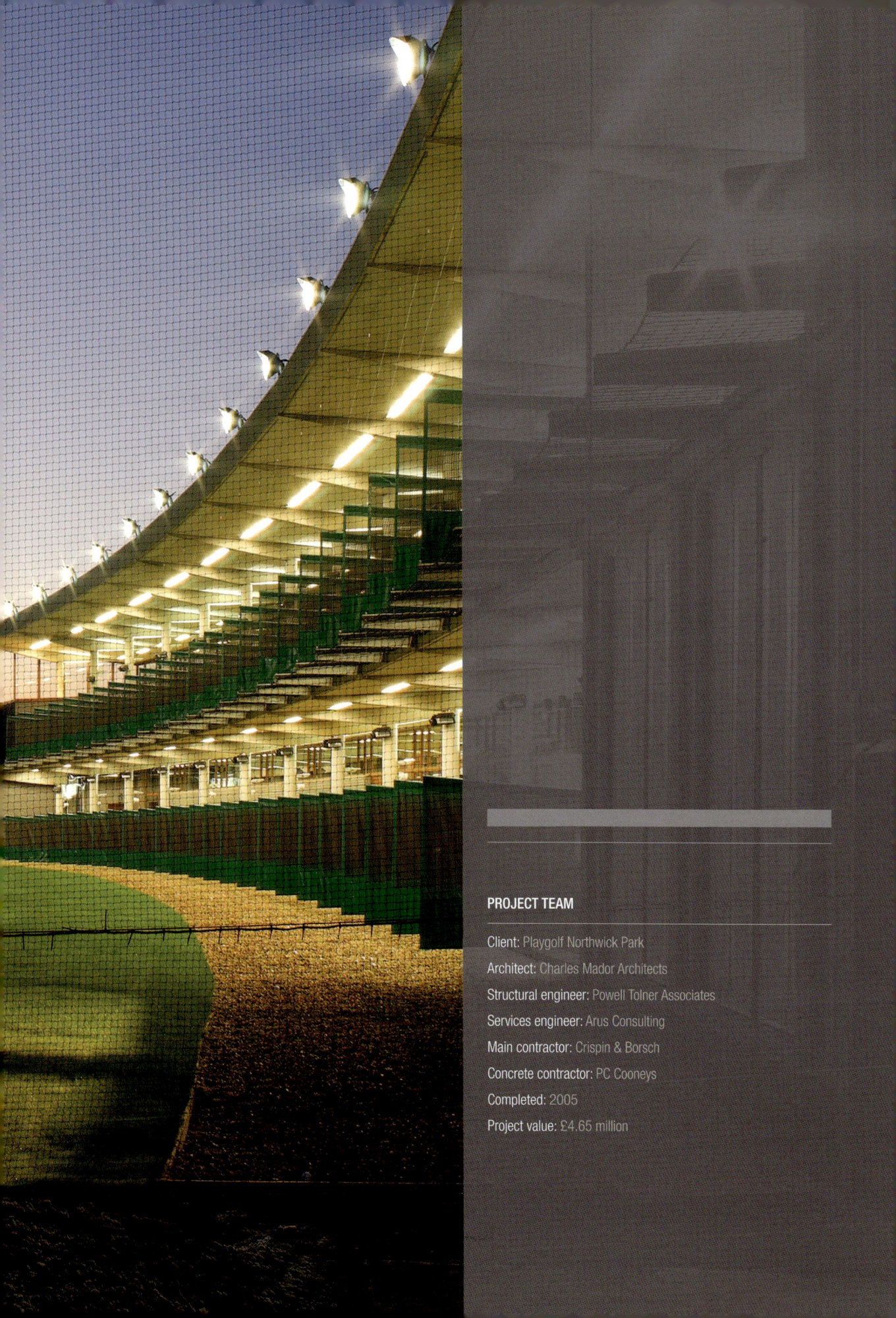

Client: Playgolf Northwick Park
Architect: Charles Mador Architects
Structural engineer: Powell Tolner Associates
Services engineer: Arus Consulting
Main contractor: Crispin & Borsch
Concrete contractor: PC Cooneys
Completed: 2005
Project value: £4.65 million

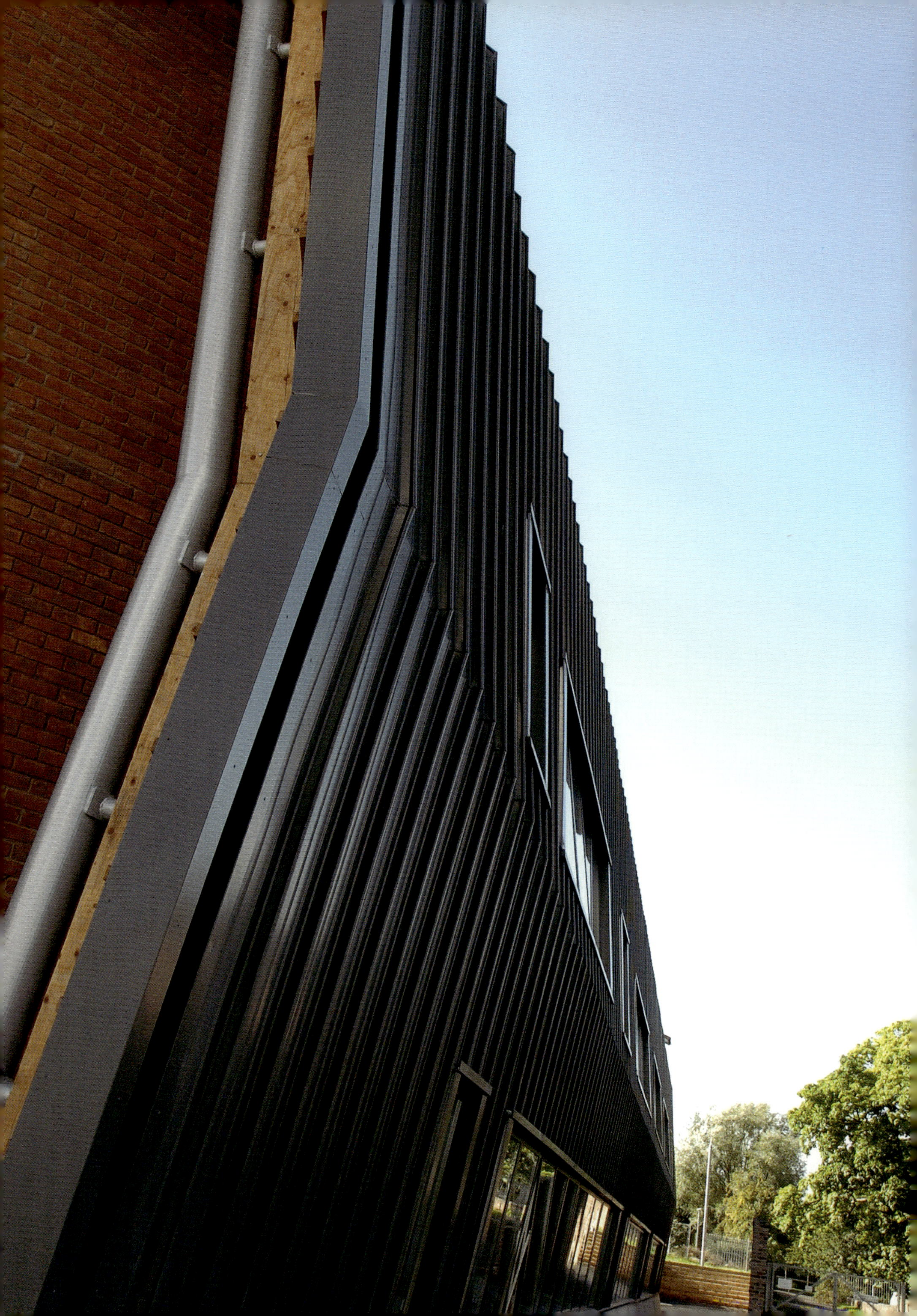

E-INNOVATION CENTRE
UNIVERSITY OF WOLVERHAMPTON, TELFORD

Architect: Building Design Partnership, Manchester

Location

The centre is located just off junction 4 of the M54 in the prestigious Priorslee area of Telford. By car, leave the M54 at junction 4 and take the fourth exit from the roundabout, signposted B5060 – Priorslee and University. Continue for approximately 1 mile then turn left onto Priorslee Avenue. Take the first exit at the next roundabout, Shifnal Road, which leads to the university campus. It is a good mile and a half walk from Telford Central Railway station.

Introduction

The building has been constructed on a sensitive site adjacent to Priorslee Hall, a Grade II* listed Georgian manor house. The University of Wolverhampton, with funding from Advantage West Midlands, paid for the construction of this facility on its Telford campus.

The aim of the E-Innovation Centre is to increase the formation of new businesses in Telford. The centre can accommodate up to 36 new enterprises at any one time. These start-up units are 20 m² in size, with an additional space of 700 m² available for expansion.

All floors have 12 start-up units, located along the north-facing side of the building. The grow-on space is open plan but there is the option of partitioning smaller areas. The ground floor is set below natural surface level, and this floor has 350 m² of grow-on space and a hospitality and meeting area of 90 m². The first floor is the entrance level where the reception area is located. It has 350 m² of grow-on space and a meeting facility of 34 m². The top floor has 200 m² of grow-on space with access to a balcony. There is a 250 mm-high raised access floor with carpet or rubber stud covering to give the building maximum flexibility for future development.

The E-Innovation Centre provides reception facilities, heating and lighting, building maintenance, cleaning, security, parking and some free use of meeting rooms. Our on-site technical support staff maintain the internal networks in the building, including access to an external broadband connection.

The architecture responds to the building's climatic control with its wrap-over aluminium roof. The internal plan is divided in two, with a dramatic three-storey high corridor slot, which contains the access staircase. At either end of the corridor slot, floating 'pods' supported on chunky columns, provide formal and informal meeting places. The building's high thermal mass is used to regulate internal temperatures and normal opening windows provide natural ventilation.

ARCHITECTURAL INSITU CONCRETE - CASE STUDIES

Architecture discussion
Ian Palmer

The scheme, which is run by the University of Wolverhampton, gives opportunities to graduates with good ideas to start up in e-business, with time to develop a network before moving on. The site was long and sloping and was next to a Grade II* listed brick building and a residential hamlet. Priorslee Hall, built in the 18th century, was once the headquarters of the Lilleshall Company, and then the Telford Development Corporation. It is now the centre of the University of Wolverhampton Telford campus.

We had scale issues and access problems in trying not to overlook the residential back gardens of the houses of Priorslee village. At the same time we did not want to compete with the monumental elevation of Priorslee Hall, whose views of the surrounding area are screened by the new build. We basically set up camp in their front garden.

We designed a simple, curved plan building which is a counterpoint to the rectilinear formality of the listed building. We formed an aluminium wrap-over roof to create a calm backdrop, reinforcing the garden wall link with the adjacent residential estate. The roof was also instrumental in controlling the climate within the building. On the north elevation, facing the village, it was curved all the way to the ground with punched holes for the windows. The roof is a standing seam and is supported on timber kirto beams.

The south elevation, facing the listed building, is fully glazed with a sun terrace. Here the roof acts as a visor and shade canopy. The structural frame is concrete as are the two internal pod structures, one at each end of the central corridor.

The building is approached from the west elevation and entered at the first floor as the base is sunk into the natural slope of the land. The building has three floors and is 12 m high at the apex of the curved roof. It is 45 m long and 20 m at it widest point. The

E-INNOVATION CENTRE

structural frame is exposed insitu concrete, with a timber roof lining and timber inner cladding.

Every joint detail and connection is well crafted and expressed. The joints between the column and floor soffit and the wall panel layout were carefully detailed. Some aspects have worked better than others but we are pleased with the overall quality.

We have a central spine within the building, which is the main circulation corridor that divides the incubation units from the more traditional office space on the south elevation. There are bridge links at every floor across the corridor that leads to the front door of every unit. The open treads of the staircase rising up from the ground floor ensure a light-filled space. The staircase treads are black precast concrete elements that cantilever from the high corridor wall. There is a metal plate on the reverse side with two bars with threaded ends that connect with bars cast in the wall. The staircase was precast by Envex with a black lacquered surface coating to match the black dimpled rubber of the ground floor.

Externally, we wanted a building material that was affordable and which was in context with the brickwork of Priorslee Hall. The standing seam roof has a dark tone which is associated with barn colours. Internally, timber was used because it is tactile and friendly and concrete for fabric energy storage. The building is naturally ventilated.

The columns are all circular and made with cardboard column formers, the suspended floors are rib beams and one-way slabs. We did consider having a coffered slab but that was omitted on cost grounds.

Concrete notes

We used the structural concrete specification to define the surface finishes and control the appearance. We ensured that the concrete subcontractor provided formwork layout drawings showing the sheet joints and the position of construction joints. There were sample panels to be cast and the shear wall which was not exposed was used as the trial panel. We stipulated a plain, smooth finish and did not specify the type of ply to achieve it. We were not aware at the time that there were so many choices. In the original design we wanted something that was textured and not smooth and liked board marking, but that proved too expensive and was substituted.

top: West elevation and main entrance
bottom: Pod structure and meeting room

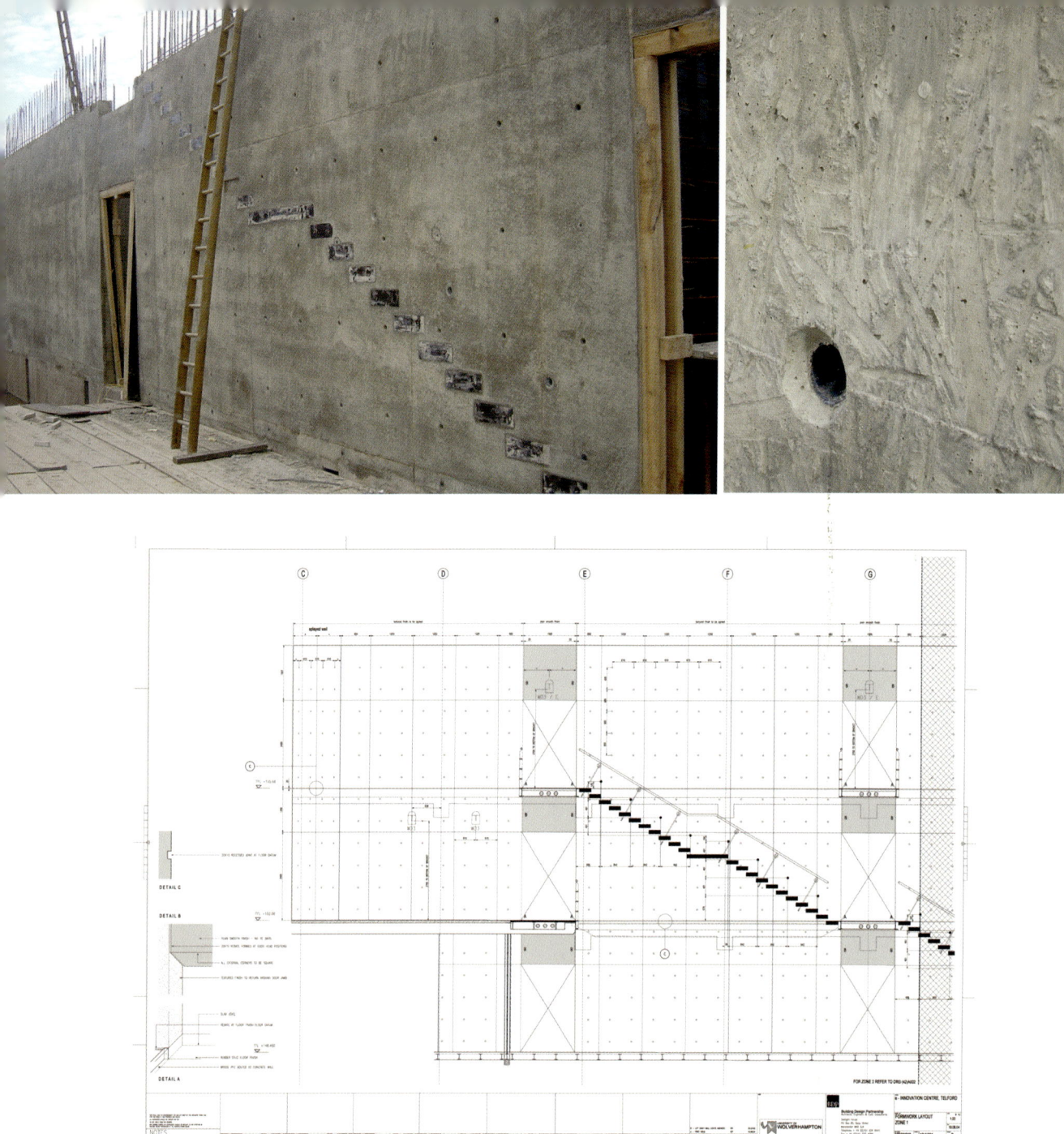

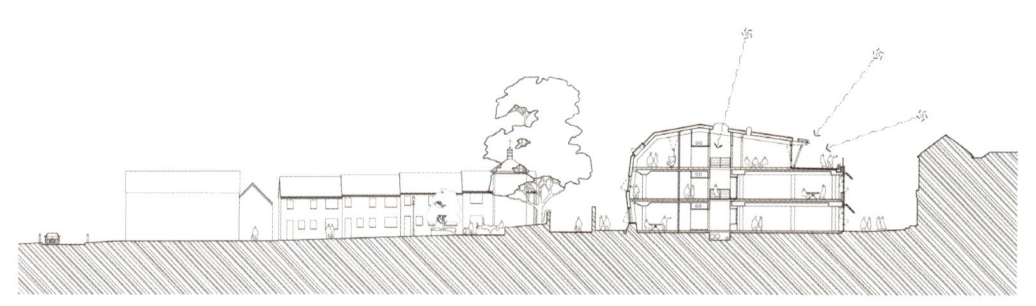

top: Corridor wall construction
middle: Wall panel layout drawing
bottom: Transverse section

192

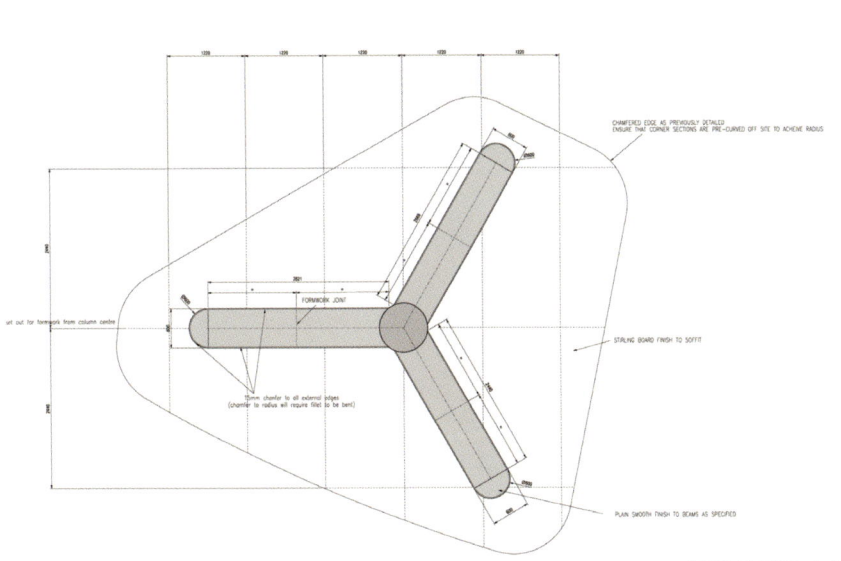

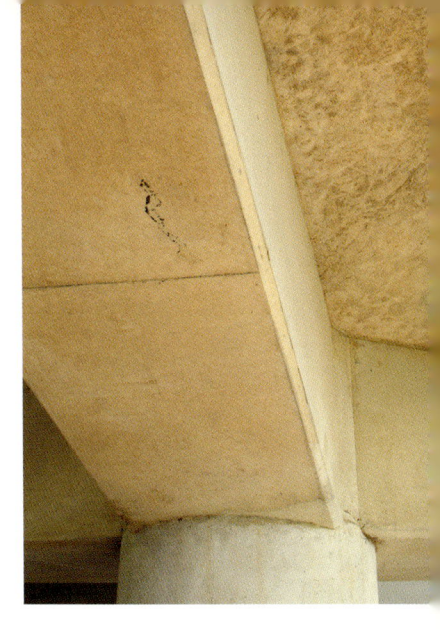

FORMWORK LAYOUT

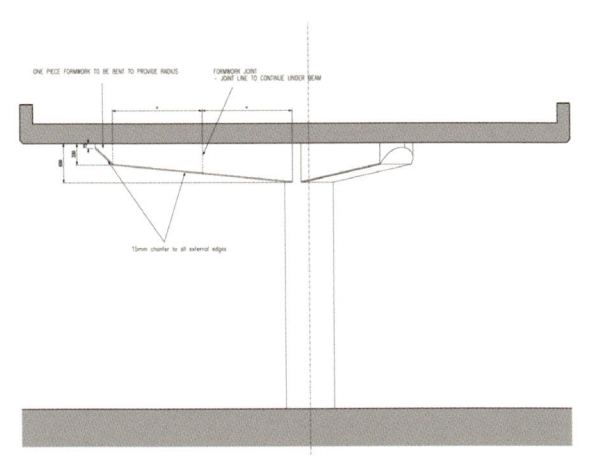

TRUE ELEVATION

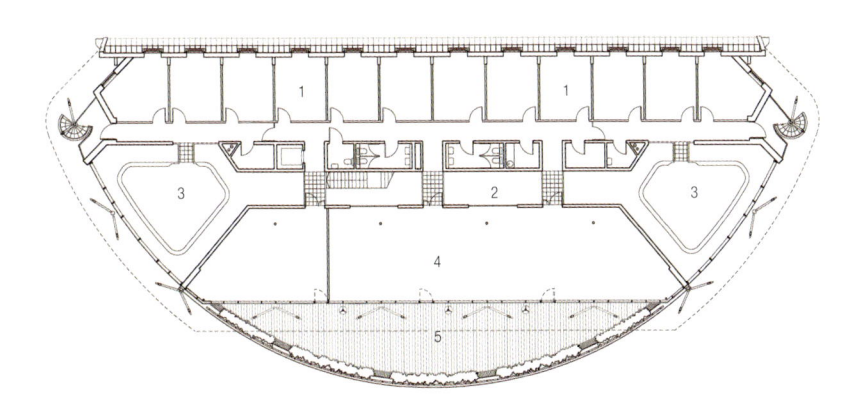

Pod structure
top left: Formwork layout
top right: Sterling board finish to soffit
middle: Elevation

bottom: First floor plan
1 Incubator units
2 Central space
3 Breakout space
4 Commercial space
5 Sun terrace

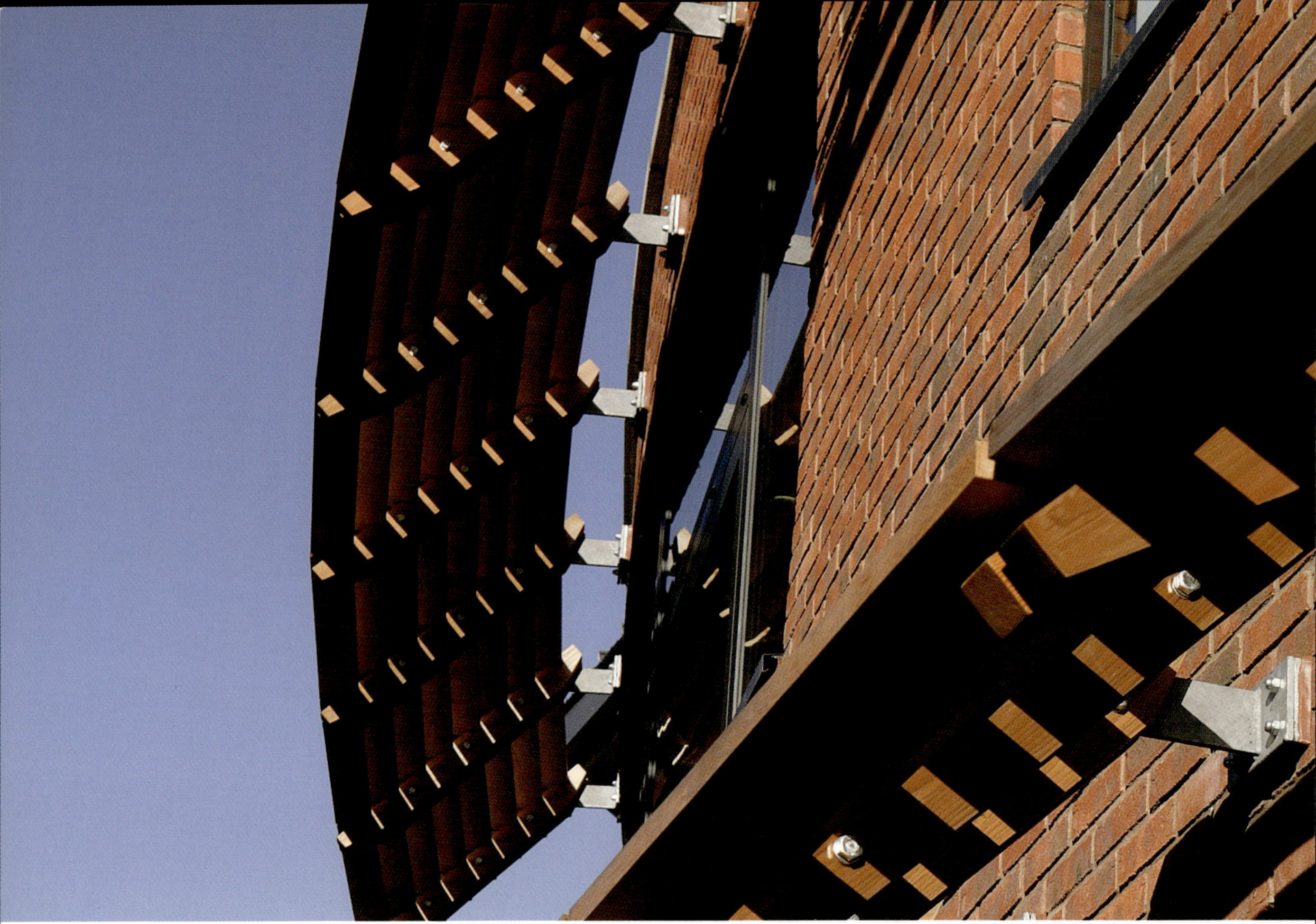

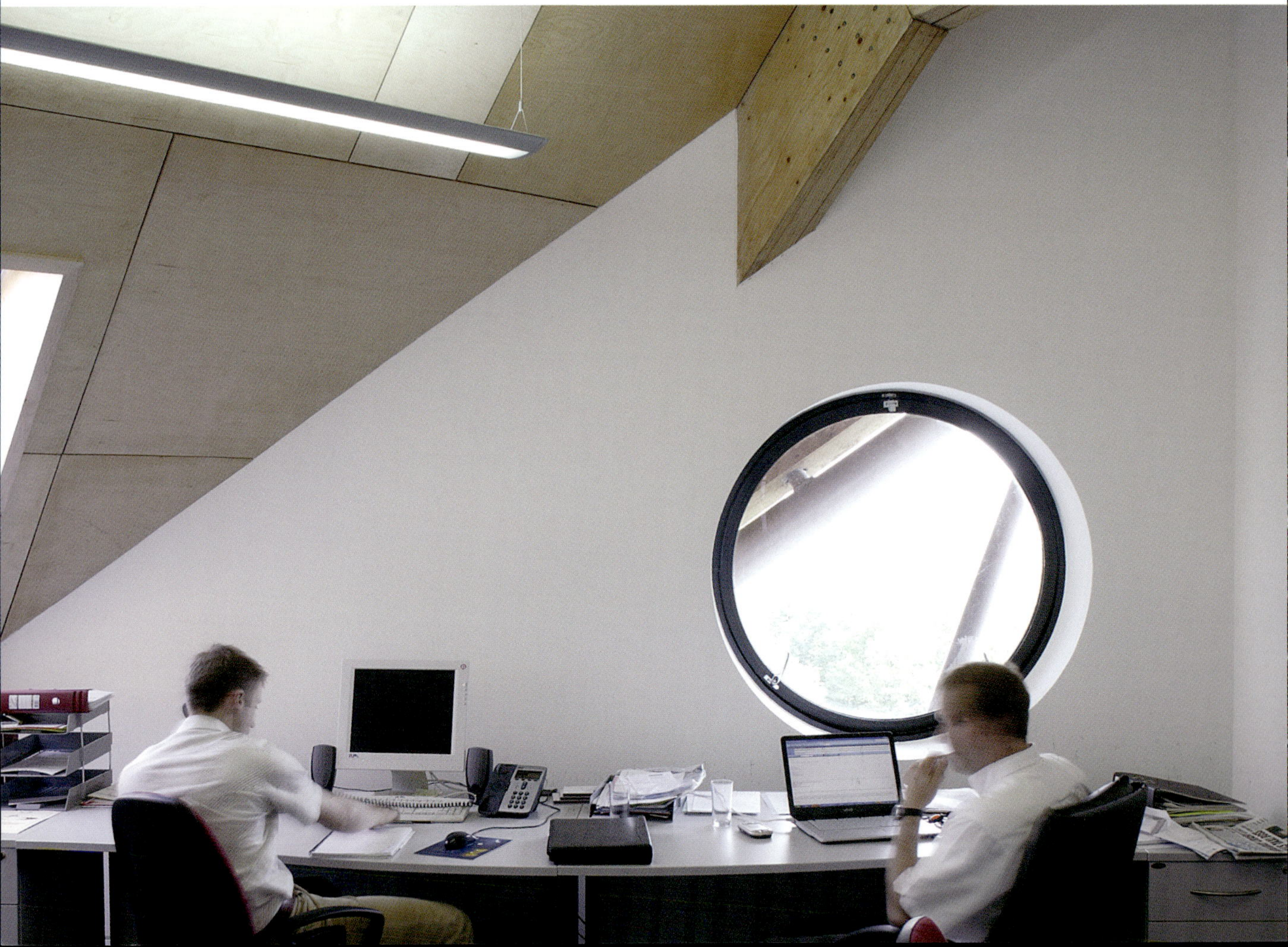

The contractor used a 'Sterling board' which is made from softwood strands, compressed and glued together with exterior grade, water-resistant resins. It is easily recognisable because of the irregular pattern of flattened, softwood strands that make up the surface. It can be primed and top coated with oil-based timber sealant. Water-based products should not be used as they may cause some swelling of the surface wafers. It gave a good finish.

CONSTRUCTION NOTES
Liam McGilloway, Danson Construction

We are based in Birmingham and carry out ground and civil engineering contracts. We also have a ready mixed operation based in West Bromwich called Any Time Concrete or ATC which supplies concrete to our own jobs and to other customers. We are trying to build our construction company with an in-house supply of materials. The ready mixed operation was started up 2 years ago.

Ashford was the main contractor and we were their concrete subcontractor for the main frame. We had not worked with Ashford before and tend to choose jobs that suit our operation. The information that appeared in the specification on concrete finishes was quite open-ended and relied on the contractor to produce a sample panel to set the standard. We priced from the bill of quantities and the set of drawings in the tender. There was mention of board-marked, plain smooth and glass reinforced plastic (GRP) formwork in the specification. The board-marked and GRP formwork was removed when the costs were reviewed. When we came to provide the sample panel we were not sure what the architect had in mind. We normally use Pourform 107 for a smooth finish so we proposed to use it. Before we poured the panel we had some discussion about the panel and sheet arrangement with the architect.

We were then asked by the architect to use Sterling board as the finish. We advised them that we would only get one use out of it because the water in the mix would make it swell and delaminate. Sterling board was not strong enough to resist the formwork pressures on its own; it had to have a backing ply. We recommended that the Sterling board was fixed to the Pourform for strength. We stripped away the Sterling board and disposed of it after each use and refixed new boards to the Pourform backing sheets. For every 60 whole sheets of ply supplied we cut at least 20 of them.

The concrete mix was selected using a crushed and graded limestone fine and coarse aggregate supplied by Enstone Quarry. We ensured that the material was from the same stockpile so that the colour would not vary and it was readily available. Panels were nailed from the form face to an agreed plan. There are a few areas where the nail has slightly split the timber; otherwise the surface finish was good.

The benefit of having our own concrete plant is continuity of supply and delivery to site when we wanted it. We asked SGB to design the formwork using their Logik 50 manhandled units, which we hired. We wanted to pour the walls to the underside of the floor slab, pour the slab and then cast the next lift of the wall on top of the floor slab, showing the construction joint lines on the corridor wall. The structural engineer had designed the rebar to do it this way. Aesthetically, however, it would not look good so we poured the walls 250 mm above slab level and cast pull-out bars to tie into the floor slab later. The joint lined up with the raised floor, so was not seen. Vertical stop ends were cast with grout checks and positioned at doorway openings.

For the wall lifts, an elephant trunk hose 4 m long was used at the end of the skip. Concrete was pumped into the skip and tremied down the rubber hose. We had to pour with the trunk fixed to the skip and moved it from one position to the next. We placed and compacted the concrete in 600 mm layers and used a high-frequency electrical internal vibrator. Besides being more efficient, it is much safer than a petrol-driven compactor, which has to be mounted on the scaffold. Two carpenters remained with the pour to check whether the tie bolts were loose and to watch for grout loss at corners and the kicker joint.

top: Timber solar shading
bottom: Incubation unit, second floor

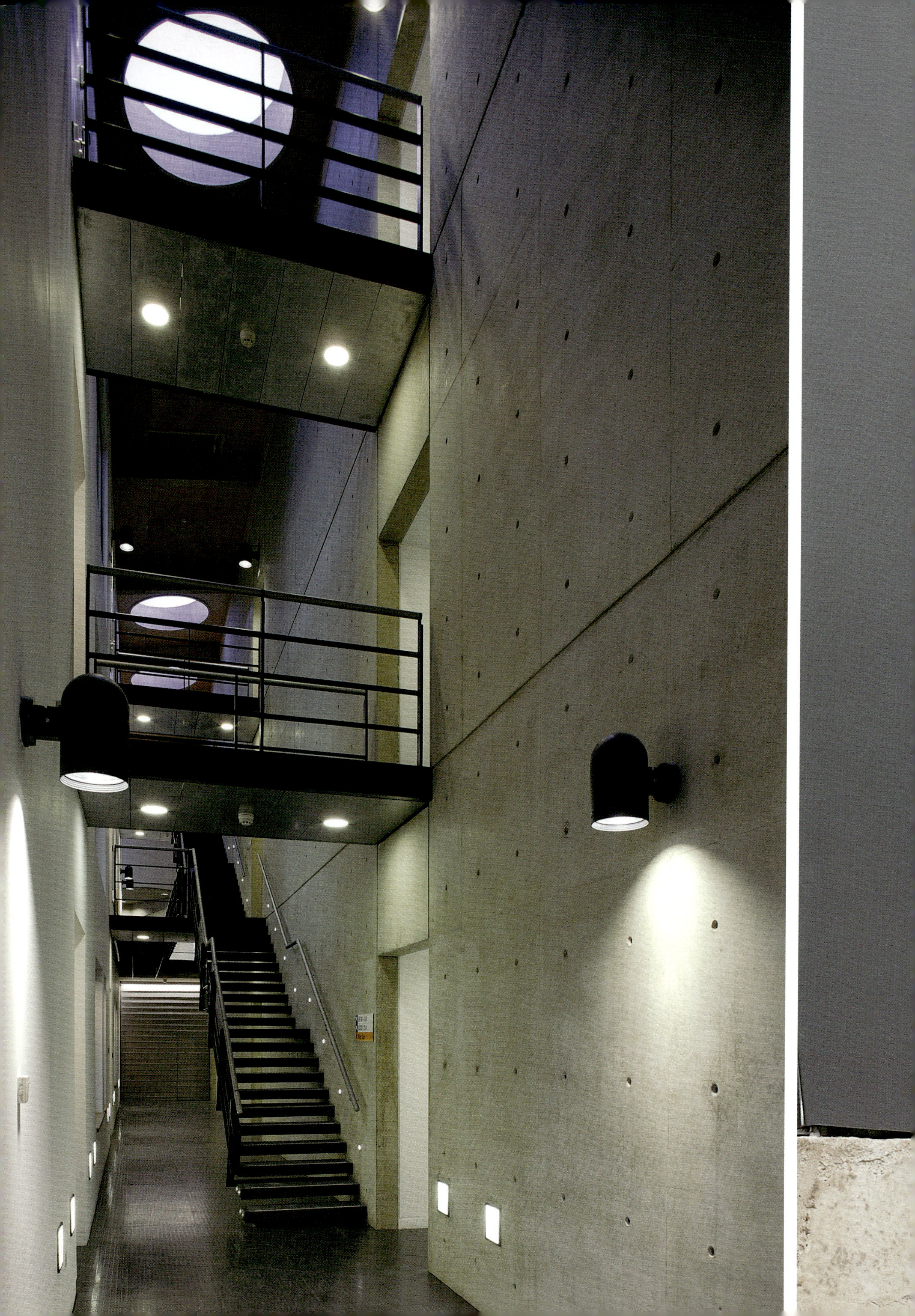

PROJECT TEAM

Client: University of Wolverhampton
Architect–engineer–QS: BDP Manchester
Main contractor: Ashfords
Concrete subcontractor: Danson Construction
Completed: October 2005
Project value: £3.85 million

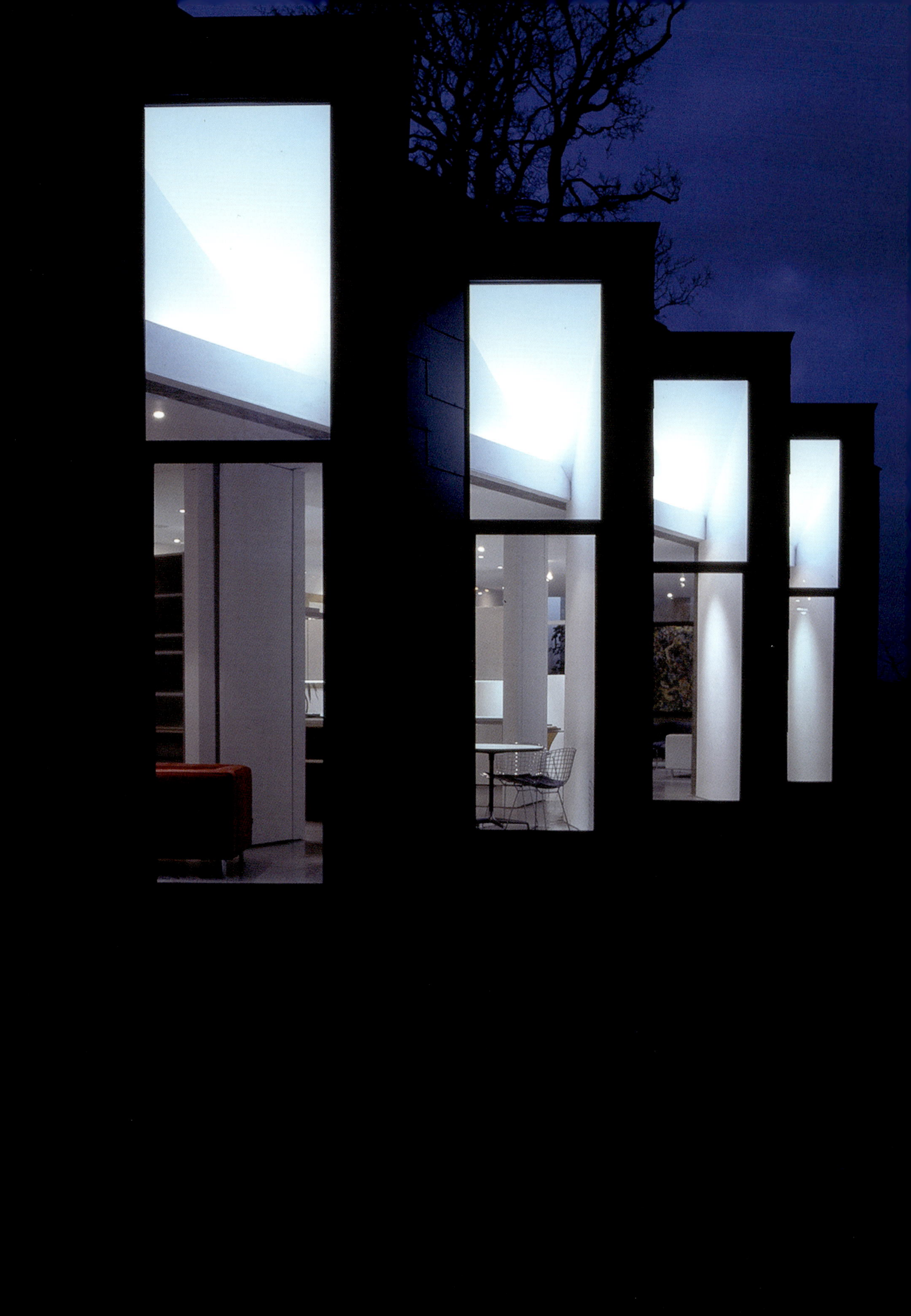

THE JONES HOUSE
RANDALSTOWN, ANTRIM

Architect: Alan Jones Architects

Location
The house is at 6 Portglenone Road in Randalstown, Antrim. Randalstown is a small town near the edge of Lough Neagh and for 150 years was a centre for the Ulster textile industry.

Client statement
Laura Jones

We have created a new house for ourselves in a small town 20 miles north of Belfast. I am a dentist, my husband an architect and together with our two boys, Isaac and Gideon, we have begun to enjoy these new surroundings.

The 0.8 acre empty site, with outline planning permission for a single house, was advertised in the local paper. We were interested in the site because of its location. We had been used to urban living with our 7 years in Camden, London and this site, next to a series of public buildings, is as urban as a small Ulster town can be. It is close to a church, a school, the shops and the local rugby club and scout hall. Many in Ulster aspire to country living, but for us the site presented a great opportunity for town living with all the advantages that brings. We still have our countryside, the adjacent stream, the extensive mature boundary planting. The views towards the Antrim Hills and Slemish Mountain ensure that nature is here also. So our new house had to acknowledge both – the urban and the natural.

When viewed from the road and the churchyard, we wanted a house that fitted in with the surroundings. The design has been so successful in this regard that the locals could not decide if it was a hotel, a church or a church hall, but didn't think of it as a house.

Inside our house, the light and views and the textured concrete are in contrast to the dark, smooth exterior. Each morning, as soon as the sun rises over the churchyard wall, our house has sunshine inside, which stays with us until the sun sets. By the tall windows facing south is where we sit to read, have breakfast or just daydream, looking out over the stream, churchyard and sky. I take pleasure in the big open ground floor space, which suits our frequent parties, and how the hidden sliding and hinged doors can be closed down to make it into a series of small, more private rooms. We have open plan and we have privacy when we want it.

The house has created much discussion locally, with strangers asking us how we are enjoying it. I am told that schoolchildren sit in buses deciding which of them like it and which do not. Our visitor's book has such comments as 'your new house appears to have always been here' and 'it is the right answer'. I particularly remember and agree with one comment that the design is 'strangely familiar' – in that different people see different things in the design, but cannot quite put their finger on what it is. I suppose that is what architecture is – different things to different people, evoking memories and past images.

ARCHITECTURAL INSITU CONCRETE - CASE STUDIES

Architect statement
Alan Jones

The old Congregational Presbyterian church, a well-known local landmark, dominates the context of the site. One of only two oval churches in Ireland, this Grade A listed church is an architectural jewel, along with other civic buildings and a British Legion War Memorial Garden to the front, all forming an immediate and challenging context for this new house. The conservation boundary runs along the stream, along the side of the site and next to the listed church. The listed, oval church is built of black stone with white window frames and is floodlit at night. The immediate aim was to create a dwelling that was visually quiet and recessive, dark with no light-coloured components that would distract a viewer from the listed church. The listed building is impressive at night, with a lantern-type image when the internal lights are on. Observing this encouraged us to ensure that our design was the opposite – mute, dark and visually quiet with only a single tall gable window illuminated at night.

The new house presents a gable to the public road, in the same way as the other buildings next to the listed church. This is partially a response to the narrow nature of the site and partially a response to its context. The tall, thin window on the public gable is of a civic scale; an architectural move enforces the ambiguous nature of the design – is it private or public, house or hall?

From the churchyard to the side and rear of listed church, the house presents tall one-sided bay windows. When leaving the listed church from the adjacent side door, the blank sides of these bay windows are visible, but not the window and view into the house. Churchgoers are not distracted by views into a domestic property and only see a dark, sombre form with no illumination.

The gradual slope of the site made it seem obvious to lift the house slightly off the site at the front, to slip a basement garage and utility room underneath and line through the ground-floor level with the rear, external south-facing deck. It was also fortuitous

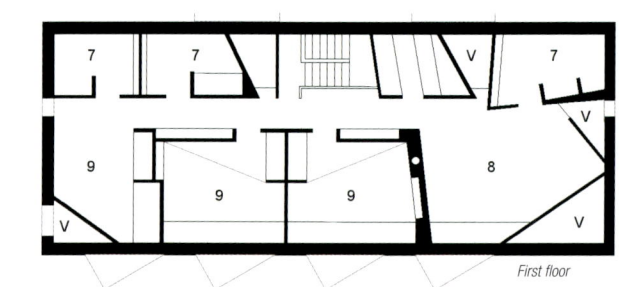

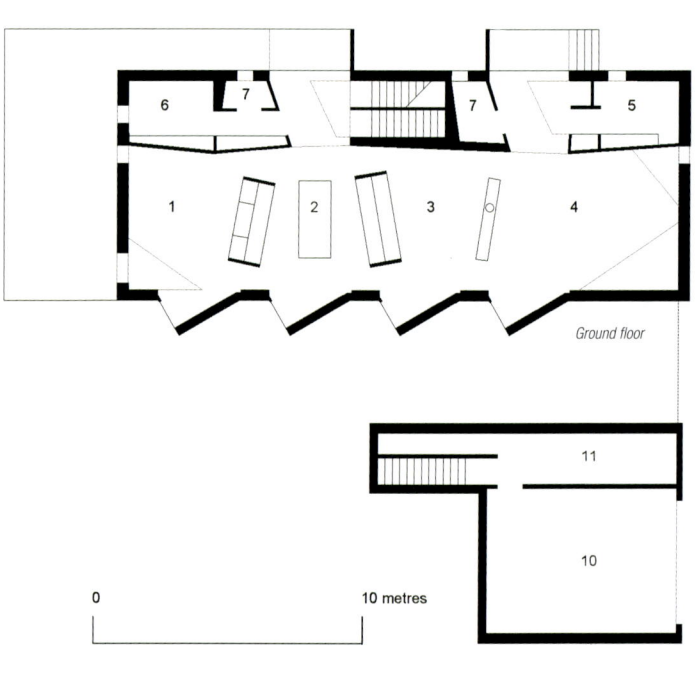

top: Rear elevation
middle/bottom: Floor plans

1 living room
2 kitchen
3 dining room
4 reception room
5 study
6 den
7 wc/shower/bath
8 master bedroom
9 bedroom
10 garage
11 plant/utility
v void

200

that the south side was away from the public road, it is only when one moves around to the private rear garden and views the south elevation that the house begins to take on a domestic scale. This is the view that the only neighbouring house has of our new dwelling. A large timber deck is placed next to the rear elevation, taking advantage of the sun from early morning through to evening – and is set at internal floor level to encourage informal garden and outside living in the summer months.

Internal arrangement

The first entrance door one approaches has an external screen that masks the second and more private entrance further along the side elevation. Immediately inside the dwelling, small spaces and the main stair are placed along the side of the house through which one enters. The front entrance leads to a study and main reception room and dining room. The stairs from the basement and to the first-floor bedroom accommodation are off the rear, more private, second entrance hall.

With low ceiling heights in proportion to the plan size of the rooms, this gave a clear zone under the first floor structure for servicing the bathroom and en-suites above. Each of the taller, bigger spaces on the ground floor has one of the side bay windows, offering views and light. Throughout the house there are voids to the roof, acknowledging that trees and banks to the sides of the site restrict the extent of natural light at certain times – and the voids also allow the extent of the concrete perimeter walls to be understood and appreciated.

Like the church next door, the interior cannot be read from the exterior. The church reveals a wonderful ox-bowl balcony and the house reveals brightly lit, textured concrete perimeter walls and plain plastered walls forming low, medium and tall spaces.

The front room of the house feels like the scale of a hotel lobby, continuing the ambiguity from outside – of public scale but for a private dwelling – the ideal space to receive guests, and a visiting minister.

The spaces within the house feel dramatic and spatial, yet relaxed and informal. Looking at the plan it appears as if the house has been ruffled, shaken, to remove any staid formalism, creating a spatial dynamism that continues to reveal itself as you walk through and sit to enjoy the interior. Above the basement garage and utility, the ground floor has the small, intimate spaces placed to

top: Graveyard elevation

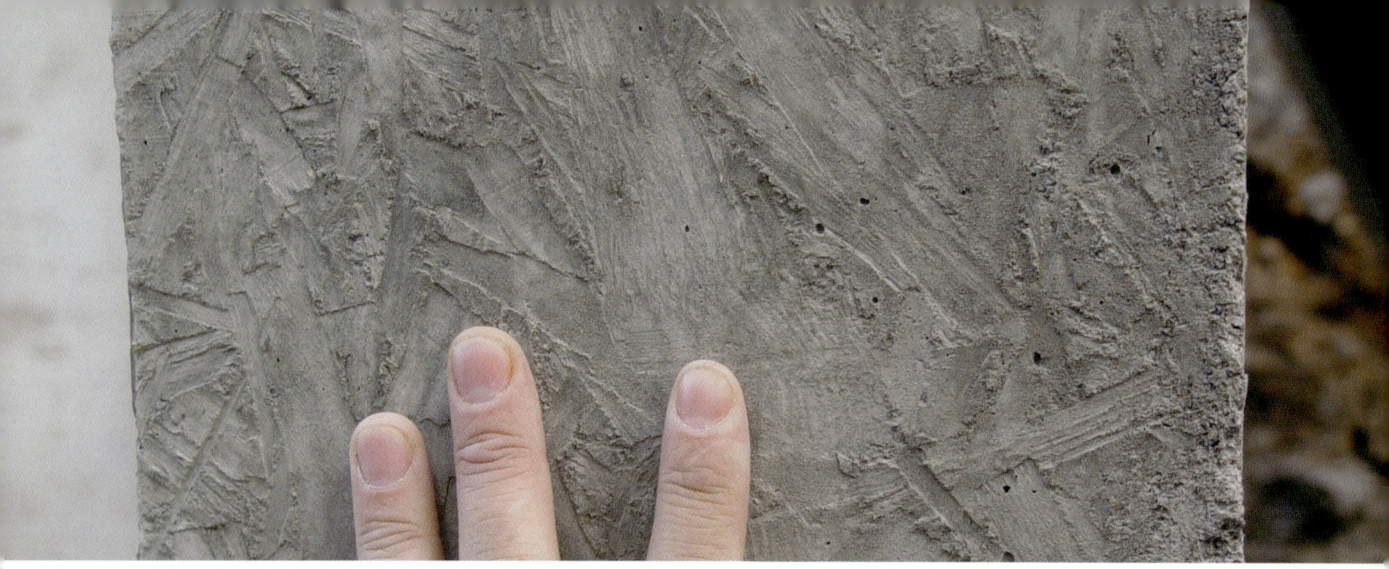

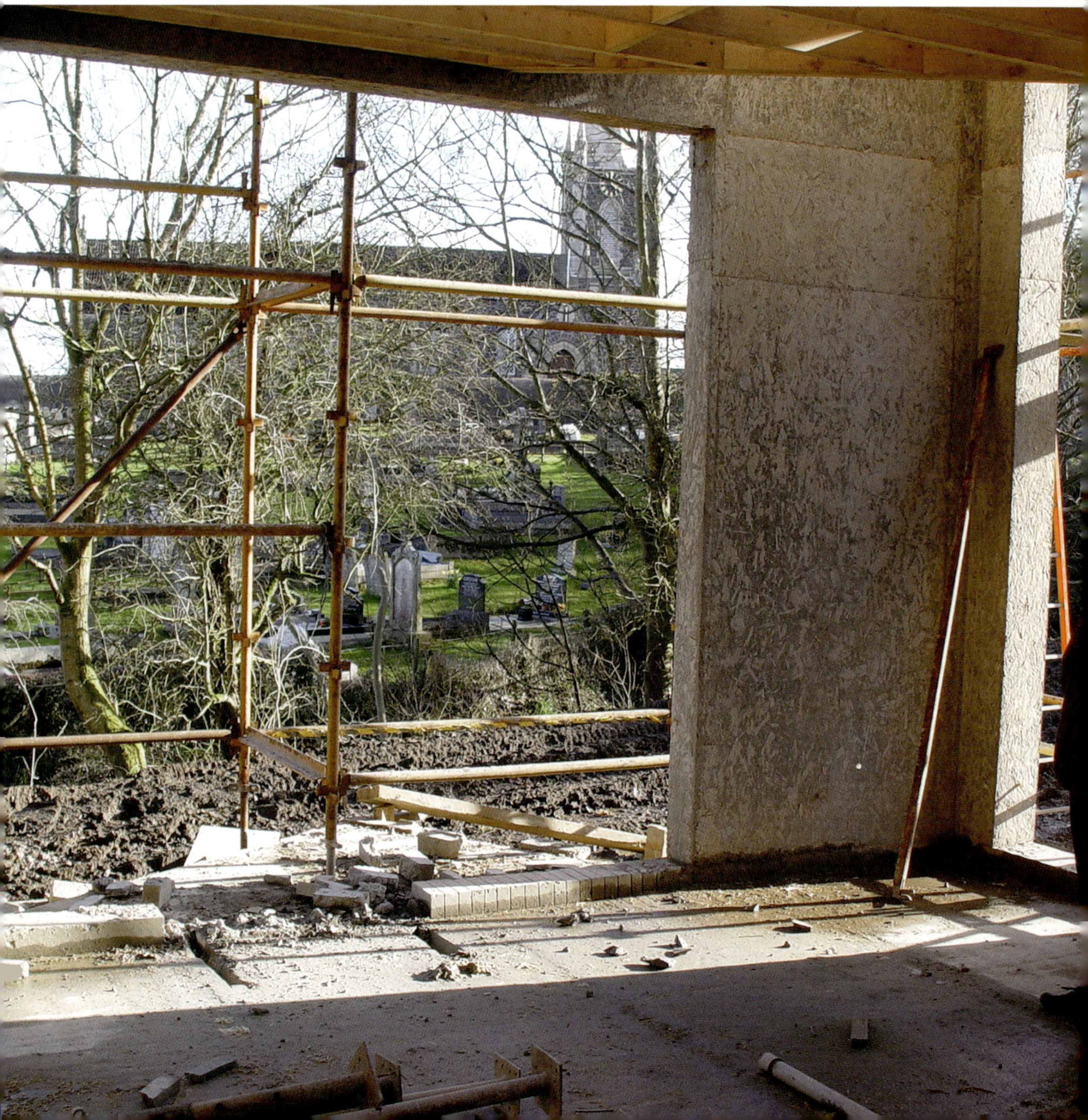

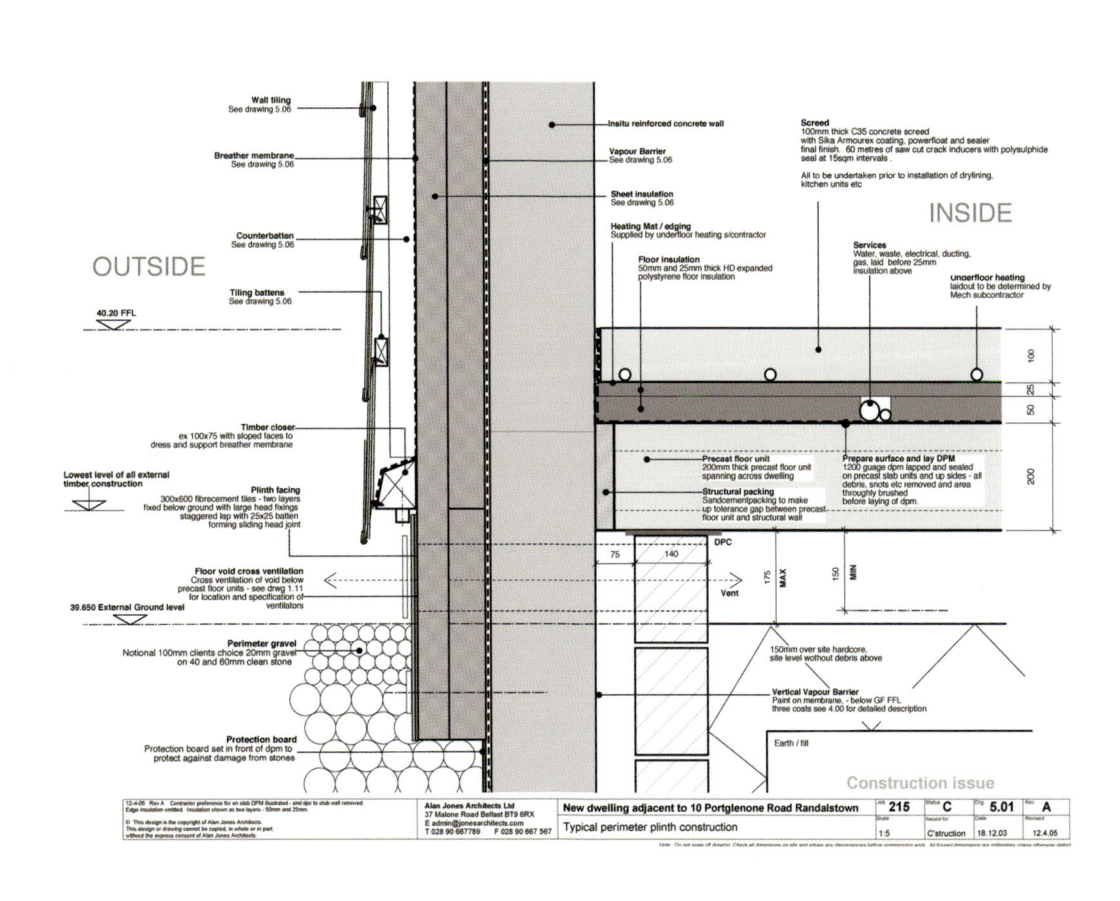

top left: OSB finish
top: OSB ply
left: Construction progress
above: Construction detail

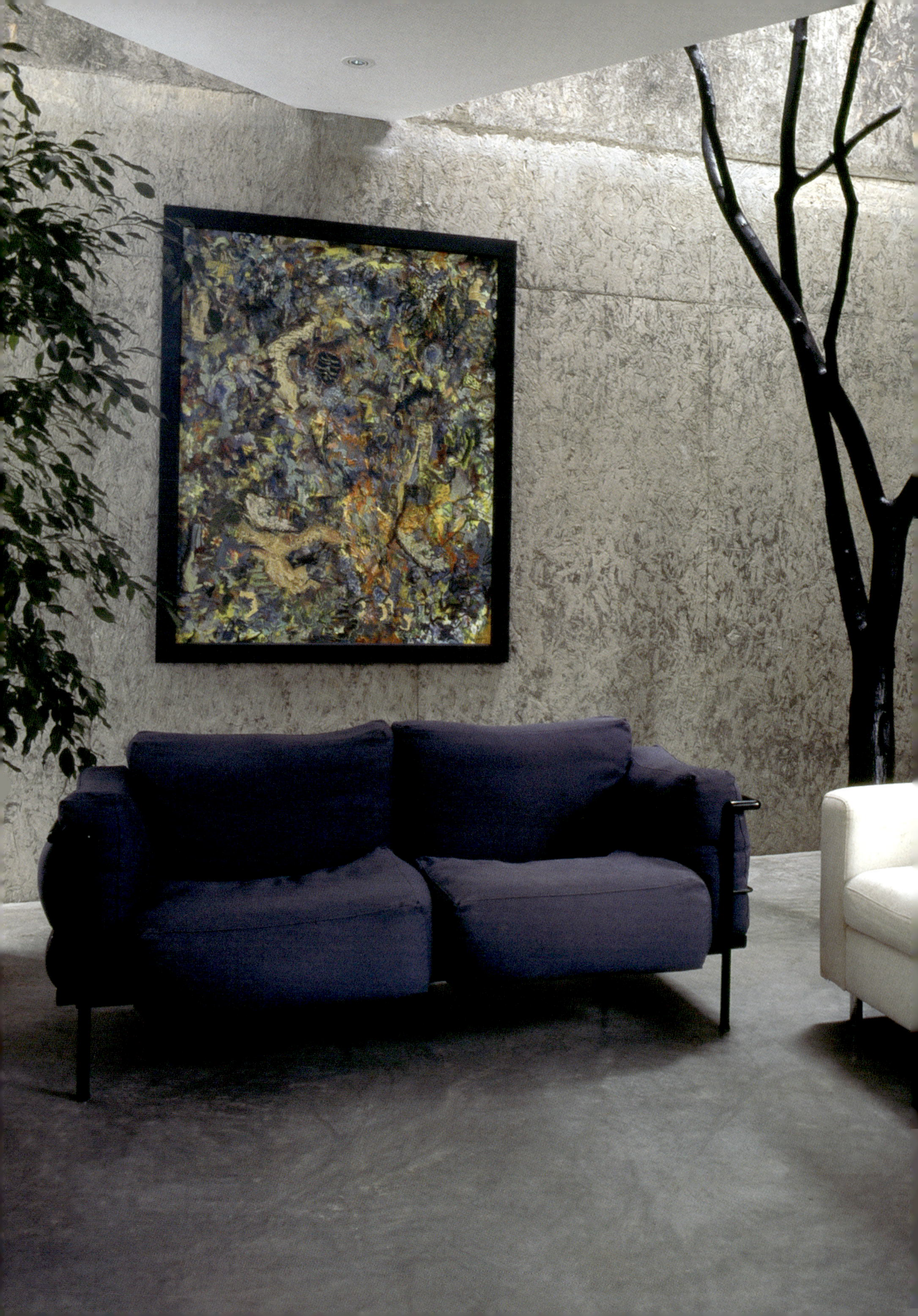

one side with the main spaces set together forming one long space over 65 ft long and 18 ft wide – an ideal party space. As necessary, doors appear from walls to segregate spaces but glass over panels continue the visual connection from gable to gable.

CONSTRUCTION NOTES

Having designed and procured a number of private one-off houses we have become increasingly frustrated with the common mix of masonry and steel structures that subsequently require plastering – with the mess, cost and subsequent cracking that can occur. Looking to the Irish countryside and the remnants of old buildings and structures, a permanent perimeter structure of insitu concrete seemed the appropriate solution to provide the internal structural stability and flexibility with floors spanning wall to wall. Concrete provided an extensive exposed thermal mass, which was in keeping with our energy strategy for the building. Externally insulating the form and covering it in a rainscreen construction allowed a relaxed approach to detailing the building.

The concrete perimeter walls were formed with cheap oriented strand board (OSB) boarding, the same boarding used as hoardings, but here the impression left is soft and semi-natural, linking interior to the natural exterior beyond. OSB is a wood-based panel comprised of wood strands from forest thinnings combined with liquid resins and wax binder. The sheets were 2,440 mm × 1,220 mm × 18 mm thick. They were used twice and afterwards the sheets were used to form hoardings for security and protection during the later stages of the project.

Forming a skeletal shell, the exposed concrete walls and the polished concrete floor create an immense thermal store, which, once warmed by the geothermal heating system give a stable thermal environment, warm in winter and cool in summer.

The project was tendered in the traditional way and Barnish Construction Ltd was appointed as main contractor. The contractor encouraged subcontracts to be placed directly with the client and consequently glazing, mechanical and electrical services, kitchen, built-in furniture, finishes and sanitary ware were omitted and direct orders placed. We were familiar with management contracting on larger projects. A close relationship between architect, subcontractor and main contractor meant that there were very few technical queries and additional costs.

Living room

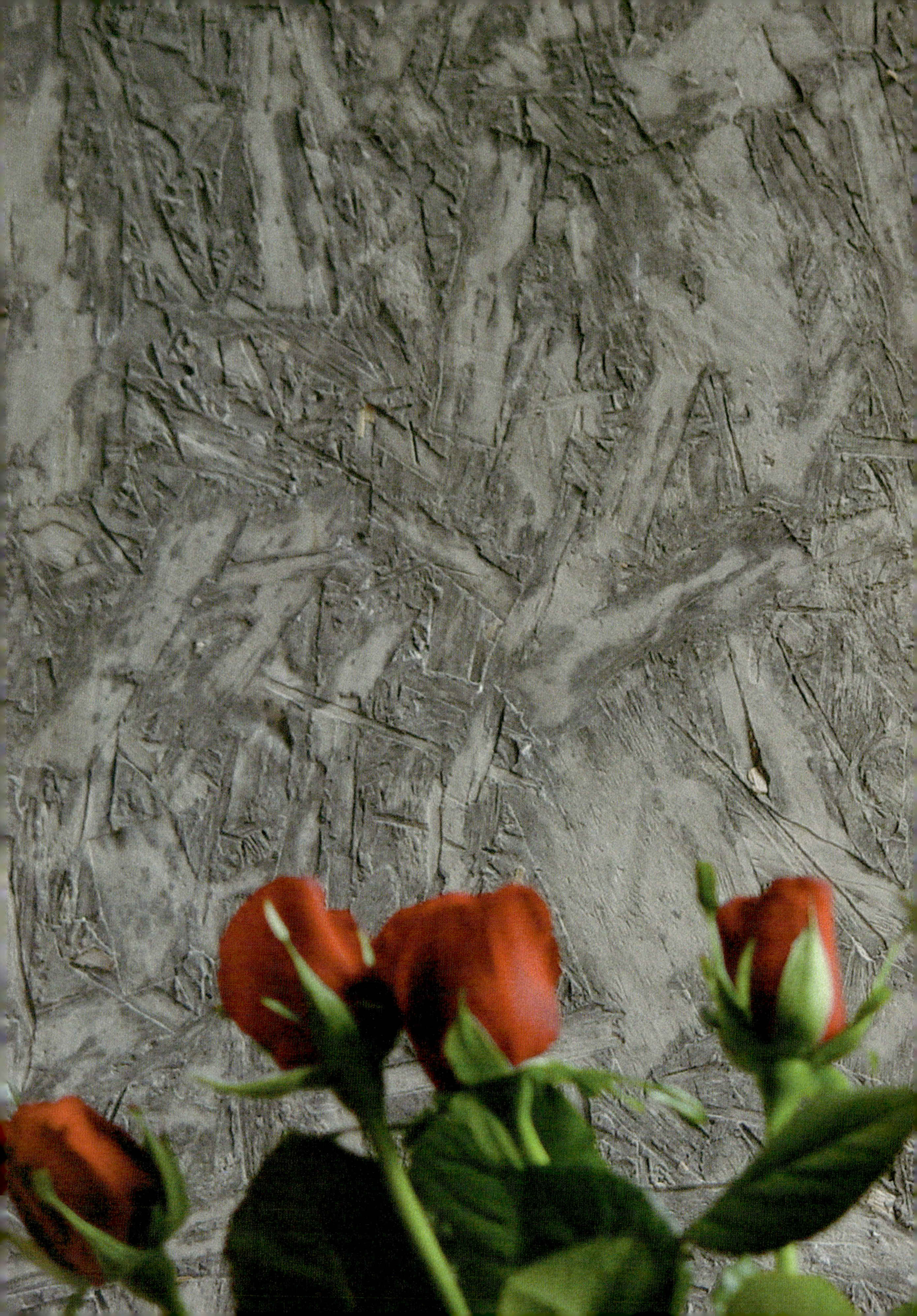

PROJECT TEAM

Architect/designer: Alan Jones Architects
Structural engineering: Doran Consulting
Cost consultant: WH McEvoy
Contractor: Barnish Construction Ltd
Completed: October 2005
Project value: £225,000

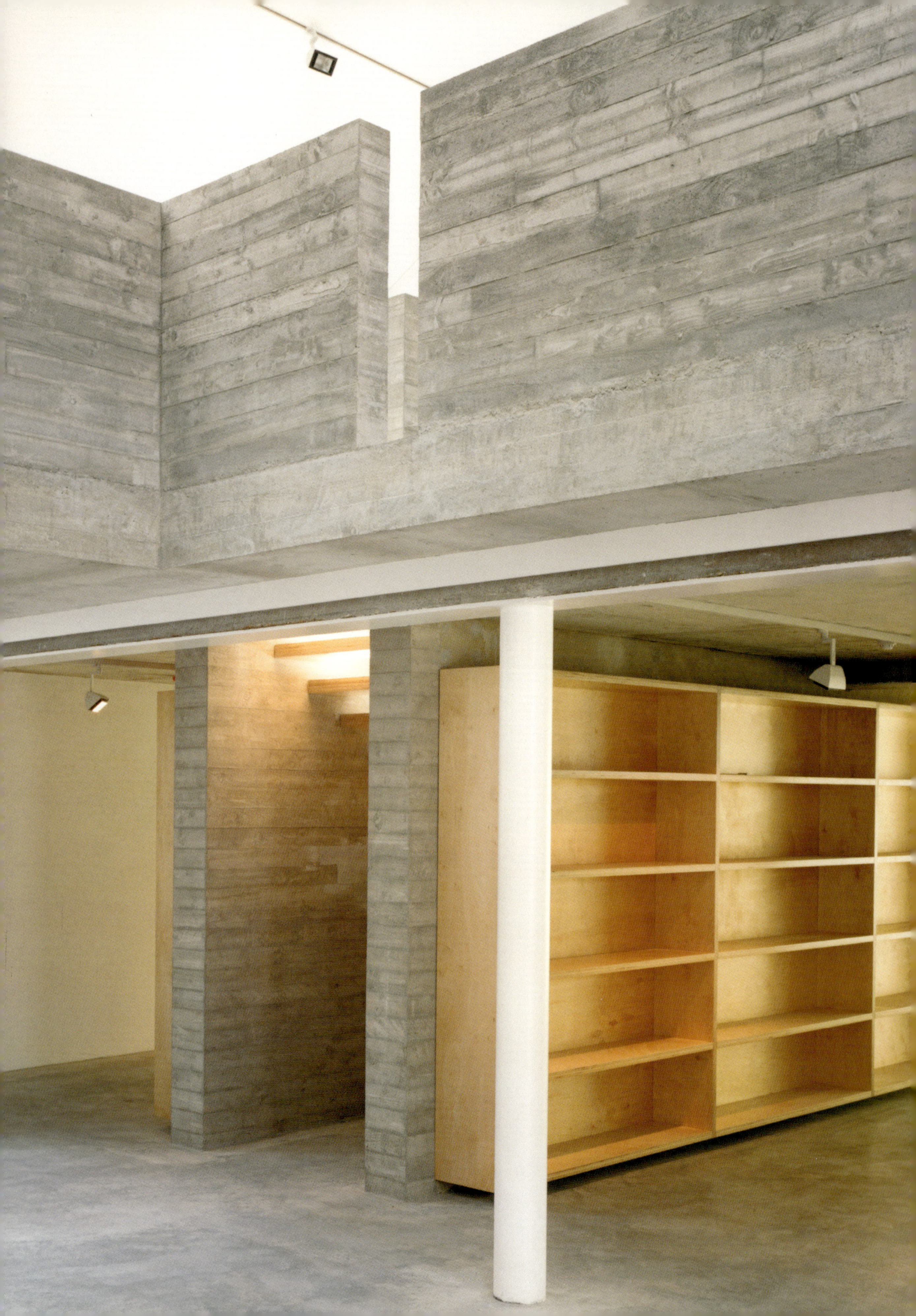

SPEDANT WORKS
PARK ROYAL

Architect: Greenway and Lee Architects

Location

Spedant Works is on an industrial estate road off Park Royal Road on the B4492 in North Acton. It's a 15-minute walk from North Acton tube station on the Central line; a walk that overlooks the tube lines and the graveyards of North Acton cemetery.

Introduction

An existing artist's studio has been transformed from a cramped and dusty warehouse into a sleek modern studio with open plan working spaces on two levels, a vast display area and room for storage. It's located in Park Royal among a cluster of industrial buildings and sheds of various shapes and sizes without much subtlety about them. They look hard and grimy, the roads and the pavement are one grey smear of tarmac coated with brake dust and chewing gum; there are cars, vans and lorries parked everywhere while the occasional waft of burning bacon, stale fat and diesel perfumes the air. Then you see Spedant Works – it's smart, it's orderly and it's architecture – and it's a splash of cologne to the nostrils at first sight.

Architecture discussion
Nick Lee

The project was to create a new workspace for stained-glass artist Brian Clarke, incorporating an office, a design studio and a display area for making presentations to clients. The initial brief was to keep the existing studio and refurbish it, but after a fuller appraisal we convinced him that it would be better to knock down the studio and rebuild a new one more suited to his current and future needs. The large space he now occupies was formerly being used for storage only, as asbestos dust had been found in the roof panels. After removing the roof sheets the new roof was raised by 1 m to create the additional working space required. The building occupies a footprint of 4,700 ft^2 or 11 × 20 m and to the ridge has a height of 9 m.

During the course of the project the existing office building at one end was gutted out and refurbished, providing space for the admin staff – offices, meeting rooms, toilets and a shower room. The client and staff moved out during the period of construction to a rented space in central London.

We introduced natural light boxes through the roof and side windows above on the first floor and fluorescent lighting and a light box on the ground floor. There was a security issue on this industrial estate, so whatever glazing we used had to be toughened and windows had to be slotted. It was quite clear that the ground floor was best for the presentations, so we installed a giant light box that illuminates the entire end wall for displaying the stained glass pieces. The first floor naturally became the studio because it has a lot of natural light. We created a double height void within the studio with a hoist, which acts as the loading bay. This set out the vertical arrangement of the space.

As it is a working studio we decided that the floors should be power floated concrete with a hard-wearing industrial finish. It would take a lot of wear and tear and could be cleaned quite easily. We then introduced a contrasting palette of materials, with

Front elevation

smooth floors, board-marked concrete walls and painted white plasterboard. The board-marked idea came from our many visits to Switzerland where my family live and where concrete is used to great effect, and from Denys Lasduns architecture which I studied as an undergraduate.

When we proposed our ideas to the client he was very excited. On the main elevation we designed a protruding window box at first floor level, which breaks out from the façade line. We cast it in the same board-marked concrete to give a hint at what's happening inside the building. By introducing full height glass panels to the sides we created a projecting bay in which to sit and reflect and which also allows the artist to view the entrance from the first floor.

A steel frame of circular columns and beams was chosen for the first floor for speed and accessibility. The cast insitu concrete floor slab that sits on the frame has exposed soffits with a birch ply finish and mitred returns for the slab edge corners. The top surface has a power trowelled finish. Lighting is surface mounted, white linear track for flexible, movable lighting spots.

We specified solid pieces of 18 mm thick Douglas fir slats for the board marking from a timber merchant in South London. The width was chosen so that it worked to the height of the wall without any half cuts. Douglas fir has knots and a graininess to it that make a wonderful imprint on the concrete surface; the effect is sublime. We specified a high-quality birch ply for the soffits which we recycled for back shuttering in forming the board-marked balustrade walls and the external wall on the first floor. We also used the panels for making the underside of the tabletops and worktops and for backing to the birch ply staircase. The Douglas fir slats were salvaged and used for garden decking, so there was no wastage.

Materials were kept to a basic palette of concrete, painted plasterboard and birch ply. The ceilings, brickwork and the blockwork walls were rendered and painted white, as was the exposed steelwork to the first floor. Furniture units – shelving, worktops, benches and storage – were made from birch ply. The staircase design idea was to have solid board-marked concrete walls and between them lightweight birchwood stair treads and no riser so you can see through them. An adjoining office to the first floor is separated by the loading bay so the client wanted a bridge link at first floor level across the loading bay and into the admin side. The bridge was created in line with the staircase walls and we

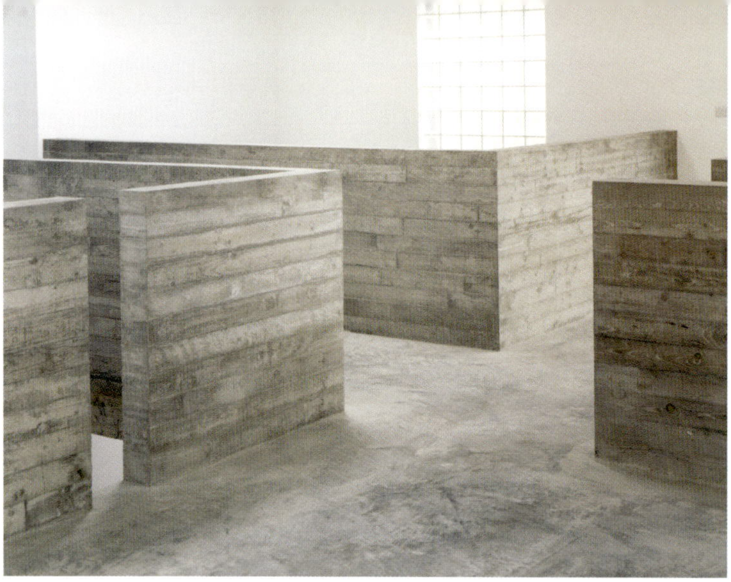
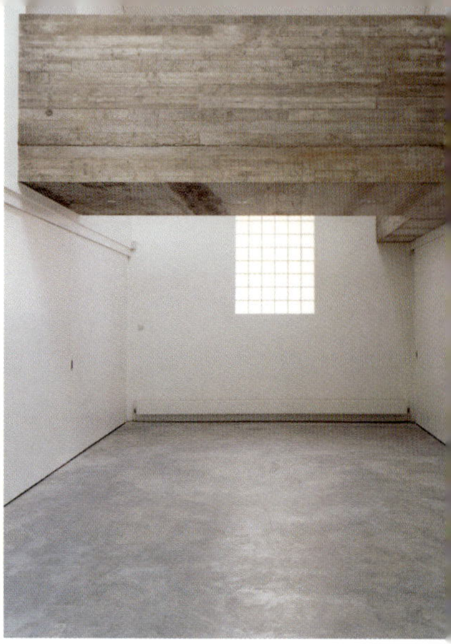
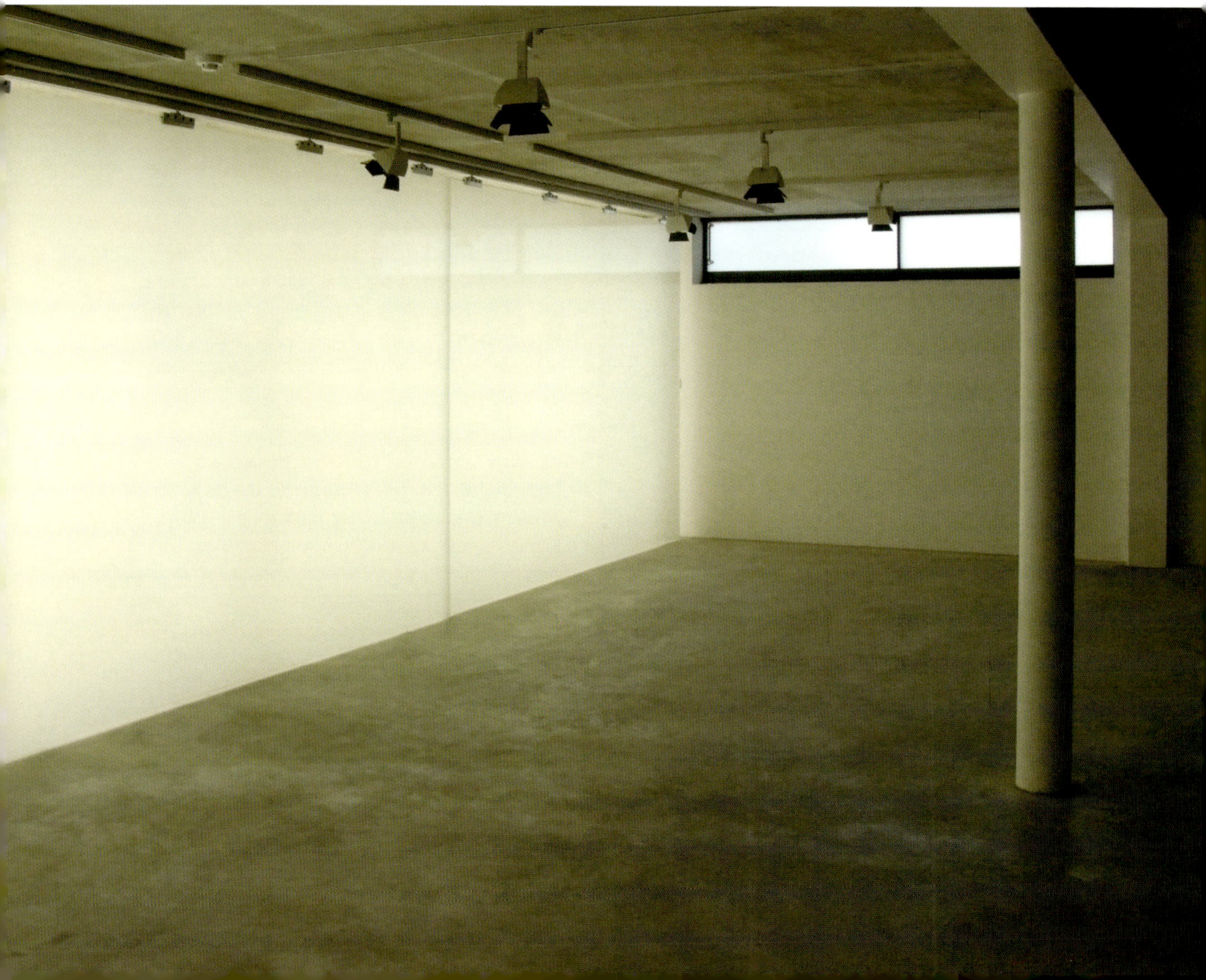

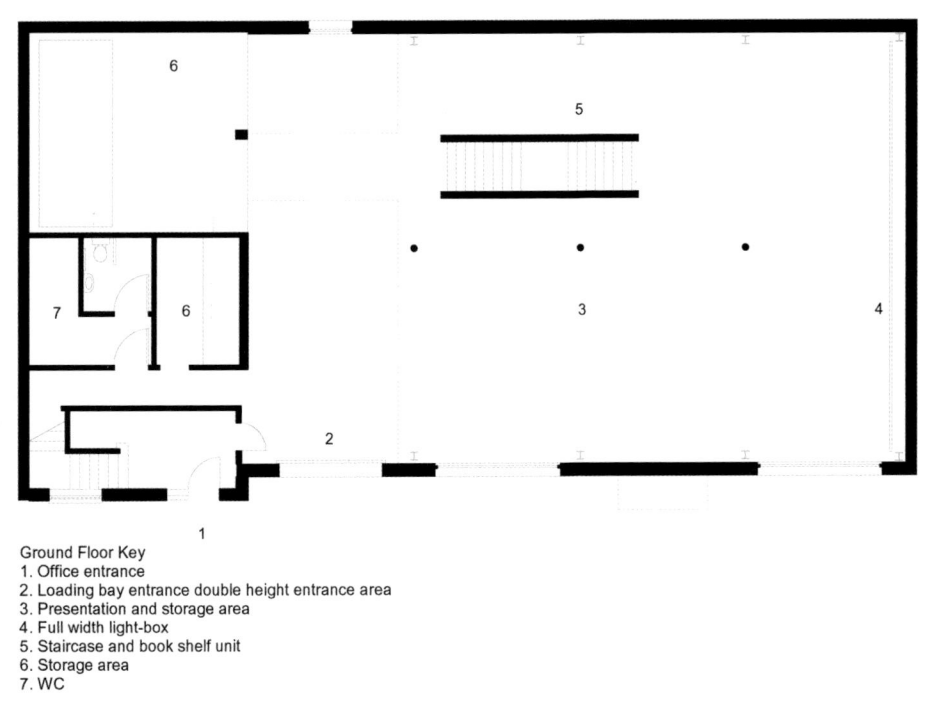

Ground Floor Key
1. Office entrance
2. Loading bay entrance double height entrance area
3. Presentation and storage area
4. Full width light-box
5. Staircase and book shelf unit
6. Storage area
7. WC

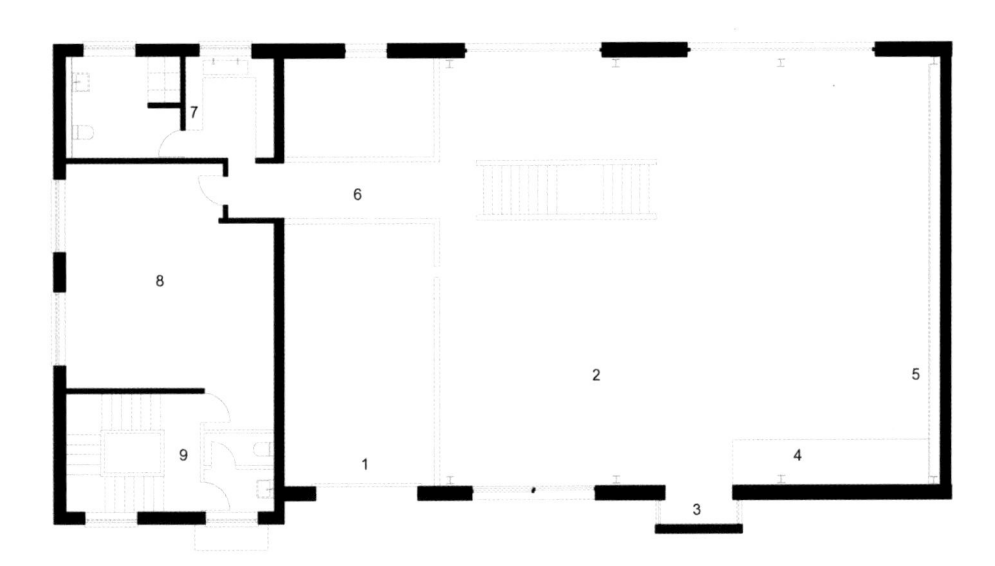

First Floor Key
1. Loading bay double height entrance area
2. Studio space
3. Bay window
4. Studio desk
5. Full width display wall
6. Concrete bridge connecting studio to office
7. Bathroom and wet room
8. Office space
9. Office circulation

top left: Mezzanine floor and board marked walls
top right: Bridge link
left: The light wall

Mezzanine floor and bay window

felt that the balustrade walls to the bridge link should also be cast in board-marked concrete, forming a cruciform shape at the top of the stairs.

On the ground floor adjoining the loading bay there is a floating partition which is closed off to keep the work area separate from the loading dock and at other times it is open for material handling. Part of the sliding partition also hides a storage area behind the staircase so that it is not visible when you come in via the main entrance. The sliding panels are 3.5 m high and are all painted white for use in presentations.

Because the client needs light to present his full-size stained glass panels to his clients, we created a giant light box that is 2.6 m high by 11 m wide along the far end of the building. This was contained behind three pieces of translucent white sliding glass. The lights are a series of Philips fluorescent tubes that are on two settings. In front of the light box is a series of brackets and chains mounted into the ceiling for the client to move the stained glass panels to check their design and quality. On the first floor there is an 11 m long by 1 m high magnetic pin-board for the client to use for displaying sketches, monitoring work in progress, mounting progress sheets, etc.

Underfloor trunk heating on the first floor is recessed in the floor thickness with surface-mounted grills to negate the need for radiators. Heat is pumped through the conduits by hot air. On the flank walls on the ground floor we have mounted some low rise conventional radiators.

CONSTRUCTION NOTES

We had a slight problem during construction of the stair wall, which was cast without the proper support system. This resulted in the wall being out of plumb and bulging, although the concrete itself was beautiful and showed no grout leakages. It was so out of alignment that the contractor had to scabble the surface to remove up to 50 mm of concrete in places to ensure that we could get the staircase in. It's a pity because the contractor spent so much time preparing the formwork and joinery, placing the board-mark planks and back fixing them. It was a very costly exercise to recast a new face using a flowing grout proposed by David Bennett which resulted in the same board-marked surface finish and maintained

top: Top of staircase
bottom: Birch ply tread detail

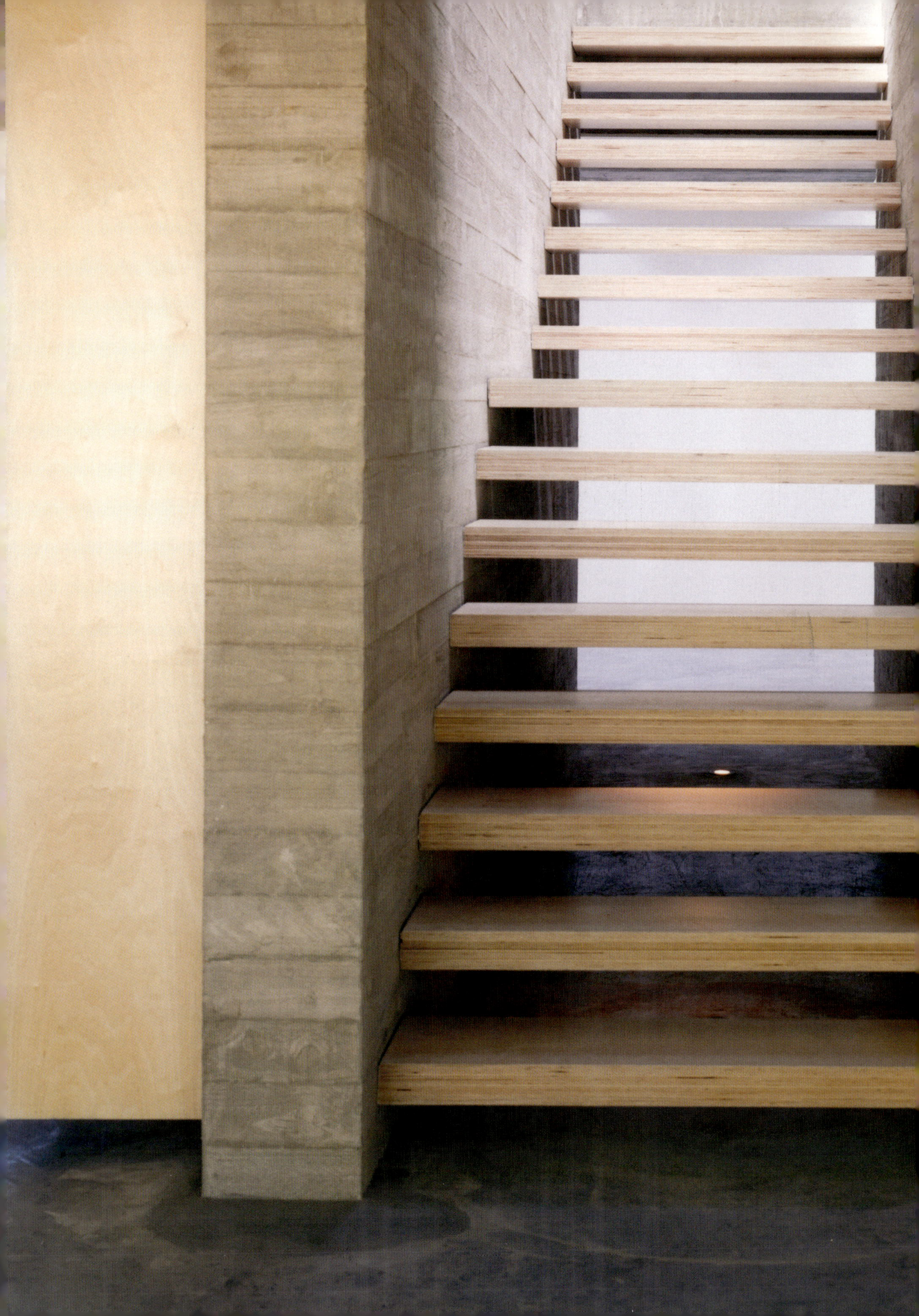

the wall in the correct alignment. This time the formwork was supported by metal props, strong backs and walings and there was no movement. The balustrades worked very well using the proprietary formwork system supplied by A-Plant.

The contractor had to do the same grouted skin to the board-marked wall of the external elevation. Minor grout marks and runs were cleaned up using hydrochloric acid cleaner. The difference in the colour of the grouted concrete section and the upper section of the stair wall is being harmonised by coating the surface with a proprietary cement wash manufactured by Keim Mineral Paints.

The wall surfaces have been coated with a clear siloxane solution to make the concrete waterproof and stain resistant.

PROJECT TEAM

Architect: Greenway and Lee Architects
Client: Transilluminate Ltd
Contractor: Zenon Builders
Structural engineer: Bolton Priestley
Completed: June 2006
Project value: undisclosed

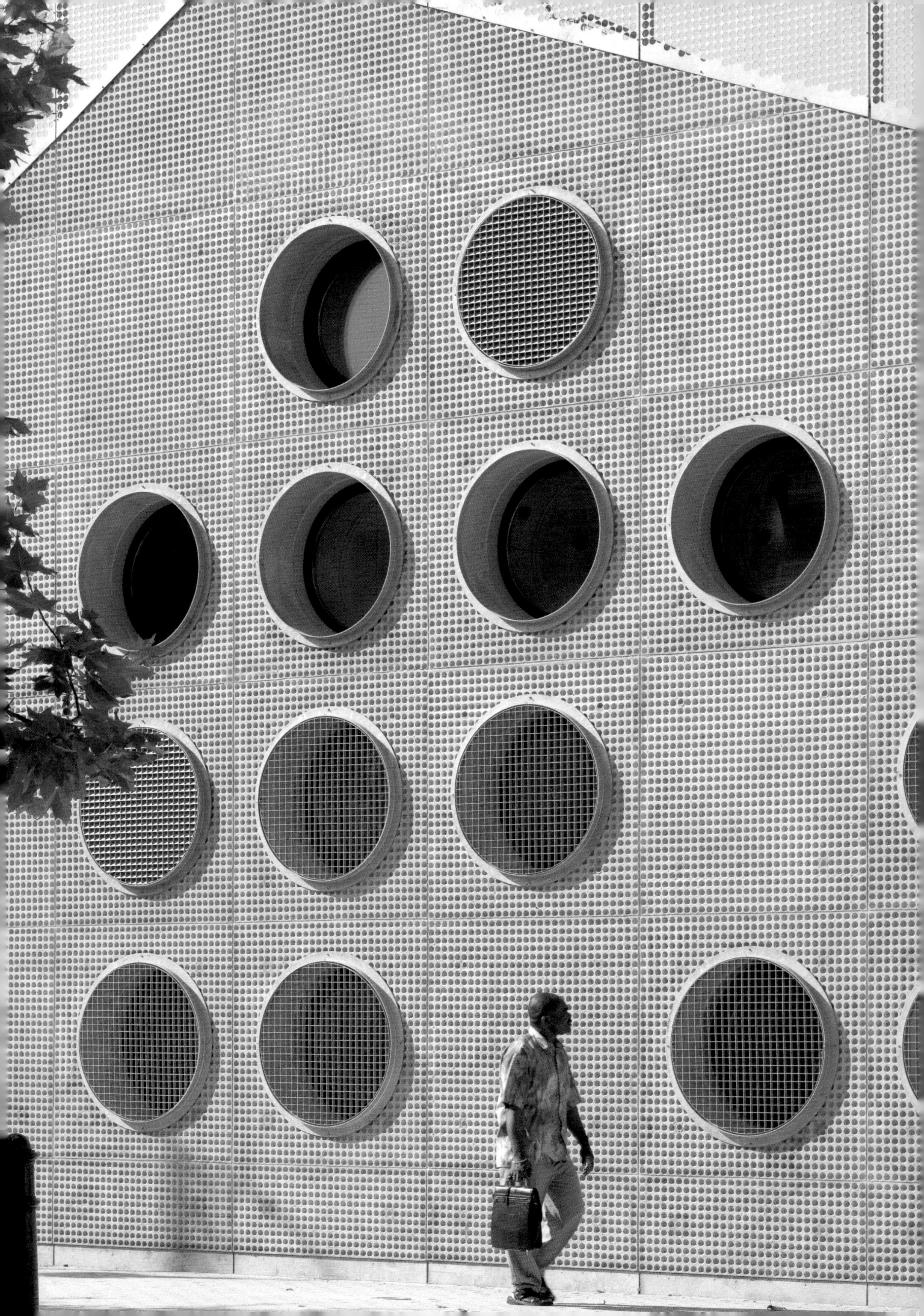

CENTRAL VENTURE PARK
COMMERCIAL WAY, PECKHAM

Architect: Eger Architects

Location

The park is situated on Commercial Way in Peckham, South London and is a good 15-minute walk from Peckham Rye train station. When the new tramway is built, there will be a stop on Commercial Way. A few minutes' walk away is Will Alsop's Peckham Library.

Introduction

The park is part of Southwark Council's Peckham Partnership urban redevelopment programme. Its concept of a green oasis for young people has evolved during an extensive process of local consultation.

The topography of the park is derived from the way dunes are formed around an oasis, encouraging imaginative play and providing natural separations between the different activity areas.

The park is extensively planted, bringing a new greenness to the area. In contrast, the perimeter is clearly defined by a wall, which rises and sinks in relation to the pavement, creating a hard urban edge.

The park will deliver state-of-the-art sports and leisure facilities, which will include an adventure play area, a toddlers' play area, a skateboard park and a youth centre. This is fully integrated into the park's landscape, with a hall housing an indoor climbing wall, offices for local youth groups, toilets and a corner shop, which may become training offices.

Architecture discussion

Selina Dix Hamilton

In 2004 we were one of five firms of architects invited to submit a portfolio of our work and to attend a selection panel (with two children on it) to design a new park for young people in South London.

It's a very exciting project because as far as we know nothing quite like it has been done in England before. We consulted very closely with many young people who lived in the area to find out what would attract them to a playground and what they liked or disliked about parks and playgrounds in general. The area has a demographic profile showing twice as many people below 20 years of age compared with the rest of the country. We went to local schools, to the Peckham Youth Forum and local Rap festivals and summer fêtes to gather information. We engaged a sculptor who worked with groups of young people to encourage them to express themselves, to share their ideas with us and to comment on our design concepts. This method of communicating was far better than getting them to answer a questionnaire or talk about their opinions as it involved them in the process by actually making something. From this came the key requirements of what should be in the park.

The young people wanted a park that was like nobody else's. They always used palm trees in the models they made and after further discussion with them we hit on the idea of creating a dunescape. The site was completely level so we had the opportunity with cut and fill to make dune-like contours over certain areas. Young people like climbing on roofs and that's usually where all the trouble begins so we thought we should make the roof of the indoor sports hall accessible to them. Therefore, the building was integrated into the landscape and was evolved into a free-form shape which particularly suited the use of concrete.

Another reason for the use of concrete was to build a very robust building. One side of the building is alongside a public pavement

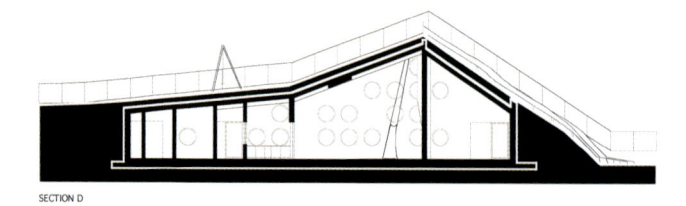

SECTION D

SECTION C

top: Site plan in CAD
above: Sections – indoor sports hall and main building
opposite:
top: Giant slide
bottom: Skating bowl

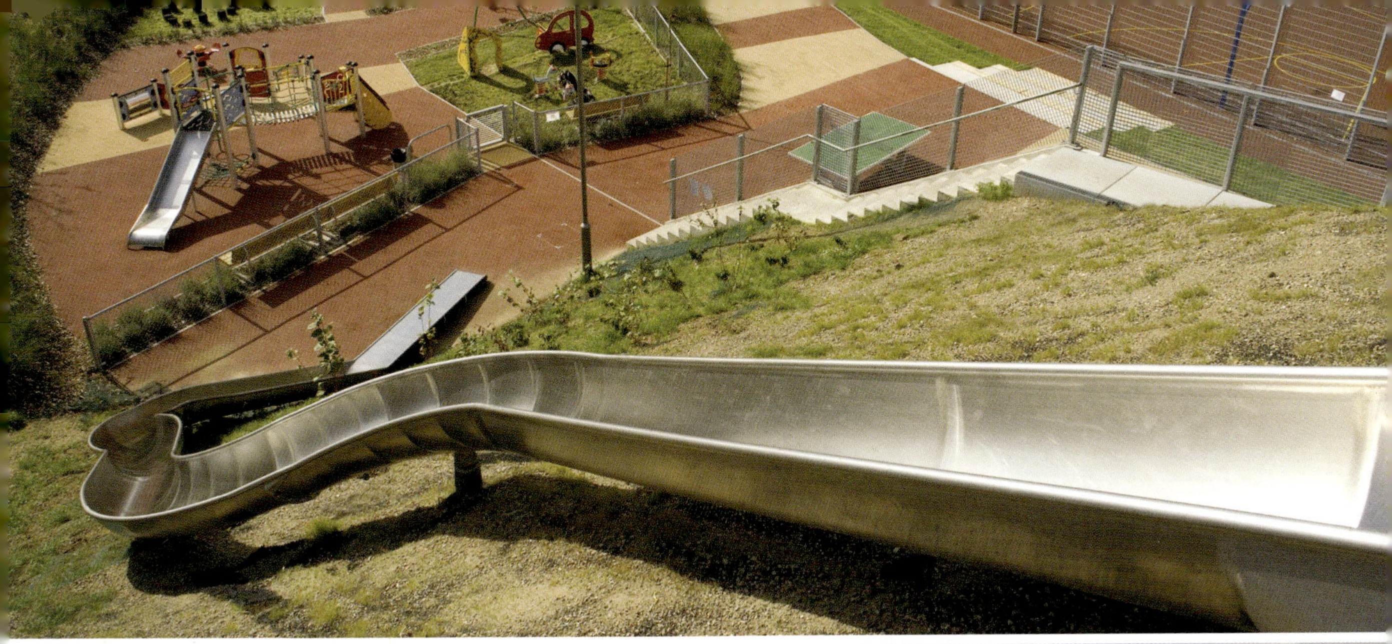
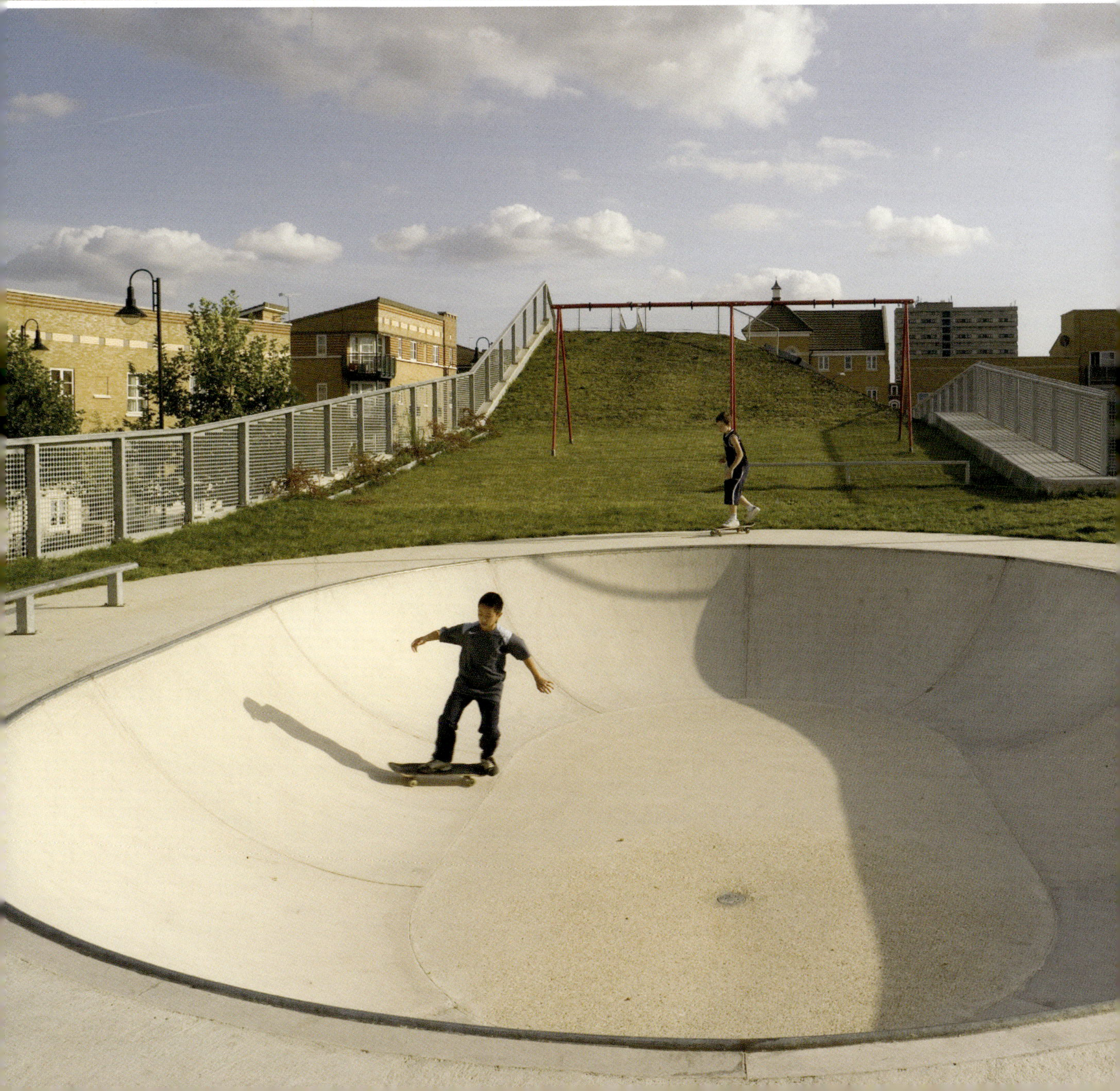

and will undoubtedly receive quite a lot of abuse. The interior of the building has to be durable and hard wearing. The park houses various play areas for different activities and different age groups, some features are for less able people and there is a small garden with play equipment for toddlers who are less than 5 years of age. On the roof of the building there are swings and a giant slide that goes down the slope of the roof, which is a spectacular attraction, visible from far away.

Inside the building we have a climbing wall 7.5 m high and that gives the form to our building structure. The climbing wall was introduced by the client partway through the project so the hill of our building became considerably steeper and the slide considerably more exciting.

The redevelopment of Peckham commenced with the use of an existing children's playground. There was a huge public outcry and the final stage of the programme was to build a new park on the last vacated plot. It was left as a level site covered with building rubble and on a square surrounded by housing association apartments.

The plot size is 50 × 100 m and the building occupies a rectangular space of 300 m^2 along one side of the park. The building is 25 m long and has a triangular elevation with the apex 8 m above ground level. The giant slope that the slide is supported on, and in part the surround to the slate bowl, are resting on soil banked up at either side of the building. Inside the building is a hall with a climbing wall and floor area for general use by local young people. It has a stainless steel curtain that protects and covers the climbing wall when it is not in use. The curtain can either be pegged out to fixings in the floor to divide the space or fixed to the climbing wall to protect it. There was to be a shop which we thought might sell snacks and drinks and sports equipment but now it is likely to be used as a training room for those persons who will run and manage the facility. There is an office for the chief play officer and a bigger office which can take up eight young people, in which they can create a magazine and organise events. The toilets are open to the public from the park side. The main door on the street side will allow the public to use the hall at night or allow it to be let to local organisations. The young people always wished for an adult to be on hand to have the playground supervised. Adults will only be allowed into the facility if they have a young person with them.

The young people wanted the park to be surrounded by a secure fence to let everyone know it was their space and to keep away

top left: Trial panel with window former
top right: Construction of main building

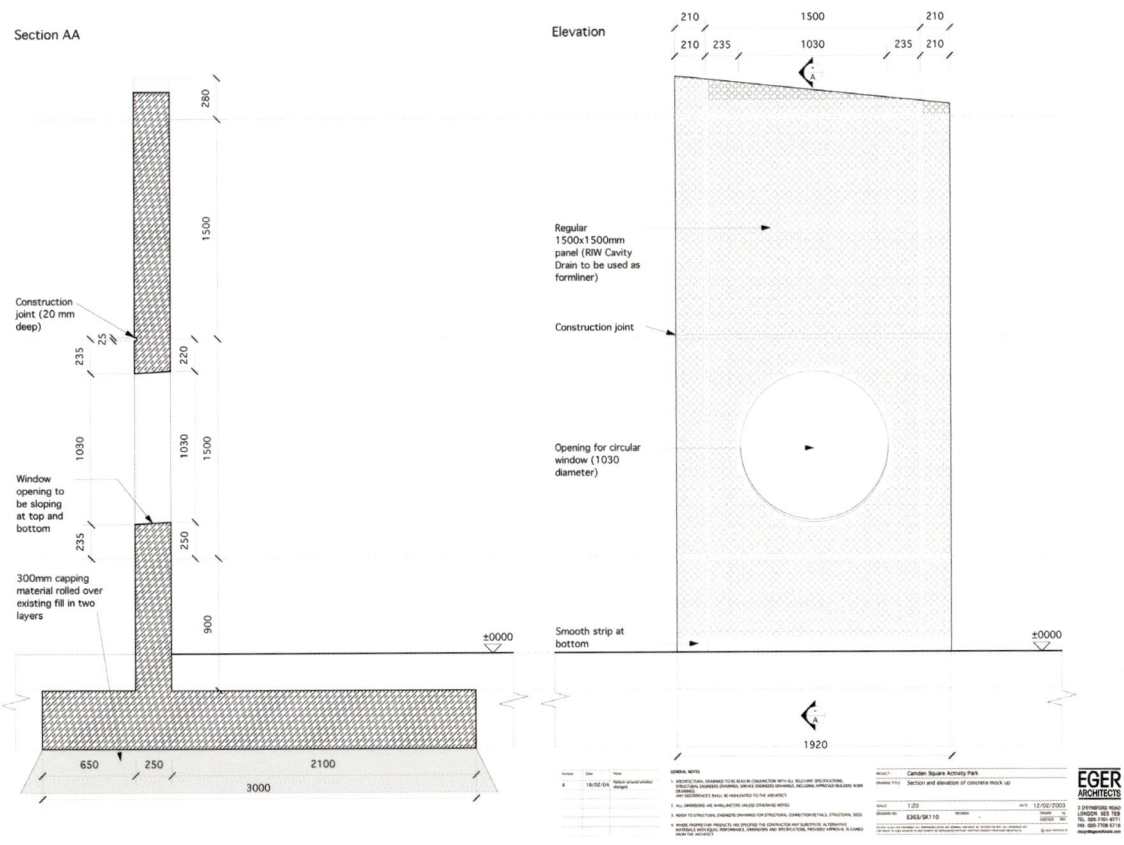

top right: Giant slide atop the main building roof
bottom: Trial panel 'mock up' drawings

223

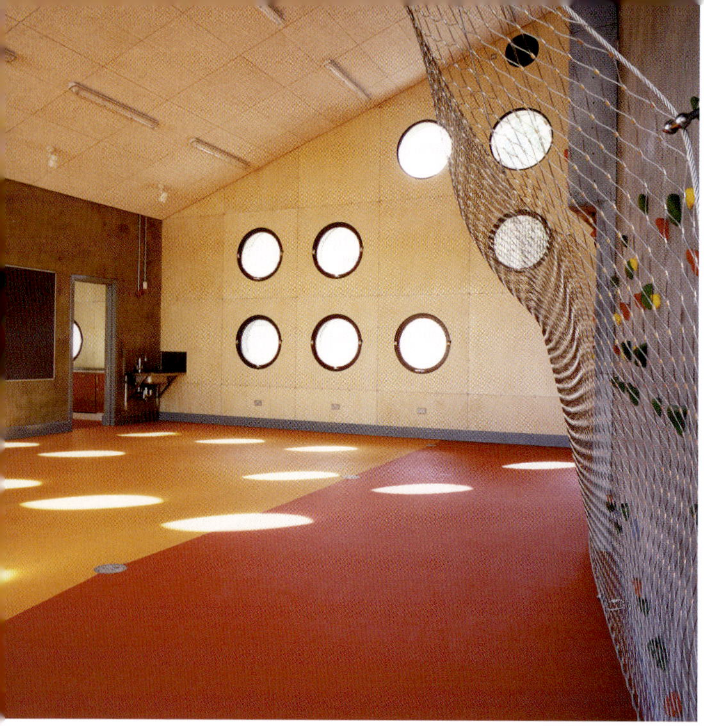
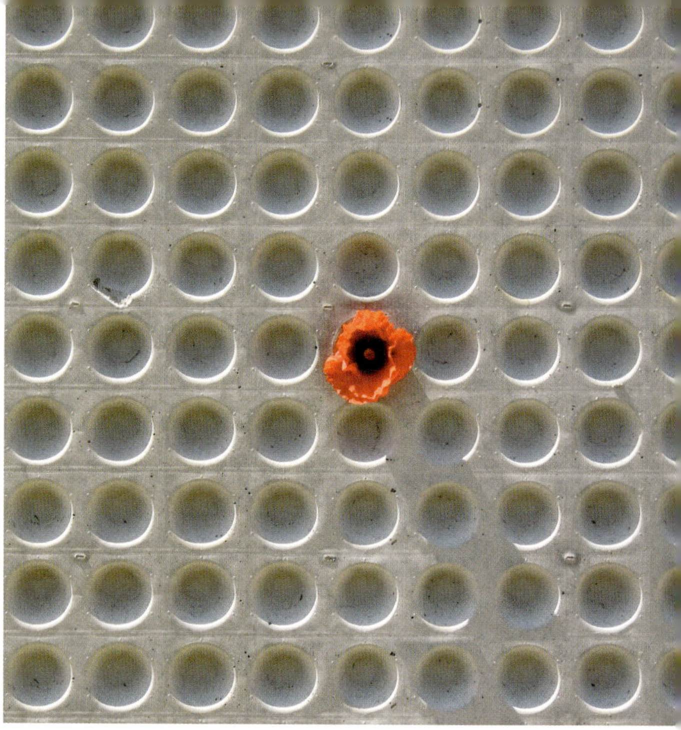
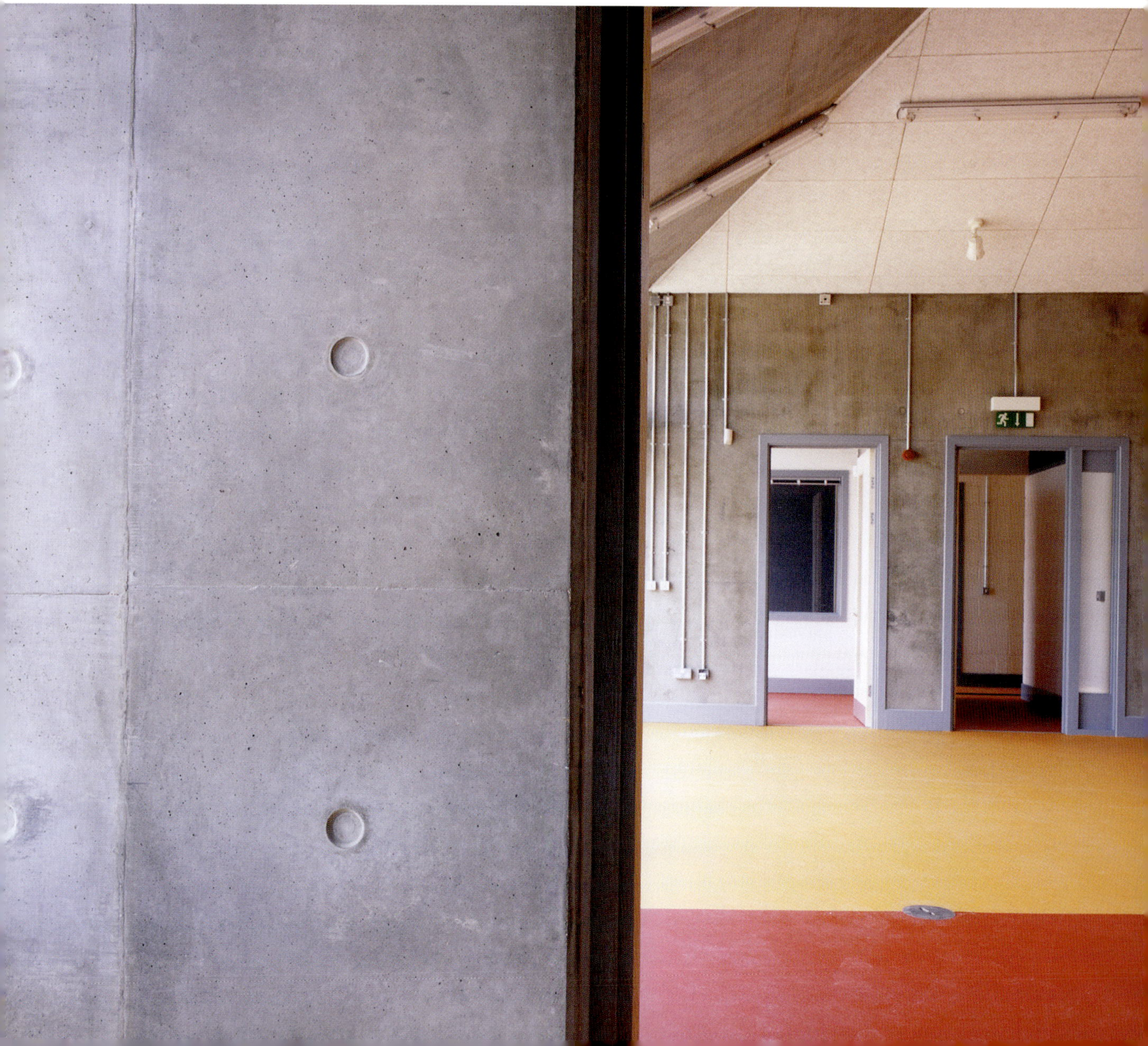

unless you were a young person. We worked closely with the local police safety officer who suggested that the fence should be quite transparent and that all points of the play area should be seen from the surroundings flats, the administrative centre or by surveillance cameras.

There is a lot of bold planting, as well as seven large palm trees. The landscape design concept is that the dunes naturally separate the zoned areas and eliminate the need to screen the skate board bowl from the main pathway. The stripes on the grass simulate movement and wind, which create dunes. We carried that idea to the resin floor inside the building and coloured it red and orange, as young people like bright colours. We thought about having permanent stripes on the grass and hardscape through the park. We contacted STRI who suggested a black and yellow grass which are both sterile (because if you plant two grasses next to each other, they intermingle after a few years and become one colour). That fits in with the hard surface which is striped with red and yellow tarmac. The striped tarmac runs into the brightly coloured floor in the building, chosen to contrast with the grey concrete walls and soffits.

The terracing around the five-a-side pitch alternates grass and concrete slopes so that on wet days the spectators don't have to get their feet wet standing on the grass.

The toddler play area has different surfaces and there are ornamental gardens with rosemary, lavender and roses for the parents and carers to sit in while they keep an eye on the children. It was also important to keep the toddler play areas away from the busy main road. Between the two main dunes is an adventure play area with an artificial rock face to scramble over, which serves as a storage space and has a door. On the high banks we have two different types of long grass mixed with bamboo. The high grass, which grows to 1 m, is supposed to discourage people from climbing the steep slopes. There are steps where they can climb the steep hill.

The long rising armadillo concrete wall of the building elevation fronts onto Kelly Avenue and hides the play area from this side of the road. This elevation, which is on the east side of the park, contains the main entrance into the building.

On top of the building you can see all round Peckham and this also gives you a bird's eye view of the park. By positioning the building along Kelly Avenue we have provided a wider pavement area for the main access. At the time we started working on the scheme we thought there was going to be a future tram station on Kelly Avenue just where we were thinking of widening the pavement so the widening would serve two purposes. In the end the tram stops are likely to be positioned on Commercial Way at the north end of the park; although the tram will still trundle along Kelly Avenue on its way to and from Elephant and Castle.

In terms of locating the building on the plot, we also wanted to maximise the daylight for the planting which all faces south and south-west.

Concrete matters

We chose concrete because it was functional and robust. The main building has a long concrete wall and we wanted it to have some surface texture or treatment for interest and to minimise the impact of potential vandalism. We had some plastic drainage panels in our office left over from another job and thought we could use them as a liner for casting the wall. As a conceptual thought we saw the park as a lump taken out of the desert and placed in this space, as if the ground had heaved and the site risen up from the pavement, with a vertical 'slice' all around it.

This wall was a large surface and we wanted to pattern the concrete in such a way that it created an interesting surface that would be difficult to spoil with graffiti. The RIW plastic cavity tray comes in a roll 20 m long and 2 m wide. To make the plastic liner rigid we stapled it to the ply and made the panels into 1.5 × 1.5 m squares that were fixed to the formwork support system via a backing ply. They would be reused a number of times.

We had hoped that the gaps between the panels would be quite wide but when the trial panels were made the gaps had to narrow to 15 mm to work out. The concrete colour which we liked used an addition of GGBS, which gives a pale, warm grey when blended with grey OP. When the trial panels were cast, and the sample approved, the contractor began building the internal walls. But instead of using the GGBS they used a PFA-blended concrete which was much darker. We ended up with darker concrete walls internally and a paler grey wall externally.

We had three full-scale sample panels made, but before that we had small-scale samples tested to check that the cavity tray did not break down or distort when it was cast against the concrete. The full-scale sample showed up the joint details and gaps between

top left: Interior of sports hall wall
top right: External RIW cavity tray finish
bottom: Internal concrete finish

the panels. It was disappointing that the concrete subcontractor didn't do the trial panels in accordance with our drawings. We had to progress the external wall without having concluded a trial panel with the full reinforcement. The rebated finish between the cavity tray panels was a problem. We tried a plywood strip, a taped joint, a tapered timber batten which got stuck in the concrete and a plastic strip which gave the best finish. The work was carried out in winter so we could not anticipate that in the summer under full sun the plastic strip would expand and distort and cause grout loss problems. The edges of the plastic strips had a chamfer and if they were put in the wrong way round they would be difficult to remove without spalling the concrete. The contractor had placed a number of the strips the wrong way round but, fortunately, we spotted that before the main wall was cast.

When the subcontractor had cast the first section of the wall they stopped work, requesting that the concrete mix was changed, suggesting that it should be self-compacting. That was very expensive and not appropriate for the wall. The dense impermeable plastic would increase the risk of air bubbles forming on the surface so it was agreed that the slump would increase from 90 mm to 150 mm, which seemed to work much better.

We prepared the entire formwork panel layout and tie bolt holes, but allowed the contractor to position the construction joints on regular spacing and to agree this with us before the work started. After the first pour, the contractor asked David Bennett to comment and help the concrete contractor with the finer points of compaction and placing technique. The concrete finish was never as good as the first pour, but was acceptable.

However, when they cast the top level pours to the wall there was noticeable grout loss and the cavity indents were disfigured and unsightly. It was agreed that a concrete finisher would be employed to carry out surface repairs, working strictly in accordance with guidance notes that had been prepared. The key was to find a person who had the right touch, was careful and knew how to work with small dabs of mortar.

They had a steady worker carrying out the repair and we were happy with the results, but partway through he left site following a dispute with the contractor. The replacement was fine initially, but some areas were left patchy and in need of further attention. The surface in the end was given a light sand blast to soften the patchy areas but not remove the grout skin to reveal the coarse

Triangular elevation of the main building

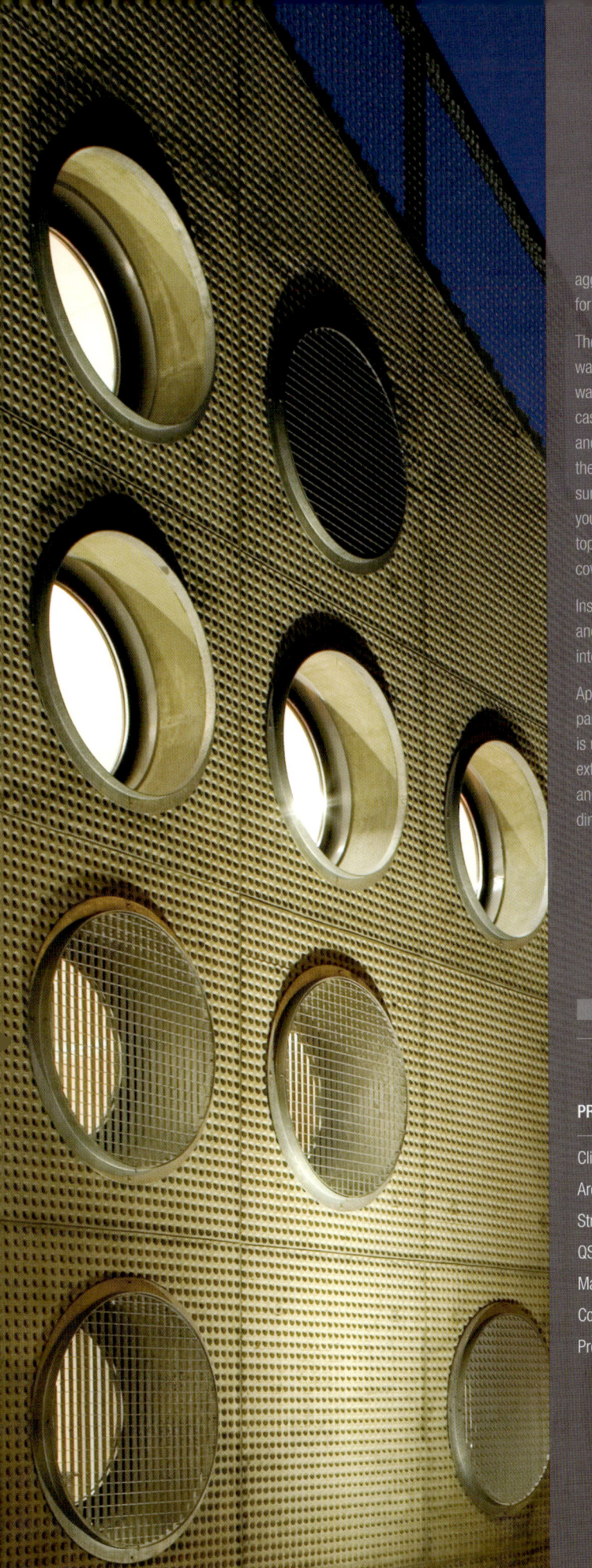

aggregates. The concrete surface was sealed with a silane solution for waterproofing and as an anti-graffiti coating.

The internal walls and soffits are fine. Casting a sloping roof of 35° was quite a challenge. A top shutter was very expensive, gunite was considered but was outside the budget price. Eventually, they cast the slope without a top shutter using a low slump concrete and fixing Expamet mesh at intervals across the slope to contain the compacted concrete as it was placed. The finish on the top surface was quite rough, but the soffit was fairly good, although you can see the pour lines. We had to trowel a screed on the top surface to create a smooth face for the insulation and roof covering.

Inside the main building the dimpled wall is covered with insulation and faced with birch ply, chosen because we did not want the internal space to look too much like a concrete bunker.

Apart from fair face concrete internally, there is a blockwork partition wall painted white. The door frames are painted, there is underfloor heating and the lighting is ceiling mounted. There is external lighting to the concrete steps, the pathways to the exits and along the pavement we have uplighters to illuminate the dimpled concrete wall.

PROJECT TEAM

Client: London Borough of Southwark
Architect: Eger Architects
Structural engineer: Elliot Wood Partnership
QS: Playle and Partners
Main contractor: Albert Soden Ltd
Completed: August 2006
Project value: £1.5 million

GLOSSARY OF CONCRETE TERMINOLOGY

Air entrained concrete
A concrete which has a small amount of air bubbles introduced into the mix to enhance and improve its frost resistance. It is used for making roads

Anti-crack reinforcement
A close mesh of reinforcement which is placed close to the surface of concrete to reduce surface cracking

Batching plant
A set of storage silos, mechanical conveyors and dispensers for automatically measuring by weight the quantities of different materials for making concrete

Binder
The cement in the concrete that glues all the constituent materials together to form a hard durable rock

Bleeding
A tendency for water to separate out of the concrete on compaction and rise to the surface

Blowhole
Air pockets on the concrete surface formed by air bubbles that are trapped on the formwork when concrete is being placed and compacted in wall and column sections

Bulk density
The weight of a material per unit volume including voids and water contained in it. Normal, well-compacted concrete has a bulk density of 2.3 t/m^3

Bulking
The increase in the volume of dry sand when its moisture content increases. This may be as much as 40 per cent when 5 per cent water is added. This increase disappears when the water content is raised above 20 per cent

Bush hammer
Percussive tooling of the concrete face to expose the lower layers and reveal the aggregate, leaving a rough surface

Calcine
The process of heating a mineral or an ore to a high temperature to drive off carbon dioxide and water

Cantilever
A beam or slab structure that is fixed at one end by the support structure and is allowed to hang freely over the rest of its length

Carborundum
The trade name for silicon carbide, which is an abrasive material that is harder than quartz, used for grinding the surface of hardened concrete

Cast insitu or cast in place
The term given to a concrete when placed into formwork on the construction site rather than in a precast factory

Chair
A bar that is bent in such a way that it holds up the top layer of reinforcement, used commonly in floor slab or beam construction

Climbing formwork (a.k.a. jump forming)
Wall formwork system that is self-supporting, being held in place by clamping to the hardened concrete lift below using bolts cast into the concrete and/or the tie bolts holes

Coarse aggregate
The biggest particles in the concrete, larger than sand particles and usually composed of gravel or crushed rock and usually 4–20 mm in diameter

Coefficient of expansion
The expansion per unit length of a material for each degree rise in temperature

Column
A vertical strut or post that supports a floor, roof or beam

Compaction
The action of vibrating concrete to fluidise the mix to release entrapped air; to make the concrete denser and improve its strength and durability

Concrete mix
The mix constituents of concrete expressed in terms of their constituent weight per cubic metre

Concrete mixer
A rotating drum or pan in which the concrete ingredients are mixed to make concrete

Concrete pump
Usually a lorry-mounted pump that has a hydraulically driven extending boom or a static line that conveys concrete deposited in its hopper via a pipeline to the point of placement

Contraction joint or shrinkage joint
A break or joint in the concrete to allow for drying and thermal contraction to minimise the risk of uncontrolled or indiscriminate cracks forming in undesirable places

Cover
The minimum thickness of concrete between the reinforcement and the concrete surface

Cube test
A test carried out on a cube of concrete made and cured to standards set out, which is then crushed to determine the compressive strength of the concrete

Curing
Keeping moisture retained within the concrete for the first week so that the concrete is allowed to harden fully

Curing compound
A sprayed-on liquid membrane that is applied to the fresh concrete surface to prevent evaporation and moisture loss

Dry batched concrete
The constituent materials are introduced separately into the drum of the truck mixer and then mixed

Drying shrinkage
The shrinkage of concrete during the hardening and drying out process

External vibrator
A vibrator that is clamped and fixed on the outside of the formwork to compact concrete

Falsework
Another name for the support system for formwork

Fine aggregate
The sand or crushed rock particles smaller than 4 mm in a concrete mix

Flow table test
A standard test for assessing the fluidity of flowing concrete with a very high slump, by measuring its spread

Formliner
A membrane or sheet of elastomeric polymer, latex rubber or glass reinforced plastic that is attached to a backing plywood or metal form to create a surface profile and/or a synthetic textured finish

Formwork
Boarding or sheeting of timber, plywood and steel that is erected to contain the concrete during placing and until it hardens. The face texture can greatly improve the finished concrete surface

GGBS
Ground granulated blastfurnace slag, a by-product in the manufacture of steel which can be used as a replacement cement for Portland cement

Grading curve
A curve made by plotting the percentage of aggregate particles (by weight) that are smaller than and can pass a range of sieve sizes

Gravel
Naturally occurring river bed or estuarine pebbles that are well rounded and usually larger than 5 mm in diameter

GRC
An abbreviation of glassfibre reinforced concrete: a lightweight precast concrete reinforced with alkali-resistant glassfibre strands

Grit blaster
A compressed air line that forces grit through a nozzle abrading the surface of smooth concrete to remove the laitance and reveal the aggregate

Grit blasting (a.k.a. shot blasting)
The equivalent of sand blasting but using grit, which is much safer to use than sand because it does not create silica dust

GLOSSARY

Grout loss
This occurs when there are gaps in the form face panels, when the joints and arrises are not watertight, allowing the mixture of cement, water and sand to leak through

Gunite
A cement, sand and aggregate mix that is forced onto a wall at high speed using compressed air. The gunite sticks to the surface due to the cohesiveness and stickiness of the mix. The low water content forms a dense, high-strength concrete

Hard standing
Any hard surface that is suitable for parking vehicles

Hardcore
Hard lumps of stone, brick, crushed concrete, etc. suitable for filling soft ground

Honeycombing
This appears on the concrete surface as a cluster of aggregates with voids between them where there is no cement and fines

Hydrostatic pressure
This is equal to the depth of the liquid multiplied by its density

Laitance
The skin on the surface of concrete formed of cement, sand and water

Lap
The length by which one reinforcing bar must overlap the other to maintain continuity

Layout
A drawing showing the general arrangement of a proposed building and its construction

Lightweight aggregate concrete
A concrete made with low-density, lightweight aggregate, usually an expanded shale or clay aggregate

Mass concrete
Concrete without reinforcement

Mix design
See Concrete mix

Mortar
A paste of cement, sand and water

Mould
A temporary structure built to hold and confine the concrete until it sets. It is usually associated with precast concrete

Particle size distribution
The proportions by weight of different particle sizes in sand determined by the percentage passing various sieve sizes

PFA
Pulverised fuel ash is a by-product of coal-burning power stations which can be used as a replacement cement for Portland cement

Plasticiser
An admixture added to the concrete mix to improve workability

Post-tensioning
A method of prestressing concrete in which the cable strands are stressed after the concrete has hardened

Pozzolana
A volcanic dust for Pozzouli or pulverised fuel ash used as a hydraulic cement which, mixed with Portland cement, becomes a hydraulic cement and will set

Precast concrete
Concrete that is cast in a precast factory

Ready mixed concrete
Concrete that is mixed at a remote batching plant and carried to the site by a truck mixer

Rebar
An abbreviated term for reinforcement

Reinforced concrete
Concrete containing more than 0.6 per cent by volume of reinforcement. Reinforcement bars are rods of steel that are embedded in concrete to provide tensile strength to prevent the concrete cracking

Relative density
The weight of a material divided by the weight of the same volume of water

Release agent
A mould oil or chemical liquid applied to formwork to prevent concrete hardening onto it

Retarder
An admixture that is added to concrete to slow or delay the setting time of the concrete

Sand blasting
A compressed air jet throws sand and flint through a nozzle directed at the concrete surface to abrade it. The depth of abrasion can vary according to the coarseness of the sand and number of passes with the air nozzle

Self-compacting
A very fluid viscous concrete with high fines content which can be placed without segregation and without the need for compaction

Shot blasting
Small pellets of metal are fired from a wheeled machine at the concrete surface at close quarters using compressed air

Shuttering (a.k.a. formwork)
Another term for formwork that supports and confines the wet concrete. Sometimes defined as the formwork in contact with the concrete

Skip
A large hoisting bucket lifted by a crane for placing fresh concrete

Slag
See GGBS

Slump test
A test for the workability of concrete which is carried out on site. A conical mould is held firmly in place on a level surface to ensure there is no grout leakage. The mould is filled with concrete in three layers, each layer is tamped 25 times with a tamping rod. The cone is lifted and the difference in height of the mould and the highest point of the concrete is the slump

Soldier
An upright support in a formwork support system that takes the tie bolt

Sprayed concrete
See Gunite

Starter bar
A reinforcing bar projecting from the construction joint to splice or lap with the adjoining bar to knit the joint together for continuity

Substructure
Part of the structure that is below ground

Superstructure
Part of the structure that is above ground

Tie bolt or tie rod
A high tensile steel rod that is threaded at both ends to receive anchor plates. It acts as a tension strut to restrain the soldiers and prevent them from moving when fresh concrete is placed against the formwork

Tremie
A hopper with a pipe leading out at the bottom that reaches down to the base of the formwork into the foundation for placing concrete underwater or in deep formwork out of the water. Sometimes referred to as trunking where a pipeline is inserted into vertical formwork to deposit concrete with the minimum of free fall and segregation

Truck mixer
A lorry with a tilting drum mixer to mix and carry concrete

Vibrated concrete
Concrete that has been compacted by using an internal or immersion poker vibrator

Viscosity
The resistance of a fluid or material to flow

Waling
A horizontal support that spans between the soldiers in a formwork system to which the sheets of formwork are attached

Workability
The ease with which concrete can be placed. Workability can be measured by slump test. For certain types of concrete a flow table or compaction factor tests are used

WRA
Stands for water reducing admixture which could be a plasticiser or superplasticiser

Further reading

Blackledge, G. F., *Concrete Practice (third edition)*, British Cement Association, 2002

Kind-Barkauskas, F., Kaushen, B., Polonyi, S. and Brandt, J. *Concrete Construction Manual*, Edition Detail, Birkhauser, 2002

Formwork: A Guide to Good Practice (second edition), Concrete Society, 1995

Herzog, T., Naterrer, J., Schweitzer, R., Voltz, M. and Winter, W., *Timber Construction Manual*, Edition Detail, Birkhauser, 2004

Lindon, K. and Sear, A., *The Properties and Use of Coal Fly Ash*, Thomas Telford, 2001

Chapman, S. and Fidler, J., *The English Directory of Building Sands & Aggregates*, Donhead, 2000

Bennett, David, *Innovations in Concrete*, Thomas Telford, 2002

Bennett, David, *Exploring Concrete Architecture*, Birkhauser, 2001

Neville, A. M., *Properties of Concrete (third edition)*, Pitman, 1981

Scott, John S., *The Penguin Dictionary of Civil Engineering (third edition)*, Penguin Books, 1980

Useful contacts

Form face materials
UPM Kymenne – www.wisa.com
Kronoply – www.kronoply.de
Thomasi – www.thomasi.com.br

Cement
Lafarge Cement – www.lafarge-cement-uk.co.uk
Castle Cement – www.castlecement.co.uk
Civil and Marine (GGBS) – www.civilmarine.co.uk
Rugby Cement – www.rugbycement.com
PFA – www.ukqaa.org.uk

Compaction equipment
Wacker – www.wacker.group.com

Aggregates
Quarry Products Association – www.qpa.org.uk
British Marine Aggregates Product Association – www.bmapa.org.uk
English Heritage – www.english-heritage.org.uk

Specialist formwork systems
A-Plant Acrow – www.aplant.com
Peri UK – www.peri.ltd.uk
SGB – www.sgb.co.uk
Ischebek Titan – www.ischebek-titan.co.uk
Doka – www.doka.com
RMD – www.rmdkwikform.net

Concrete advisory
The Concrete Centre – www.concretecentre.com
The Concrete Society – www.concrete.org.uk
The British Cement Association – www.bca.org.uk
British Ready Mix Concrete Association (BRMCA) – www.brmca.org.uk

Pigments
Lanxess (Bayer) – www.lanxess.com
Hatfields – www.royhatfield.com

Architectural concrete consultant
David Bennett Associates (tel: 01279 439562) – www.concretebennett.com

Picture credits

Part 1: Technology
Figures:
Adrian Howley, 92–95
A Plant Acrow Ltd, 69
British Cement Association (BCA), 18, 20, 22–24, 38, 40, 75, 88–91
Castle Cement, 1, 2, 4–14
Civil and Marine, 15
Creteco Ltd, 86, 87
David Bennett, 3, 19, 21, 25, 28–36, 39, 55, 59–64, 67, 68, 77, 78, 79 (left), 84
David Tucker, 53, 54, 85
Eldridge Smerin Architects, 70–72
George Dawes, 41–43, 74
Kronoply GmbH, 48, 51
Lanxess GmbH, 37
Linden Sear (UKQAA), 16, 17
PERI Ltd UK, 56–58
SGB Ltd, 73
The Concrete Society, 65, 66
UPM-Kymmene Wood Ltd, 44–47, 49, 50, 52
Wacker (UK) Ltd, 76, 79 (right), 80–83

Part 2: Case studies
Drawings and plans are reproduced with the kind permission of the architectural practices.
Thames Barrier Park: pages 74–83
Patel Taylor Architects, 76, 79; Martin Charles, 74, 77, 78, 83.

Persistence Works: pages 84–95
David Grandorge, 85, 87, 88, 90 (top); Martine Hamilton Knight, 90 (bottom); Stuart Blackwood@YASS, 92; Mandy Reynolds, 95.

The Art House: pages 96–105
Hélène Binet

The Anderson House: pages 106–115
Sue Barr, 106, 112, 115; David Grandorge, 109; Jamie Fobert Architects, 111

Aberdeen Lane: pages 116–125
Keith Collie

One Centaur Street: pages 126–135
M. Mack/A. de Rijke

85 Southwark Street: pages 136–145
Dennis Gilbert

The Bannerman Centre: pages 146–155
David Butler

The Brick House: pages 156–165
Hélène Binet

The Collection: pages 166–177
Hélène Binet, 169, 172 (bottom), 175, 177; Peter Hudspith Architects, 166, 170, 171, 172 (top)

Playgolf, Northwick Park: pages 178–187
Andy Spain, 178, 181, 185, 186; Charles Mador Architects, 182, 183

E-Innovation Centre: pages 188–197
David Barbour, Building Design Partnership

The Jones House: pages 198–207
Alan Jones Architects

Spedant Works: pages 208–217
Mathew Hawkins, 208, 210, 212 (top), 214, 215 (top), 216; Nick Lee, 212 (bottom), 215 (bottom)

Central Venture Park: pages 218–227
Matt Chisnall, 218, 221, 226; Eger Architects, 220, 222, 223; Robert Teed, 224

We would like to thank all of the architects and photographers who provided the numerous excellent photographs for this book.

INDEX

acid washing, 35
acoustics, case studies, 113
additions, cement *see* blended cements
admixtures, 27, 31
aggregate/cement ratio, 40
aggregates, 14–21
   concrete mix, 20–1, 39–40
   material quality, 19–20
   selection, 18–19, 20–1
   self-compacting concrete (SCC), 26
   surface finish, 34
      case studies, 163
   types, 16–20
alkali–silica reaction (ASR), 20
architectural drawings, formwork, 55–9
ASR (alkali–silica reaction), 20

birch ply formwork, 43–4
   case studies, 123, 211
blastfurnace slag (GGBS) additions *see* ground granulated blastfurnace slag (GGBS) additions
blended cements, 5, 11–13 (*see also* ground granulated blastfurnace slag (GGBS) additions; pulverised fuel ash (PFA); silica fume additions)
blowholes, 67, 70
board-marked finish, 42
   case studies, 134, 173, 174, 176
brickwork with insitu concrete,
   case studies, 161–4
BS 197-1, 3
BS 882, 20
BS 1014, 36
BS 4027, 3
BS 5268, 46
BS 6566, 44
BS 6699, 12
BS 8500, 5
BS EN 196, 12
BS EN 206-1, 5
BS EN 313/314, 46
BS EN 413, 3
BS EN 450-1, 13
BS EN 635/636, 46
BS EN 1992 (Eurocode 2), 17
BS EN 12620, 20
BS EN 12878, 37

carbonation, 6
'CEM' designation, 3
cement, 2–13
   blends *see* blended cements
   colour *see* colour, cement
   manufacture, 2–3, 6–10
   mineral composition, 3, 11
   properties, 5–6
   raw ingredients, 2, 6, 10–11
   types, 3–4
cement content, 26, 39–40
cement replacements *see* blended cements
china clay additions, 5
chipboard formwork, 47–8
   case studies, 109
chlorides, 19–20
circular column formwork, 50

cladding, case studies, 80, 151
cleaning, surface, 70
colour, 34–8
   aggregate selection, 19
   case studies, 87, 113
   cement, 4, 5–6, 10–11
      GGBS, 12–13
      PFA blends, 13
   discoloration, 71
   pigments, 36–8
column formwork, 50
combination cements *see* blended cements
compaction, 25, 62–4
   case studies, 82
concrete terrazzo, 35
consolidation, 62–4
construction joints, 54
   case studies, 80
   grout loss, 68
corrosion protection, reinforcement, 65
cracking, case studies, 93
crushed rock aggregates, 17
   concrete finish, 20–1
curing, 67
cutting formwork, 51–2

discoloration, 68–70
disposable column formers, 50
Douglas fir formwork, case studies, 211
dry batch plant, 24, 29–31
dry shake surface colours, 38
durability, 5

efflorescence, 37–8, 68
elastomeric formliners, 49
energy performance
   case studies, 173
   cement production, 2
environmental precautions, 33
Eurocode 2 (BS EN 1992) :
   Design of concrete structures, 17
expanded shale aggregate, 18–19
exposed aggregates, 34
   case studies, 163

fibre-cement panels, case studies, 132
fibreglass formliners, 48–9
film faced plywood formwork, 44–6
   case studies, 80, 82, 91, 94, 134, 144, 164, 181
fineness, cement, 10, 11
finish *see* surface finish
fixing formwork, 51 (*see also* tie bolts)
   chipboard, 48
   oriented strand board (OSB), 46–7
flakiness index, 19
floor slabs, 63, 64
   case studies, 92–3, 143, 153, 163, 184, 186
flyash *see* pulverised fuel ash (PFA)
formliners, 48–9
formwork, 41–59
   architectural drawings, 55–9
   chipboard, 47–8
      case studies, 109
   design and assembly, 51–3, 54, 56, 68
      case studies, 82, 91, 94, 102, 113, 143–4

disposable column formers, 49–50
formliners, 48–9
oriented strand board (OSB), 46–7
   case studies, 205
plywood, 43–6, 56 (*see also* film faced plywood formwork)
   case studies, 114, 123, 195, 211
release agents, 55
reuse, 51, 54
   case studies, 83
sawn board timber formwork, 42–3
   (*see also* board-marked finish)
steel faced, 48
striking times, 54–5

geotextiles, 49
glass reinforced plastic (GRP) formliners, 49
gravels, 16–17
   concrete finish, 20–1
green roofs, case studies, 181
grey cement, 5, 10–11
grit blasting, 35, 70
   case studies, 163
ground granulated blastfurnace slag (GGBS) additions, 5, 11, 12–13
   case studies, 82, 87, 89, 114
grout loss, 68
grout runs, 70
GRP (glass reinforced plastic) formliners, 49

hand–arm vibration syndrome (HAVS), 63
hardened self-compacting concrete (SCC), 26
HAVS (hand–arm vibration syndrome), 63
HDO (high-density overlay) formwork, 45, 46
high density chipboard formwork, 47–8
high density overlay (HDO) formwork, 45, 46
honeycombing, 70–1

impurities, aggregates, 19–20
internal poker vibrators, 63

joints, construction *see* construction joints

lay-flat hoses, 62
   case studies, 176
Liapor, 18–19
lighting (*see also* service layouts)
   case studies, 153, 210
lightweight aggregates, 14, 17–19 (*see also* Liapor; pulverised fuel ash (PFA))
lime bloom, 37–8, 68
Lytag, 18
   case studies, 99, 102

manufacture
   cement, 2–3, 6–10
      effect on colour, 5–6, 10–12
   concrete, 22–33
marine aggregates, 17
masonry cement, 3, 10
MDO (medium density overlay) formwork *see* medium density overlay (MDO) formwork

INDEX

mechanical properties, aggregates, 19
medium density overlay (MDO) formwork, 45, 46
    case studies, 82, 134, 144
metakaolin additions, 5
metal formwork, 48
microsilica, case studies, 102
milling, cement, 10
mineral composition, cement, 3
mixes, 33, 39–40
    aggregates, 20–1
    self-compacting concrete (SCC), 26
mixing, 23–4
    self-compacting concrete (SCC), 25
models, case studies, 168
    'Multi Cem', 10

oriented strand board (OSB) formwork, 46–7
    case studies, 205

panel layout (formwork), 54, 56
    case studies, 82, 91, 113, 174, 176
paper faced panels (formwork), 45, 46
    case studies, 164, 181
PFA *see* pulverised fuel ash (PFA)
phenolic film faced (PFF) formwork, 44–5, 45–6
    case studies, 91, 94, 134
    oriented strand board (OSB), 47
pigments, 36–8
plastic cavity tray formliners, case studies, 225–6
plastic sheeting formliners, 50
    case studies, 109, 113, 114
plywood formwork, 43–6, 56 (*see also* film faced plywood formwork)
    case studies, 114, 123, 195, 211
pollution precautions, 33
polythene sheeting formliners, 50
    case studies, 109, 113, 114
polyurethane formliners, 49
polyvinylchloride (PVC) formliners, 48–9
'Portland' cement, 3 (*see also* cement)
pour planes, 64
pressure assessment, 53–4
properties
    aggregates, 19
    cement, 5–6
    self-compacting concrete (SCC), 26
pulverised fuel ash (PFA)
    aggregates, 17–18 (*see also* Lytag)
    cement additions, 5, 11–12, 13
pumping concrete, 61–2
    case studies, 82–3, 114
PVC formliners, 48–9

quality control (*see also* testing)
    case studies, 81
    concrete mix, 30–1, 40
    pressure assessment, 53–4
    stains and blemishes, 68–70

rapid hardening cement, 10, 11
ready mixed concrete, 23–33
    admixtures, 27
    dry batch plant, 24, 30–1
    self-compacting concrete (SCC), 24–7

    wet mix/batch plant, 23–4, 32–3
rebar, 64–5
recycled aggregates, 14, 16
reinforcement, 64–5
release agents (formwork), 55
    film faced plywood, 45
    formliners, 49
repairs, surface *see* surface repairs
revibration (compaction), 64
rigid formliners, 49
roofing, case studies, 164, 174, 184

sand/cement ratio, 40
sands, 16–17, 20 (*see also* aggregates)
sawn board timber formwork, 42–3 (*see also* board-marked finish)
segregation, 25, 39
self-compacting concrete (SCC), 24–7
    case studies, 173–4
semi-dry batching, 32–3
service layouts (*see also* lighting)
    case studies, 114, 184
silica fume additions, 5
    case studies, 102
skip discharge, 60
    case studies, 102
slump flow test, 26
slump mix, 26, 39, 40
    admixtures, 27
slump testing, 40
sound insulation, case studies, 113
spacers, reinforcement, 65, 66
SRPC (sulphate-resisting Portland cement), 3
stains, 68
    case studies, 82, 91, 92
Sterling board, case studies, 195
stonework with insitu concrete, case studies, 173
storage
    aggregates, 29
    formwork, 50–1
    reinforcement, 64
striking times (formwork), 54–5
sulphate-resisting Portland cement (SRPC), 3
superplasticisers, 27
surface coatings
    case studies, 91
    formwork, 44–6
    water repellent, 67–8
        case studies, 92
surface colours, 38
    case studies, 87, 91
surface discoloration, 68–70
surface finish, 34–5, 67
    case studies, 81, 143–4, 163, 184
    concrete mix, 20–1
    formwork, 41–2
surface repairs, 70–1
    case studies, 144, 226–7
surface retarders, 35
surface stains and blemishes, 68–70
synthetic oxide pigments, 37

terrazzo, 35
    case studies, 134, 164, 184
testing, 30, 33

    aggregates, 19
    self-compacting concrete (SCC), 26
    trial panels, 67
thermal bridging, case studies, 99
thermal mass, case studies, 173
tie bolts, formwork, 52–3, 56
    case studies, 114, 144
timber formwork *see* chipboard formwork; oriented strand board (OSB) formwork; plywood formwork; sawn board timber formwork
transport, 29, 31, 60
    self-compacting concrete (SCC), 26
tremie pipes, 62
    case studies, 82–3, 195
trial panels, 67
truck mixing, 24, 28–9, 30–1, 40
    case studies, 81–2, 134
    self-compacting concrete (SCC), 26–7

vibrating screed rails, 63
vibration beams, 64
viscosity agents, 25
visqueen sheeting formliners, 50
voids, 25

water reducing admixtures (WRA), 27
water repellent coatings, 67–8
    case studies, 92
water/cement ratio, 39–40
waterproofing formwork
    chipboard, 48
    cut edges, 52
    untreated timber, 42, 44
wet batch/mix plant, 23–4, 32–3
white cement, 5–6, 11
workability, 39
    admixtures, 27
    pumping concrete, 61
    self-compacting concrete (SCC), 26
WRA (water reducing admixtures), 27

Zemdrain, 49